Low-Dimensional Materials

Low-Dimensional Materials

Bridging the Fundamental Principles to Practical Applications

HUI-MING CHENG

Faculty of Materials Science and Engineering/Institute of Technology for Carbon Neutrality, Shenzhen Institute of Advanced Technology, Chinese Academy of Sciences, Shenzhen, Guangdong, P.R. China

DAI-MING TANG

Research Center for Materials Nanoarchitectonics (MANA), National Institute for Materials Science (NIMS), 1-1 Namiki Tsukuba, Ibaraki, Japan

XIAOLONG ZOU

Shenzhen Geim Graphene Center, Tsinghua-Berkeley Shenzhen Institute & Tsinghua Shenzhen International Graduate School, Tsinghua University, Shenzhen, P.R. China

LILI ZHANG

Institute of Metal Research, Chinese Academy of Sciences, Shenyang, Shenyang National Laboratory for Materials Science, Liaoning, P.R. China

ELSEVIER SCIENCE PRESS

Elsevier
Radarweg 29, PO Box 211, 1000 AE Amsterdam, Netherlands
125 London Wall, London EC2Y 5AS, United Kingdom
50 Hampshire Street, 5th Floor, Cambridge, MA 02139, United States

ISBN: 978-0-443-29046-6

For Information on all Elsevier publications
visit our website at https://www.elsevier.com/books-and-journals

Publisher: Matthew Deans
Acquisitions Editor: Glyn Jones
Editorial Project Manager: Namrata Aggarwal
Production Project Manager: Anitha Sivaraj
Cover Designer: Vicky Pearson Esser

Typeset by MPS Limited, Chennai, India

Contents

Preface

Dimension is a measure of human beings' understanding of the universe. In ancient China, it was believed that the world is a square ground under the circular heaven. The Greeks realized that the Earth is a three-dimensional sphere by geometric measurements. In Newtonian mechanics, mechanical dynamics included three spatial dimensions coupled with one temporal dimension. And Einstein's theory of relativity told us that the four-dimensional space–time is intertwined through the Lorentz transform. In addition to space and time, hidden dimensions such as spin have been discovered in quantum mechanics.

Since last century, not only human beings have made groundbreaking discoveries, but also there have been great progresses in controlling and utilizing at inner and finer dimensions. Regarding measurement in spatial dimensions, the precision of Laser Interferometer Gravitational-Wave Observatory has reached 10^{-18} m, 1/1000th of the atomic nucleus size, enabling detection of extremely small distortion of space–time caused by merging of black holes, which occurred at a distance of billions of light-years. Regarding measurement of time, the precision of atomic clocks has reached 10^{-18}, meaning that there will be an error of less than 1 second, for the time since Big Bang. With invention of transmission electron microscopy and scanning probe microscopy, it becomes possible to see and manipulate individual atoms. In addition, transistors with feature sizes smaller than 10 nanometers could be fabricated by using extreme ultraviolet lithography. The progresses in science and technology into the finer and deeper dimensions have become foundation of modern civilizations.

On the other hand, the exponential growth of human civilization since the Industrial Revolution has brought great pressure to the ecosystem on a global scale. The population has increased from 1.65 billion in 1900 to 7.60 billion in 2021 and is expected to reach 10 billion in 2100. Along with the growth of world gross product from \$1.4 to \$84.7 trillion from the 1960s to the year 2020, the world energy consumption has grown from 52,399 TWh in 1965 to 173,340 TWh in 2019. In the meantime, the atmospheric concentration of CO_2 has dramatically increased, reaching a record high of ~300 ppm over the last 800,000 years and reaching 412.5 ppm in 2021. In 2019 an editorial article published by *Nature* discussed about the irreversible tipping points of global climate, as evidenced, and alarmed by the melting of glaciers, degradation of the Amazon rainforest, slowing down of the Atlantic circulation, massive death of Australian coral reefs, release of greenhouse gases from the Siberian permafrost, and so on.

How to extend and progress the civilization on the planet Earth sustainably with limited space and energy? We may learn from history that the key to enhance productivity and to solve grand challenges is fundamental and technological innovations. In the First Industrial Revolution, chemical energy in coal was transformed into heat and mechanical power by inventing steam engines. In the Second Industrial Revolution, electricity was generated and utilized by inventing generators and motors, where electrons became the energy carrier. In the Third Industrial Revolution or the Information Revolution, information was processed by switching weak electrical signals, stored in electron spins, and transmitted by light. We can see that the industrial revolutions were realized by understanding and controlling the matter, energy, and information in the deeper and deeper dimensions. As Richard Feynman stated in his famous 1959 speech "There's plenty of room at the bottom: an invitation to enter a new field of physics," the next industrial revolution will be started at the small scale and low-dimensional materials.

Indeed, low-dimensional materials have unique structures, extraordinary physical and chemical properties, and potentially revolutionary applications as demonstrated by the list of Nobel Prizes in related fields: integer and fractional quantum Hall effect of two-dimensional electron gas (1985, 1998), semiconductor heterojunctions in optoelectronic devices (2000), two-dimensional graphene (2010), invention of LED (2014), topological phases and transitions (2016) in physics, supramolecular chemistry (1987), zero-dimensional fullerene (1996), one-dimensional polyacetylene conducting polymers (2000), molecular machines (2016), development of lithium-ion batteries (2019), discovery and development of quantum dots (2023) in chemistry.

The authors of this book and their research teams have been researching for many years in the field of low-dimensional materials, especially in controlled growth and physical property exploration of carbon nanotubes and graphene, electrochemical energy storage and conversion devices such as lithium-ion batteries and supercapacitors, and solar energy conversion materials. Representative achievements include macroscopic growth of single-walled carbon nanotubes, controlled growth of graphene and novel two-dimensional materials, construction of three-dimensional graphene networks, design of hierarchical porous materials for efficient energy storage, development of red anatase TiO_2 photocatalysts for solar energy conversion, and development of in situ electron microscopy for growth mechanisms, intrinsic properties, and device performance of low-dimensional materials. Through years of dedicated and diverse research, we become convinced that we must go back to the fundamentals to explore the full potential of low-dimensional materials to realize the "zero to one" destructive innovation. Starting from fundamental theories to explore the deeper dimensions and establish the relationship between microscopic structures and functions, we may design, control, and fabricate novel devices based on new principles to solve the grand challenges. In this book, we review the fundamental theory, summarize the progresses, reflect on the problems, and share our thoughts on low-dimensional materials with the research community. The book included seven chapters as follows:

In Chapter 1, theoretical foundations of low-dimensional materials will be introduced, with emphasis on the effects of symmetry, dimension, and confinements to energy forms, distribution, transport, and conversion between energy and information carriers, such as electrons, photons, and phonons.

Chapter 2 is about the unique structures of low-dimensional materials. The influence of dimension, length scale, symmetry, surface, and strain energy will be discussed. Examples include magic numbers for the stability of zero-dimensional clusters, chirality of carbon nanotubes, long-term order in two-dimensional materials, and heterostructures with low-dimensional building blocks.

In Chapter 3, we will introduce bottom-up and top-down strategies to control the structures of low-dimensional materials by growth, assembly, and processing. Growth and assembly will be discussed in a similar manner on different length scales, based on thermodynamics and kinetics of nucleation and growth. And various interactions and forces, including optical and particle beams and mechanical forces, will be introduced for precisely processing and fabricating functional devices.

In Chapter 4, structural characterization methods will be introduced, to establish correlations between structure and properties, in real, reciprocal, momentum, and energy spaces. Emphasis will be placed on advanced in situ transmission electron microscope and scanning probe microscope for observation, manipulation, and measurement, along with spectroscopy for analysis of composition, defect, and interfaces at atomic resolution and on multiple length scales.

Extraordinary properties of low-dimensional materials will be discussed in Chapter 5, including mechanical, electrical, thermal, chemical, optical, and magnetic properties. Particularly, influence of length scale, dimension, shape, interface to the physical and chemical properties, and coupling of different properties in the functional devices will be presented.

Chapter 6 is about applications of low-dimensional materials in information technology, clean energy, and health care. Dependence of device performance on the structures will be discussed, emphasizing on the common fundamental working principles for different devices.

In the last chapter, Chapter 7, we will give a summary of the progresses, challenges, and an outlook for the low-dimensional materials. After reviewing discoveries and inventions during the last several decades, we will examine the achievements in practical applications, existing problems, and challenges. Finally, we will share our thoughts on the atomic limit and possible directions for the future breakthroughs.

We would like to acknowledge the great support and kind help from colleagues in the Shenzhen Institute of Advanced Technology (SIAT) and Institute of Metal Research (IMR) of the Chinese Academy of Sciences (CAS), the Tsinghua–Berkeley Shenzhen Institute (TBSI), and the National Institute for Materials Science (NIMS). We would like to thank many collaborators and peers for valuable discussions and for providing many materials for this book. Low-dimensional material is a vastly expanding and rapidly developing field, it is inevitable to miss out many important works. We hope this book will initiate thoughts and discussions and welcome any comments from the readers.

Hui-Ming Cheng
Dai-Ming Tang
Xiaolong Zou
Lili Zhang

CHAPTER ONE

Theoretical basis of low-dimensional materials

Low-dimensional materials are defined as those with at least one dimension in the nanoscale range. They mainly include zero-dimensional (0D), one-dimensional (1D), and two-dimensional (2D) materials, as well as composite structures, assemblies, and functional devices constructed by low-dimensional units as building blocks. The science of low-dimensional materials is based on quantum mechanics to explore dimensionality and scale effects in these systems ranging from the atomic to the macroscopic scales. It aims to establish structure—property correlations, design and synthesize low-dimensional materials and structures, fabricate functional devices, and explore their applications in energy, environment, information, health, among many other fields (Fig. 1.1).

The dimensionality and scale effects manifest themselves in the energy distribution and transport processes. The characteristic length determining the occurrence of quantum effects in a system is the de Broglie wavelength, denoted as $\lambda = h/P = h/\sqrt{2mE}$, which is inversely proportional to the particle mass, $\sqrt{m}$. For example, the photon, a massless particle, exhibits quantum wave characteristics at any temperature and scale. For electrons near

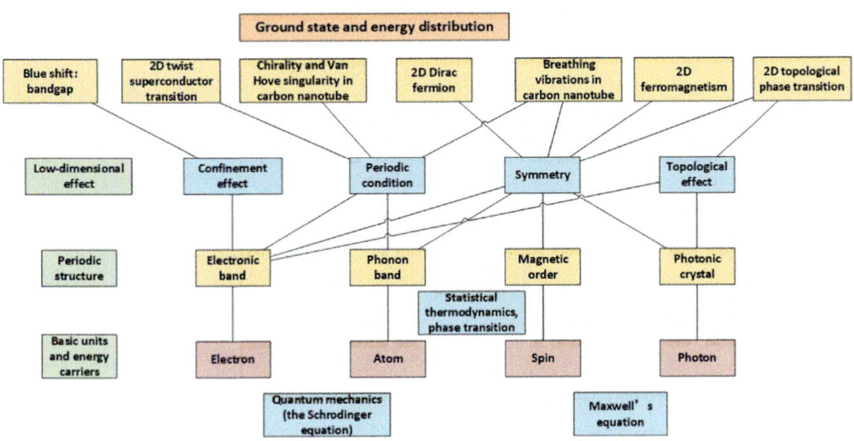

Figure 1.1 Theoretical foundations of low-dimensional materials.

Low-Dimensional Materials.
DOI: https://doi.org/10.1016/B978-0-443-29046-6.00007-9

the Fermi surface in materials, the corresponding de Broglie wavelength is $\lambda_F = h/\sqrt{2m^* E_F}$, which is inversely proportional to the electron's effective mass, $\sqrt{m^*}$. In metals and semiconductors, the de Broglie wavelength typically falls within the nanometer scale. Other characteristic lengths include the mean free path and phase relaxation length governing transport behaviors, corresponding to the lengths over which particles lose their initial momentum and phase information, respectively. These characteristic lengths are also typically in nanometer scale. Therefore low-dimensional materials tend to exhibit quantum effects, such as discrete energy levels, ballistic transport, and coherent transmission.

1.1 Basis of quantum mechanics

Quantum mechanics, known as the "first principles", is the fundamental principle governing the structures, properties, and functionalities of materials. This section will begin with the Schrödinger equation and introduce the solutions of the Schrödinger equation for electrons in various systems, ranging from free electrons to low-dimensional systems. The boundary conditions in each case determine the quantization under the confinement and dimensionality effects.

1.1.1 The Schrödinger equation

The Schrödinger equation describes quantum states and their time evolution:

$$i\hbar \frac{\partial \Psi}{\partial t} = H\Psi$$

where Ψ, i, $\hbar$, $\frac{\partial}{\partial t}$, H are the wave function of the system, the imaginary unit, the reduced Planck constant, the time derivative operator, and the system's Hamiltonian, respectively. The wave function Ψ is a complex function of the system's degrees of freedom (spatial coordinates, time, spin, etc.) and must satisfy the symmetry requirements corresponding to either fermions or bosons. The wave function itself is not a physically measurable quantity and does not have a direct classical correspondence. From the wave function Ψ, the probability distribution can be calculated as the square of the wave function's amplitude $(\Psi^* \cdot \Psi)$, representing the probability of the system being in a certain state at a specific time. Thus the wave function can be understood as a probability wave.

If Ψ is a solution to the equation, any other solution with different phases $\Psi\exp(i\varphi)$ is also a solution, and it does not change the probability distribution or the measurement results of the system's physical quantities. The phase of the wave function is the origin of quantum interference. Furthermore, the Schrödinger equation is a linear equation, so if Ψ_1 and Ψ_2 are both solutions, any linear combination of them is also a solution. This brings about wave-like effects in quantum mechanics, such as quantum superposition and interference.

The Hamiltonian, H, describes the interactions within the system and its coupling with the environment. Typically, the Hamiltonian includes the kinetic energy (T) and potential energy (V) of each particle in the system, that is, $H = T + V$. Under nonrelativistic conditions, the Hamiltonian can be generally expressed as $H = -\frac{\hbar^2}{2m}\nabla^2 + V(\vec{r}, t)$. Different quantum systems are distinguished by their specific potentials. Charged particles in an electric field have electric potential energy, while particles in a magnetic field experience the electromagnetic vector potential. Coulomb interaction is important for any multiparticle system, whereas spin—orbit interaction is critical for materials containing heavy atoms. In principle, once the system's Hamiltonian is determined, solving the Schrödinger equation can yield all the information about the system. The practical difficulty lies in the understanding of the system's interactions and determination of the form of the Hamiltonian, as well as the exponential increase in difficulty to solve the equation with an increasing number of particles in the considered system.

The Schrödinger equation can be separated into spatial and temporal components through variable separation if the potential does not depend on time, with the spatial part yielding the time-independent Schrödinger equation:

$$H\Psi(\vec{r}) = E\psi$$

where H represents the Hamiltonian operator, $\Psi(\vec{r})$ is the wave function of the system, and E denotes the eigenenergy. Solving this equation yields a set of wave functions, $\Psi(\vec{r}, t) = \Psi(\vec{r})e^{-iEt/\hbar}$, referred to as stationary states. As the eigenenergy E is constant, the corresponding physical observables in the stationary state are also constant. The determination of energy eigenvalues and their distribution through the solution of the time-independent Schrödinger equation forms the foundation for the understanding the physical properties of materials. For instance, the analysis of molecular energy level distribution enables the study of characteristic features in optical absorption spectra. Likewise, semiconductor band structures provide rich information, from band types and gaps to the topology-related electronic properties.

In addition to the stationary states, there are nonequilibrium processes such as electron scattering in microscopic characterization, electron transport in semiconductor devices, and photo-generated carriers or excitons in photocatalytic semiconductor materials. It is necessary to consider the time evolution of quantum states, which is determined by the time-dependent Schrödinger equation:

$$H\Psi(\vec{r},t) = -\frac{\hbar^2}{2m}\nabla^2\Psi(\vec{r},t) + V(\vec{r},t)\Psi(\vec{r},t) = i\hbar\frac{\partial}{\partial t}\Psi(\vec{r},t)$$

It is important to note that the Schrödinger equation is not Lorentz-invariant and cannot describe the relativistic quantum mechanics. Under relativistic conditions for spin-1/2 particles, the well-known Dirac equation is applicable, with its corresponding Hamiltonian being $H = c\vec{\alpha}\cdot\vec{p} + mc^2\beta$. In this equation, the coefficient matrices are denoted by $\alpha_i = \begin{pmatrix} 0 & \sigma_i \\ \sigma_i & 0 \end{pmatrix}$, $\beta = \begin{pmatrix} I & 0 \\ 0 & I \end{pmatrix}$, where σ_i is the Pauli matrix, and I is the 2×2 identity matrix. Generally, relativistic effects need to be considered only for systems with particle velocities close to the speed of light. However, in 2D graphene, due to its unique crystal symmetry, the equations describing the behaviors of electrons take the form of the zero-mass Dirac equation.

1.1.2 Quantum confinement and dimensionality effects

The physical and chemical properties of materials are mainly determined by their electronic structures. The interactions among electrons give rise to different chemical bonds, determining the chemical properties of materials. These chemical bonds are further combined to form molecules and crystals, influencing the mechanical and thermal properties. In a periodic potential field, electrons form energy bands, which govern electrical properties of materials. The interaction between electrons and the electromagnetic field determines optical properties.

The electronic properties of materials can be obtained by solving the Schrödinger equations with different Hamiltonians and boundary conditions. Model systems include free electrons, electrons in an infinite potential well, electrons in a spherically symmetric potential well, and electrons in a periodic potential field of crystals.

The Hamiltonian for free electrons consists of kinetic energy only. The eigenfunction solution is a plane wave, and the energy—momentum dispersion relation is simply $E = \frac{\hbar^2 k^2}{2m}$. Such a parabolic form applies approximately for nearly free electrons in metals and for electrons near the bottom of the conduction bands and the top of the valence bands in semiconductors.

The electronic structure of a hydrogen atom, under a spherically symmetric Coulomb potential, has an analytical solution $\Phi(\vec{r}) = R_{nl}(r)Y_{lm}(\theta, \phi)$, where r, θ, and ϕ are the radial, polar angle, and azimuthal angle coordinates, respectively. R is a function that decays exponentially with r, Y represents spherical harmonics, while n, l, and m are the principal, angular momentum, and magnetic quantum numbers, respectively. Different electronic orbitals correspond to different quantum number combinations.

The electronic properties of low-dimensional materials can be described by the infinite potential well model. Under such boundary conditions, the wave function at boundaries is zero, leading to the energy quantization: $E = \frac{1}{2m}\left(\frac{nh}{2d}\right)^2$, inversely proportional to the square of the width of the potential well d. The size of low-dimensional materials determines the width of the potential well, which is the fundamental reason for the quantum confinement effects in the band gap related to the featured size, diameter, and thickness of 0D nanocrystals, 1D carbon nanotubes, and 2D materials (Fig. 1.2A).

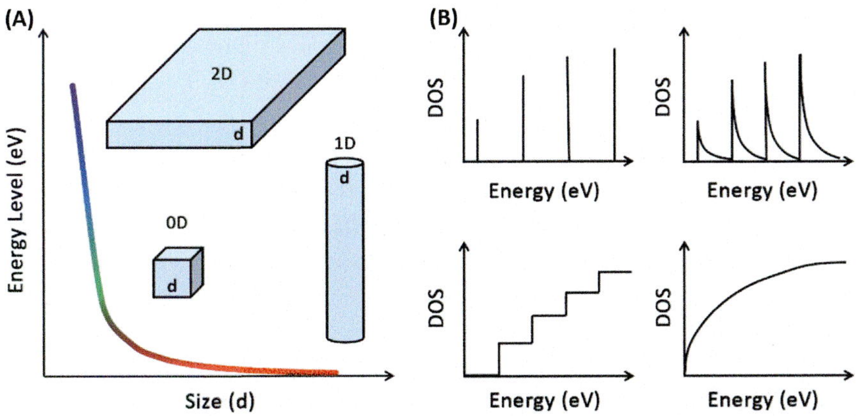

Figure 1.2 Quantum confinement and dimensionality effects in electronic structure: (A) the characteristic energy levels of low-dimensional materials, such as 0D, 1D, and 2D systems, increase inversely with their characteristic size (d^2); (B) schematic representation of the density of states for different dimensional systems, that is, discrete energy levels, van Hove singularities, step-like features, and parabolic characteristics for 0D, 1D, 2D, and 3D systems. *0D*, Zero-dimensional; *1D*, one-dimensional; *2D*, two-dimensional; *3D*, three-dimensional.

The relationship between the electronic structure of low-dimensional materials and dimensionality is reflected in the form of the density of states for nearly free electrons in different dimensions (Fig. 1.2B). For free electron gas, the density of states can be obtained based on its parabolic dispersion, which is summarized below: $g(E) \propto \sqrt{E}$ for three-dimensional (3D) case, $g(E) \propto$ constant showing a step-like feature for 2D case, $g(E) \propto \frac{1}{\sqrt{E}}$ exhibiting a series of van Hove singularities for 1D case, and $g(E) \propto \delta(E)$ representing discrete energy levels for 0D case.

1.2 Energy distribution in low-dimensional systems

The energy carriers in low-dimensional materials mainly include electrons, phonons, and photons. In a crystal's periodic potential field, the particle Hamiltonian exhibits periodicity, and the form of the ground-state wave function is determined by the Bloch theorem:

$$\psi\left(\vec{r} + \vec{R}\right) = e^{i\vec{k}\cdot\vec{R}}\psi(r)$$

where $\vec{R}$ is the lattice vector of the crystal, and $\vec{k}$ is the wave vector. According to the Bloch theorem, the electronic wave function in a periodic potential can be written as the product of a plane wave and a periodic function, $\psi(r) = e^{i\vec{k}\cdot\vec{r}}u(r)$, in which $u(\vec{r})$ satisfies $u(\vec{r}) = u(\vec{r} + \vec{R})$, and $\psi(r)$ is called the Bloch function.

The Bloch theorem is applicable for the motion of electrons in a crystal governed by the Schrödinger equation, forming the foundation of electron energy band theory. It is also applicable for the Maxwell's equations in photonic crystals, where periodic structures in dielectric materials with different refractive indices can create characteristic photonic band structures similar as electronic ones.

1.2.1 Electronic band structure

The steady-state properties of a system are determined by the time-independent Schrödinger equation, $H\psi(r) = E\psi(r)$, where $H = -\frac{\hbar^2}{2m}\nabla^2 + V_{eff}(r)$ is the Hamiltonian operator, and the many-body interactions are incorporated through the effective potential function $V_{eff}(r)$. As claimed by Dirac, the fundamental physical and chemical theories of materials, along with mathematical methods, have been established. However, the challenge lies in applying these theories to specific systems due to the complexity of the resulting equations (on the order of 10^{23}), making exact solutions impractical. Therefore various approximations must be adopted:

1. Adiabatic approximation: Due to the significant difference in mass between electrons and atomic nuclei, the influences of the electron–nucleus interaction on motion of ions are notably smaller than that for electrons. As electrons can instantaneously respond to the motion of the nuclei, the motion of electrons and nuclei can be separated. The system's Hamiltonian is then separated into the sum of the electronic and nuclear parts. The variable separation of the Schrödinger equation leads to the product form of the system's wave function, consisting of separate wave functions for electrons and nuclei, which can be solved independently.
2. Classical nucleus approximation: After the separation of nuclear and electronic motion, the Schrödinger equation for atomic nuclei remains challenging to solve due to its high-dimensional nature. Considering that the nuclear wave function is highly localized compared to the bond length or electronic wave function, the quantum effects of the atomic nuclei are not significant. Therefore the classical nucleus can be used as an approximation, and the equation of motion can be described by solving the Newtonian equations.
3. Electron–electron interaction approximation: Based upon the previous approximations, the system is simplified to the electronic Schrödinger equation under fixed atomic nuclei positions. However, the equation is a multielectron problem, which needs further approximations for electron–electron interactions. These approximations include the independent particle approximation (neglecting electron–electron Coulomb interaction), Hartree approximation (introducing Coulomb interaction classically), Hartree–Fock approximation (considering the exchange interaction among electrons using Slater determinants), and density functional theory (approximating exchange-correlation effects and the ground states determined by the electronic density).

1.2.1.1 Band structures of graphene and carbon nanotube

Graphene is a hexagonal lattice structure composed of carbon atoms with sp^2 hybridization, as shown in Fig. 1.3A, consisting of two sublattices

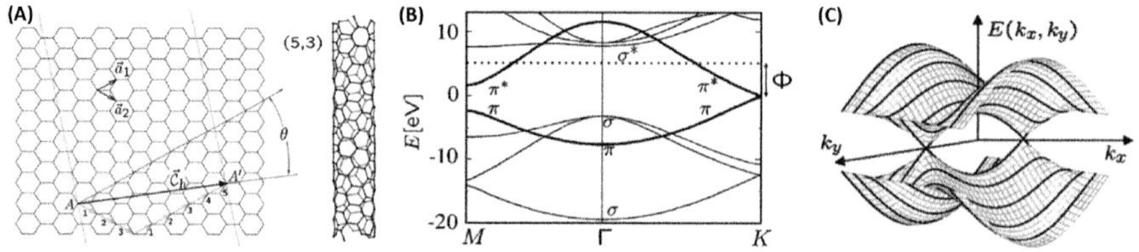

Figure 1.3 The influences of geometric boundary conditions on electronic structure: (A) structural models of graphene and carbon nanotubes, and the latter is formed by rolling up graphene along the C_h direction. (B) The band structure of graphene exhibits a linear dispersion relation at the *K* point, known as the Dirac point. (C) The band structure of carbon nanotubes shows a stripe distribution due to the periodic condition imposed along the circular circumference (Charlier et al., 2007).

(A and B positions). As early as 1947, Wallace (1947) used tight-binding approximation to calculate the band structure of graphene, and its dispersion relation can be expressed as follows:

$$E_{\pm}(k) = \pm t\sqrt{3 + f(k)} - t'f(k)$$

$$f(k) = 2\cos\left(\sqrt{3}k_y a\right) + 4\cos\left(\sqrt{3}k_y a\right)\cos\left(\frac{3}{2}k_y a\right)$$

where t (~2.8 eV) and t' (~0.1 eV) represent nearest and next-nearest neighbor interactions, while "$\pm$" corresponds to the upper (antibonding π*) and lower (bonding π) bands, respectively. Near the Fermi level, the above dispersion is expanded around the $\vec{K}$ and $\vec{K'}$ points in the reciprocal space, with the wave vector expressed as $\vec{k} = \vec{K} + \vec{q}(|\vec{q}| \ll |\vec{K}|)$, which yields

$$E_{\pm}(k) \approx \pm v_F|q|$$

where v_F is the Fermi velocity of graphene. Two linear bands cross each other near the $\vec{K}$ point forming a Dirac cone (Fig. 1.3B). The linear dispersion can be compared to the relativistic energy formula $E^2 = c^2p^2 + m^2c^4$, corresponding to a Dirac relativistic particle with zero mass ($m = 0$).

Carbon nanotubes can be regarded as quasi-1D structures formed by rolling up graphene, introducing periodic boundary conditions in the circumferential direction (Fig. 1.3A), along with restrictions on the values of the reciprocal space vector $\vec{k}$. The band structure of carbon nanotubes emerges from the discrete slicing of the graphene energy bands (Fig. 1.3C). Whether the slicing line passes through the $\vec{K}$ point determines the metallic or semiconducting nature, with the band gap of semiconducting carbon nanotubes being inversely proportional to the diameter.

1.2.2 Phonon of lattice vibration

The adiabatic approximation separates the motion of electrons and atomic nuclei, and the quantization of lattice vibrations for nuclear motion yields the concept of phonons. They serve as carriers of thermal energy, determining the thermal properties by their energy distribution, transport, and interactions such as electron–phonon coupling, and photon–phonon coupling.

The essential physics of atomic vibrations can be described using a 1D harmonic model. The potential energy of vibration is quadratic in terms of the displacement x from the equilibrium position, $V(x) = \frac{k}{2}x^2$, where k is the elastic constant. Its Hamiltonian expression is $H = \frac{1}{2}p^2 + \frac{1}{2}\omega^2x^2$, and the corresponding operator form is $H = -\frac{\hbar^2}{2}\frac{\partial^2}{\partial x^2} + \frac{1}{2}\omega^2x^2$. Accordingly, the stationary Schrödinger equation takes the following form:

$$-\frac{\hbar^2}{2}\frac{\partial^2\psi(x)}{\partial x^2} + \frac{1}{2}\omega^2x^2\psi(x) = E\psi(x)$$

Solving the above equation yields a harmonic oscillator energy spectrum $E_n = h\omega\left(n + \frac{1}{2}\right)$, where n corresponds to the nth eigenstate. Clearly, the system's energy is quantized, with an even energy difference of $\hbar\omega$. Importantly, the ground-state energy is nonzero but $\frac{\hbar\omega}{2}$, signifying zero-point vibrational energy.

The eigenstates of lattice vibrations can be viewed as linear combinations of vibrations of all atoms within the crystal. After introducing normal coordinates, vibration eigenstates can be represented as $Q_\sigma(\vec{q}, t)$, where $\vec{q}$ is the phonon wave vector and σ represents distinct collective vibration modes. The corresponding Hamiltonian is given by

$$H = \frac{1}{2}\sum_{\vec{q},\sigma}[P_\sigma^*(\vec{q}, t)P_\sigma(\vec{q}, t) + \omega_\sigma^2(q)Q_\sigma^*(\vec{q}, t)Q_\sigma(\vec{q}, t)]$$

In this context, P represents the conjugate momentum of Q, and the lattice vibration is described as a sum of classical harmonic oscillator Hamiltonians. The corresponding vibrational energy is likewise expressed as the sum of independent harmonic oscillator energies:

$$E = \frac{1}{2}\sum_{\vec{q},\sigma}\left(n_{\vec{q},\sigma} + \frac{1}{2}\right)\hbar\omega_{\vec{q},\sigma}$$

where the sum of σ runs over a total of $3N$ degrees of freedom, corresponding to the system's vibrational modes. Thus lattice vibrations can be described using quasiparticles known as phonons, which are quanta of lattice vibrations, and each distinct phonon mode possesses an energy of $\hbar\omega_{\vec{q},\sigma}$. At a given temperature, when a certain vibration is in an eigenstate $\left(n_{\vec{q},\sigma} + \frac{1}{2}\right)\hbar\omega_{\vec{q},\sigma}$, there are $n_{\vec{q},\sigma}$ number of phonons excited in that state.

From the lattice vibration Hamiltonian, eigenvalues $\omega_{\vec{q},\sigma}$ as a function of wave vector $\vec{q}$, namely, the phonon dispersion relation, can be obtained. In contrast to the infinite number of electronic eigenstates at each wave vector $\vec{k}$ in the energy band dispersion, the number of lattice vibration eigenstates is finite and is determined by the total number of atoms N in the unit cell. Phonon modes can be categorized into acoustic and optical types, in which three acoustic branches describe the overall cell motion, while the other $3(N-1)$ optical branches describe relative atomic motion within the unit cell.

1.2.2.1 Phonons of graphene and carbon nanotube

The vibrational modes of molecules and lattices are determined by their symmetries described by molecular or crystalline symmetry groups. Low-dimensional materials possess both molecule and crystal characteristics, exhibiting unique vibrational modes distinct from those of 3D bulk crystals.

Here, 2D graphene and quansi-1D carbon nanotubes are taken as examples to introduce the dimensional effects on the phonon band structures. A graphene unit cell consists of two atoms (Fig. 1.3A), yielding six phonon branches in its phonon spectrum (Fig. 1.4A), consisting of three

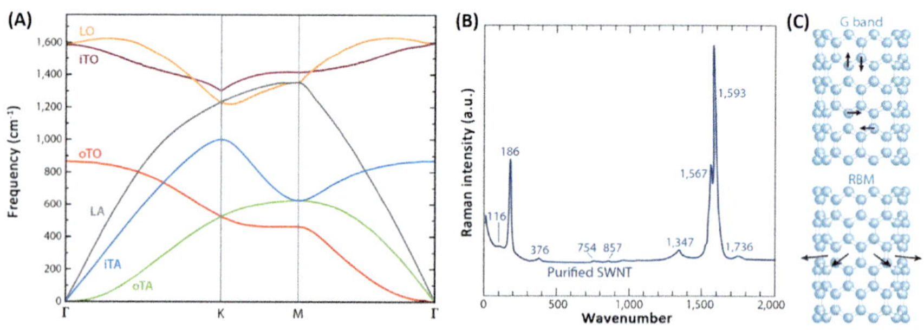

Figure 1.4 Effects of geometric symmetry on phonon structures: (A) phonon spectrum of 2D graphene, including acoustic and optical, in-plane and out-of-plane, and longitudinal and transverse-type branches. (B) Raman scattering spectrum of quasi-1D carbon nanotubes, including G, D, and breathing modes. (C) Layer-stretching vibration mode (G band) and RBM of carbon nanotubes, corresponding to 1593 and 186 cm^{-1} respectively in (B) (Dresselhaus et al., 2010). *1D*, one-dimensional; *2D*, two-dimensional; *RBM*, radial breathing mode.

acoustic (A) and three optical branches (O). These branches are further categorized into in-plane (*i*) and out-of-plane (*o*) modes based on their vibrational directions. Furthermore, they are classified as longitudinal (L) or transverse (T) types according to the relationship between vibrational and propagating directions (Jorio et al., 2011).

Using methods similar to those to calculating electronic band structures, the phonon band structure of carbon nanotubes can be approximately obtained by folding that of graphene. Constrained by periodic conditions, the phonon band structure of carbon nanotubes can be viewed as stripes cut from that of graphene. Due to dimensionality effects, the phonon density of states in carbon nanotubes also features 1D Van Hove singularities (Jorio et al., 2011).

The 1D tubular structure of carbon nanotubes gives rise to unique vibrational modes. One of these modes is the torsional mode, where the vibrational eigenstate involves the rotation of all atoms around the tube axis. Another mode corresponds to a vibrational eigenstate where all atoms simultaneously contract or expand radially, known as the radial breathing mode (Fig. 1.4B and C) (Dresselhaus et al., 2010).

1.2.3 Photonic crystal

In addition to electrons and phonons, another fundamental particle serving as an energy carrier is the photon. In classical physics, electromagnetic waves can be regarded as a superposition of numerous monochromatic plane waves with different wave vectors $\vec{k}$ and frequencies ω. Upon introducing the concept of photons, these monochromatic plane wave modes can be correlated with photons, possessing momentum $\hbar\vec{k}$ and energy $\hbar\omega$. Microscopic particles are classified into fermions and bosons based on their spin being half-integer or integer, following Fermi–Dirac and Bose–Einstein distributions, respectively. Electrons, with a spin of 1/2, are fermions, subject to the constraints of the Pauli exclusion principle. Photons and phonons are bosons, which can occupy the same state. In extreme conditions, all the bosons can stay at the same ground quantum state, which is the Bose–Einstein condensation.

From the perspective of waves, light and electrons show many similar wave-like behaviors, despite their different equations of motion. Electrons follow the Schrödinger equation, while light waves are described by the Maxwell wave equations. The electron wave function is represented as $\psi(\vec{r}, t) = e^{-iEt/h}\psi(\vec{r})$, associated with the Hermitian operator $\frac{-\hbar^2\nabla^2}{2m} + V(r)$.

The electromagnetic wave function is denoted as $H(\vec{r},t) = e^{-i\omega t}H(\vec{r})$, associated with the Hermitian operator $\nabla \times \left(\frac{1}{\varepsilon(\vec{r})} \nabla \times\right)$. Electron waves in periodic potential fields form Bloch waves and band structures. Similarly, light waves in structures with periodic dielectric constants can also develop analogous energy band structures, known as photonic crystals (Yablonovitch, 1987).

1.2.4 Dimensionality effects on density of states

Now we can generally describe the dimensionality effects on energy dispersion. For photons, the dispersion is typically linear, $E = ck$, where c is the speed of light. For phonons, different vibrational modes show distinct dispersion relations. Acoustic phonons exhibit linear dispersion, $E = vk$, where v is the speed of elastic waves. In 2D systems of graphene, flexural phonons display parabolic dispersion, $E = vk^2$. In the case of electrons, the simplest relationship is parabolic, $E = \frac{(\hbar k)^2}{2m}$, applicable for free electron gases and the band-edge electrons in semiconductors. However, in graphene or topological insulators, electrons near the Dirac points follow a linear dispersion relation, $E = v_F k$. Generally, the aforementioned dispersion relations can be expressed as $E = k^n$.

The general expression for the density of states is $g(E) = \frac{dN}{dE} = g\left(\vec{k}\right)\frac{d\vec{k}}{dE}$, where $g\left(\vec{k}\right) = \frac{dN}{d\vec{k}}$ is the density of states in k space, which is related to the dimensionality of $\vec{k}$ space. $\frac{d\vec{k}}{dE}$ can be derived from dispersion relations, demonstrating the features related to the dimensionality:

$$\text{For 3D cases, } \frac{d\,k^{\rightarrow}}{dE} = \frac{4\pi k^2 dk}{dE} = \frac{4\pi}{n}E^{(3-n)/n}$$

$$\text{For 2D cases, } \frac{d\,k^{\rightarrow}}{dE} = \frac{2\pi k dk}{dE} = \frac{2\pi}{n}E^{(2-n)/n}$$

$$\text{For 1D cases, } \frac{d\,k^{\rightarrow}}{dE} = \frac{2dk}{dE} = \frac{2}{n}E^{(1-n)/n}$$

1.3 Topological band space

Energy bands not only provide the relationship between energy and momentum but also constitute a "band space" with intriguing topological structures. Intuitive examples of real-space topological structures include a solid sphere and a torus, having 0 and 1 holes, respectively. During continuous deformations, the number of holes, which is a topological invariant, remains unchanged. Mathematically, the number of holes is defined as the genus g, and it can be expressed using the surface integral of the Gaussian curvature K. Topological structures in materials include dislocations in crystals and topological excitations like skyrmions in spin systems, etc. For reciprocal band space, the topological structures and the properties of wave functions are closely related to the crystal symmetry (Chen et al., 2013).

In 1980 Vonklitzing et al. (1980) discovered the integer quantum Hall effect (IQHE) in 2D electron gases at low temperature and under strong magnetic fields, where the Hall conductivity $\sigma_H = \frac{Ne^2}{h}$ was found to be quantized. In 1982 Thouless et al. connected mathematical topological concepts with electron wave function phases. They pointed out that the origin of the IQHE lies in the existence of gapless and topologically protected chiral edge states at the Fermi level (TKNN theory). The number of these states is a topological invariant, corresponding to the Chern number (Thouless et al., 1982) (similar to the genus as a topological invariant, as shown in Fig. 1.5).

The system's topological properties can be expressed through the characteristics of the phases of electron wave functions. Neglecting accidental degeneracies, when an electron state m is integrated along a closed path in $\vec{k}$ space, its wave function $|u_m\left(\vec{k}\right)$ accumulates an additional phase, called the Berry phase. Through the wave function, the Berry connection $\vec{A}_m = i\langle u_m|\nabla|u_m\rangle$ and Berry curvature $\vec{B}_m = \nabla \times \vec{A}_m$ can be defined. The Chern number is then given by the line integral of $\vec{A}_m$ or the surface integral of $\vec{B}_m$, that is, $n_m = \frac{1}{2\pi}\int d^{2k}\vec{B}_m$. TKNN theory demonstrated that the quantized coefficients of Hall conductivity σ_H have the same expression as the Chern number, and the integer coefficient N corresponds to the summation of n_m obtained from this expression.

Graphene lattice serves as a significant topological model system. In 1988 Haldane (1988) proposed the quantum anomalous Hall effect without an external magnetic field based on the graphene model. In 2010, Yu et al. (2010) suggested that the quantum anomalous Hall effect could potentially be realized in the Bi_2Se_3/Bi_2Te_3 system through magnetic doping with Cr and Fe. In 2013, Chang et al. (2013) successfully observed the anomalous quantum Hall effect in $Cr_{0.15}(Bi_{0.1}Sb_{0.9})_{1.85}Te_3$ thin films.

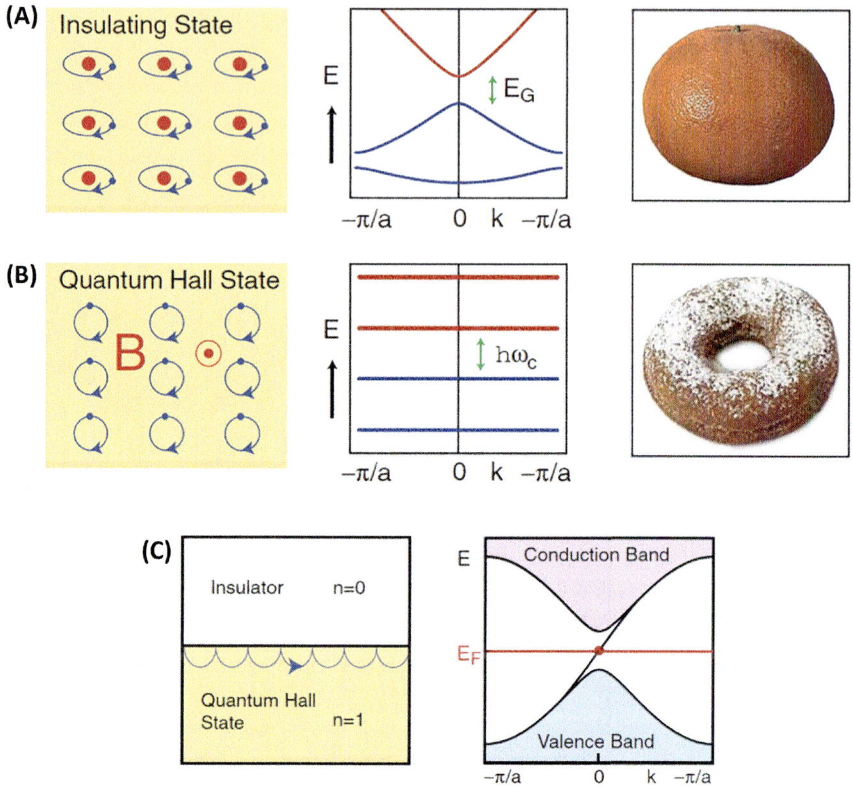

Figure 1.5 Atomic insulators and topological insulators: (A) Schematic diagram of motion of electrons in atomic insulators, band structure, and solid sphere with genus of 0. (B) Schematic of circular movement of electrons in quantum Hall states, band structure, and doughnut with genus of 1. (C) Unidirectional conductive channel at the boundary of topological phases (Hasan & Kane, 2010).

Inspired by the electronic band topology, scientists have also been exploring topological order phenomena in the realms of photonics and phononics. Topological photonic crystals were proposed by Haldane and Raghu (2008), where they identified that the key to realize chiral edge states within photonic crystals is the presence of nonreciprocal media breaking time-reversal symmetry. Wang et al. (2009) experimentally observed one-way propagation of topologically protected microwave electromagnetic waves, free from disorder scattering.

In 2009 Prodan and Prodan (2009) took microtubules as an example and showed the existence of topologically nontrivial modes in mechanical models. Later, Zhang et al. (2010) introduced the theoretical understanding of topological phonon Hall effect and its phase transition, which ignites the study of topological phonon transport. Further, Wang et al. (2015) proposed a new type of topological phononic crystals with the gyroscopic inertial effects to break time-reversal symmetry, which can realize the phononic quantum (anomalous) Hall effect, leading to the one-way elastic wave.

1.4 Thermodynamics of low-dimensional systems

Basic physical laws and equations, such as the Schrödinger equation, are time-reversible, while actual processes are practically irreversible. Moreover, macroscopic systems are composed of a large number of microscopic particles. It is almost impossible and not necessary to provide an accurate description of the states for each individual particle. The macroscopic properties of a system are statistical averages of the microscopic quantities, which forms the basis of statistical mechanics and thermodynamics.

Statistical mechanics rests on two fundamental assumptions: (1) If the state has the same total energy and number of particles, the probabilities of all possible microscopic states are equal, known as the principle of equal a priori probabilities. (2) Starting from any initial state, the system will traverse all possible microscopic states after a sufficiently long time, referred to as ergodicity. The probability of being at a macroscopic state is determined by the number of possible combinations of the microscopic states with certain distributions. For systems with a sufficiently large number of constituents, the probability of being in a particular macroscopic state becomes overwhelmingly dominant, ensuring the stability of macroscopic properties. Meanwhile, thermodynamic quantities fluctuate around their mean values, with the scale of fluctuation depending on the particle numbers of the system. In the context of low-dimensional systems, fluctuations can play a significant role, possibly leading to the presence of multiple locally metastable states.

The thermodynamic free energy consists of the contributions from both system energy and entropy. When different configurations have comparable energies, configurational entropy becomes critical in low-dimensional systems to determine the structure. Recent research has indicated that the configurational entropy at the nanoscale boundaries of single-walled carbon nanotubes can influence the stability of nanotubes of different chiralities, thus explaining the observed chirality distribution at different temperatures (Magnin et al., 2018).

An important theory of low-dimensional thermodynamics is the Mermin–Wagner theorem (Hohenberg, 1967; Mermin & Wagner, 1966), which predicts that a system with a dimension (d) smaller than 3 cannot have long-range order due to the dimension related fluctuations. According to this theory, 2D crystalline lattices or 2D long-range magnetic ordering cannot exist. However, recently stable 2D crystals with ferromagnetism have been discovered. These surprising discoveries will be further discussed in Chapters 2 and 5.

1.5 Energy transport in low-dimensional systems

Researches on equilibrium states can provide information about energy levels and energy distribution. To control and utilize different states of particles in functional devices, along with the information and energy they carry, it is necessary to consider nonequilibrium transport processes. This section will introduce energy transport in low-dimensional systems from two perspectives, that is, particle and wave transport, emphasizing the effects of ballistic transport and coherent propagation in low-dimensional systems (Fig. 1.6).

1.5.1 The Boltzmann equation

The Boltzmann equation is fundamental for calculating the transport properties of particles experiencing various interactions:

$$\frac{\partial f(\vec{k},\vec{r},t)}{\partial t} = -\frac{d\vec{k}}{dt}\cdot\nabla_k f\left(\vec{k},\vec{r},t\right) - \frac{d\vec{r}}{dt}\cdot\nabla_r f\left(\vec{k},\vec{r},t\right) + \frac{\partial f(\vec{k},\vec{r},t)}{\partial t}\Big|_{coll}$$

In this equation, the three terms on the right-hand side correspond to changes in the distribution function f in $\vec{k}$ space, $\vec{r}$ real space, and those caused by collision. Taking electrons as an example, on one hand, they undergo acceleration motion under the influence of the electric potential gradient. On the other hand, they restore to the equilibrium distribution through collision. Solving the distribution function $f(\vec{k})$ allows one to obtain the number of electrons per unit volume $d\vec{k}$ as $2f(\vec{k})d\vec{k}/(2\pi)^3$

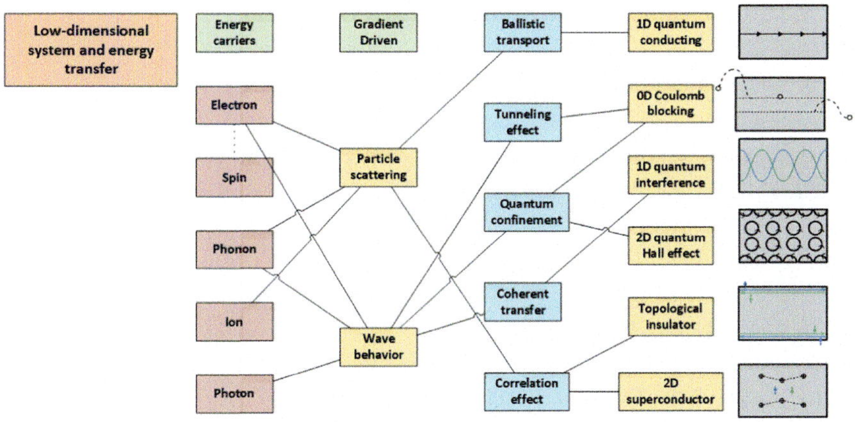

Figure 1.6 The energy transport mechanisms in low-dimensional systems involve energy carriers such as electrons, phonons, and photons. Below characteristic length scales, they show tunneling, quantum confinement, ballistic transport, coherent propagation, and correlation effects. These effects are manifested as novel transport properties, such as quantized conductivity, Coulomb blockade, coherent interference, (anomalous) quantum Hall effect, and superconductivity.

(considering the electron's spin degeneracy of 2), and the corresponding current density as $-2ef(\vec{k})\vec{v}(\vec{k})d\vec{k}/(2\pi)^3$. Based on these, various properties like electrical conductivity can be calculated.

One key in studying various transport behaviors by using the Boltzmann equation is the treatment of the collision term. The commonly used relaxation time approximation assumes that the deviation of the distribution function is caused by collision decays exponentially within the relaxation time τ:

$$\frac{\partial f(\vec{k},\vec{r},t)}{\partial t}|_{\text{coll}} = \frac{f - f_0}{\tau}$$

Under the steady-state conditions, neglecting the influences of spatial variation leads to the corresponding Boltzmann equation:

$$-\frac{d\vec{k}}{dt}\cdot\nabla_k f\left(\vec{k},t\right) = \frac{f - f_0}{\tau}$$

By performing a power series expansion of the distribution function, the general expression of Ohm's law can be obtained:

$$\vec{J} = -2e^2\int \tau\,\vec{v}\left(\vec{k}\right)\left[\vec{v}\left(\vec{k}\right)\cdot\vec{E}\right]\frac{\partial f_0}{\partial E}d\,\vec{k}\,/(2\pi)^3$$

And the electrical conductivity is expressed as

$$\sigma_{\alpha\beta} = -2e^2 \int \tau\left(\vec{k}\right) v_\alpha\left(\vec{k}\right) v_\beta\left(\vec{k}\right) \frac{\partial f_0}{\partial E} d\vec{k} / (2\pi)^3$$

For isotropic metal systems, the electrical conductivity is simplified to that for the classical electron gas:

$$\sigma = \frac{ne^2\tau}{m^*}$$

1.5.2 Scattering and diffusion of particles

The Boltzmann equation and theories could describe the transport processes of many different particles. As a result, the transport equations for these different processes show similar mathematical forms. Diffusion and drift are the common processes for the transport of scattering particles.

Particle diffusion: It is driven by concentration gradient and follows Fick's first law: $\vec{J} = -D\nabla n$, where $\vec{J}$, D, and n are the particle flux, diffusion coefficient, and concentration, respectively.

Thermal diffusion: It is driven by temperature gradient, following Fourier's law: $\vec{q}_t = -\kappa\nabla T$, where $\vec{q}_t$, κ, and T are the heat flux density, thermal conductivity, and temperature, respectively.

Electronic drift: Under the combined influences of electric-field acceleration and random scattering, electrons undergo drift, which can be described by Ohm's law: $\vec{J}_e = -\sigma\nabla\varphi = \sigma\vec{E}$, where $\vec{J}_e$, σ, φ, and $\vec{E}$ are the current density, electrical conductivity, electric potential, and electric field , respectively.

Ionic transport: While ions are driven by electric field, they also experience viscous drag. The electric field force is given by $\vec{F}_E = ze\vec{E}$, where z, e, and $\vec{E}$ are the charge number of an individual ion, elementary charge, and electric field, respectively. The viscous drag force F_f can be obtained from the Stokes relation, that is, $F_f = fv = 6\pi\eta Rv$, where η, R, and v are the dynamic viscosity coefficient, ion radius, and velocity of ions relative to the liquid. From the mechanical equilibrium conditions, the ion mobility can be derived by $\mu_{\text{ion}} = \frac{ze}{f} = \frac{ze}{6\pi\eta R}$. Apparently, ions with larger size have lower mobility.

Einstein relation: Both diffusion and drift can be viewed as scattering processes of particles. As a result, parameters like the diffusion coefficient D and mobility μ are interrelated and satisfy the Einstein relation as

$D = \mu k_B T$. For diffusion of charged particles, the equation becomes $D = \frac{\mu_q k_B T}{q}$, where q represents the charge carried by the particles.

1.5.3 Ballistic transport of particles

When the system size is smaller than the mean free path, particles will undergo ballistic transport. The mean free path l represents the average distance that a particle travels between two scattering events. It is expressed as $l = v\tau$, where v is the particle's velocity. For ballistic transport, the classical scattering theory and equations in the last section are no longer applicable, and quantum effects such as quantized conductivity emerge.

The main participants in electronic transport are electrons near the Fermi surface. Hence the mean free path is related to the Fermi velocity v_F. Due to the unique linear dispersion relation of carbon nanotubes and graphene, they have high Fermi velocities (10^6 m/s), and thus the mean free path can reach the micrometer level. In 2001 (Kong et al., 2001) observed quantized conductivity $2G_0 = 4\frac{e^2}{h}$ in metallic carbon nanotubes, reflecting ballistic transport properties of electrons. Meanwhile, they found the oscillation of conductivity with gate voltage and Fermi energy, indicating wave-like coherent transport characteristics.

In contrast to electronic transport, where the unit quantum conductivity is constant, phonon transport corresponding to a unit quantum thermal conductivity is proportional to temperature. In electrical measurements, it is possible to separately modulate the electronic chemical potential and temperature, while in thermal measurements, the temperature is normally the single variable. Moreover, the mean free path of phonon is usually much shorter than that of electron, making it more challenging for the observation of ballistic transport of phonons. Schwab et al. (2000) prepared silicon nitride nanostructures to act as 1D waveguides with good coupling between the lowest energy mode and the thermal bath and observed quantized thermal conductivity $G_{th} = \frac{\pi^2 k_B^2 T}{3h}$ at a temperature below 600mK.

1.5.4 Coherence propagation of waves

Previous discussions are about the scattering of particles. According to the wave-particle duality of quantum mechanics, energy carriers also have wave-like characteristics, showing propagation properties of waves within the coherent length.

Wave propagation is a universal phenomenon. Different types of waves and their couplings contain rich physical information. Mechanical waves and phonons are essentially mechanical vibrations in different frequency ranges. Light with different wavelengths can be classified as radio waves, microwaves, visible light, X-rays, and γ-rays. Sound waves have lower frequencies and energies compared to electromagnetic waves, but high-frequency sound waves can overlap with the photon spectrum in energy. In the GHz frequency range, the resonance between mechanical vibrations and microwaves forms the basis for quantum optomechanical devices. In the THz frequency range, the interaction between phonons and photons is fundamental for Raman and Brillouin scattering spectroscopies.

The coherent transport of particles is closely related to their wavelengths. Photons have long wavelengths, for example, visible light has wavelengths of a few hundred nanometers. It is easy for them to show the coherent behaviors. On the other hand, electrons or lattice phonons in materials have much shorter wavelengths, typically on the order of 1–10 nm. Accordingly, their wave-like effects can only be observed in small-scale low-dimensional structures and at low temperature.

Through single-electron double-slit experiment, Tonomura et al. (1989) clearly demonstrated the probabilistic wave nature of electrons and their coherent transmission in vacuum. Transmission electron microscopy is a typical example to utilize the coherent transport of electrons in materials, in which case different phases of electron waves after Bragg diffraction of the lattices interfere to form high-resolution images. In clean low-dimensional systems, the coherent transport of electrons can clearly be demonstrated. Cheianov et al. (2007) proposed to use gated graphene p–n junction to control electron interference and focusing phenomena. A Veselago prism was formed at the n–p–n junction with the Fermi momentum playing the role of refractive index in optics, being positive in the conduction band and negative in the valence band.

Compared to electrons, observing coherent transport of phonons with shorter wavelengths presents more significant challenges and requires better control over surface roughness and interface flatness of nanostructures. (Luckyanova et al., 2012) measured the thermal conductivity of GaAs/AlAs superlattices, showing that the thermal conductivity increases linearly with the increase of the superlattice period in the range of 30–150K, reflecting the coherent characteristics of thermal transport. (Anufriev et al., 2017) fabricated arrays of microholes with different sizes on silicon films, arranged them radially with increasing radii, and observed focusing and divergence effects of phonons.

1.6 Energy conversion in low-dimensional systems

The application of low-dimensional functional materials and devices involves the conversion of information, energy, and matter. These three types of conversion are often intertwined. While matter serves as the carrier of information and energy, energy carriers such as electrons and photons also serve as information carriers. For energy conversion devices, solar cells absorb light and convert it into electrical energy. Light-emitting diodes, on the other hand, emit light by injecting and recombining electrons and holes. Resistive heating wires convert electrical energy into heat. Thermoelectric devices, conversely, convert temperature difference into electrical energy. Photosynthesis converts light into chemical energy, while in fuel cells, chemical energy is converted into electrical energy (Fig. 1.7, left).

During the energy conversion processes, there is energy input and output between the systems and the environment. Not only the transport of individual energy carriers, but also the coupling between different carriers needs to be considered. Furthermore, energy carriers may be excited or absorbed, and the total number of carriers is no longer constant. This section takes the example of photoelectric effect to introduce the energy conversion in optical processes and the design principles of solar cells, based on the nature of light and the interaction between light and matter (Fig. 1.7, right).

In classical theory, Maxwell's equations are the fundamental theory of light and electromagnetism. When applied to vacuum conditions, they yield the wave equations, $\nabla^2\vec{E} - \mu_0\varepsilon_0\ddot{\vec{E}} = 0$. The plane wave solution is $\vec{E}(\vec{r}, t) = \vec{E}_0\exp[i(\vec{k}\cdot\vec{r} - \omega t)]$, where the wave vector and angular frequency are related by $\frac{c^2k^2}{\omega^2} = 1$. In quantum theory, by introducing the vector potential $\vec{A}$ and scalar potential ϕ, the Hamiltonian for electromagnetic waves can be written in a form similar to that of a harmonic oscillator as $H = \sum_{\vec{k},s}\hbar\omega_{\vec{k},s}a^{\dagger}_{\vec{k},s}a_{\vec{k},s}$, where the subscripts $\vec{k}$ and s represent the wave vector and polarization direction of the electromagnetic waves, respectively. $\omega_{\vec{k},s}$ is the frequency of the corresponding mode, while $a^{\dagger}_{\vec{k},s}$

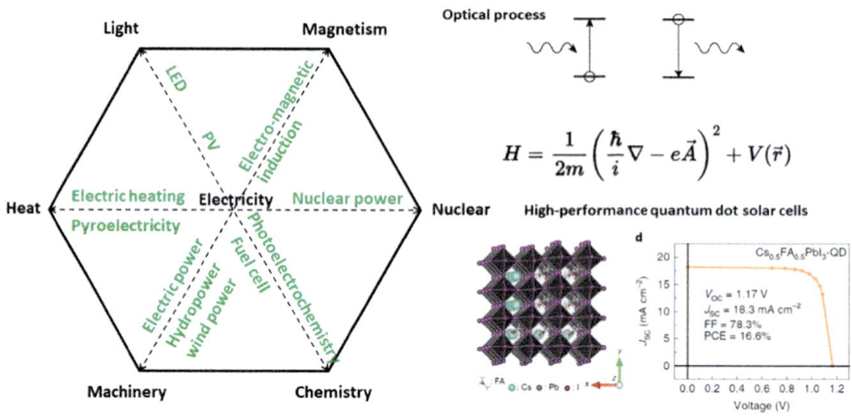

Figure 1.7 (Left) Energy conversion in low-dimensional materials and functional devices. (Right) Optical processes, Hamiltonian of electrons in electromagnetic fields, and the design of high-performance quantum dot solar cells (Hao et al., 2020).

and $a_{\vec{k},s}$ are the creation and annihilation operators. The energy associated with this Hamiltonian is $E_{\vec{k},s} = \left(n_{\vec{k},s} + \frac{1}{2}\right)\hbar\omega_{\vec{k},s}$, where $n_{\vec{k},s}$ is a nonnegative integer, and the corresponding quasiparticles are photons. Due to its nature as a quantized harmonic oscillator, the Hamiltonian and energy of photons are analogous to those of phonons discussed in Section 1.2.2.

Optical processes can be considered from both macroscopic and microscopic perspectives. On the macroscopic side, reflection, refraction, and transmission processes can be described by complex dielectric function and complex refractive index. The corresponding relation can be obtained by considering the boundary continuity conditions at different interfaces. On the microscopic side, by describing the interaction between light and matter, fundamental processes such as absorption, emission, and scattering can be understood.

Under the conditions of no free charge, no magnetization, and negligible current density in the medium, Maxwell's equation is expressed as $\nabla^2\vec{E} - \mu_0\varepsilon_0\varepsilon(\omega)\vec{E} = 0$. The solutions are the same as in vacuum, but in this case, the relationship between wave vector (k) and angular frequency (ω) is given by $\frac{c^2k^2}{\omega^2} = \varepsilon(\omega)$, where ε is the complex dielectric function of the material. The complex refractive index of the material is defined as $\tilde{n}^2 = \varepsilon(\omega)$, $\tilde{n} = n + i\kappa$. Assuming the electromagnetic wave propagates along the z-direction in the medium, the electric field wave function can be represented as $E = E_0e^{i(\omega t - \tilde{n}kz)} = E_0e^{-\kappa kz}E_0e^{i(\omega t - nkz)}$. It can be seen that the imaginary part κ in the complex refractive index represents the optical absorption rate, while the real part n represents the ordinary refractive index.

The excitation process of light and matter can be presented by using an energy level model, with E_g and E_{ex} representing the energies of the ground state and the excited state, respectively. When the energy of the incident photon $\hbar\omega$ is greater than or equal to the energy difference between the excited state and the ground state, the electron can absorb the energy of photon and undergo transition from the ground state to the excited state. The electron in the excited state can then spontaneously return to the ground state, through two possible processes, that is, radiative and nonradiative recombination. In the presence of an incident photon, an electron in the excited state can also return to the ground state and emit a photon with the same energy as the incident photon. This process is known as stimulated emission, which is the basis of laser. If the transition process is accompanied by the generation or annihilation of phonons, it corresponds to Raman scattering or Brillouin scattering, depending on the type of phonons involved.

The quantum treatment of optical processes involves the total Hamiltonian of electrons, electromagnetic fields, and their interactions, which can be expressed as $H = H_{el} + H_{light} + H_{inter}$ with the interaction term treated as a perturbation. By introducing the vector potential $(\vec{A})$, the single-particle Hamiltonian can be written as $H = \frac{1}{2m}\left(\frac{\hbar}{i}\nabla - e\vec{A}\right)^2 + V(\vec{r})$, where $V(\vec{r})$ is the electrostatic potential. In the Coulomb gauge $(\nabla \cdot \vec{A} = 0)$, the first-order approximation for the vector potential leads to

$$H = \frac{\hbar^2}{2m}\nabla^2 + V(\vec{r}) - \frac{e}{m}\vec{A}\cdot\frac{\hbar}{i}\nabla = H_{el} - \frac{e}{m}\vec{A}\cdot\frac{\hbar}{i}\nabla = H_{el} + H^{(1)}$$

with the first-order perturbation Hamiltonian denoted as $H^{(1)}$. The transition rate from the initial state i to the final state f can be calculated as $W_{if} = \frac{2\pi}{\hbar}\left|H_{if}^{(1)}\right|^2 D(E)$, where $D(E)$ is the density of final states satisfying the momentum conservation condition. The transition matrix element $H_{if}^{(1)}$ can be expressed as $H_{if}^{(1)} = \int \psi_f^*(r^{\rightarrow})H^{(1)}\psi_i(r^{\rightarrow})d\tau = \langle\psi_f|H^{(1)}|\psi_i\rangle$, and its squared modulus gives the transition probability.

The detailed derivation of the calculation process is not elaborated here. The calculation results yield the following three important conclusions:

1. The expression for the transition rate of the absorption process is $W_{if} \propto |A_0\ \psi_f|H^{(1)}|\psi_i\ |^2 D(E_i + \hbar\omega) = A_0^2|H_{if}^{(1)}|^2 D(E_i + \hbar\omega)$, which is proportional to the light intensity A_0^2.

2. Using the Hermitian property of $H^{(1)}$ and the orthogonality between the eigenwavefunctions, one obtains $\psi_f|H^{(1)}|\psi_i = \psi_i|H^{(1)}|\psi_f$. The transition probabilities for stimulated emission and absorption are equal. The difference lies only in the number of states between the initial state i and the final state f; thus the prerequisite for the stimulated generation of laser is the inversion of the number of excited and ground states.
3. Substituting $H^{(1)}$ and the wave function into the equation and integrating the exponential part, one gets $E_{\text{ex}} - E_g - \hbar\omega = 0$ and $\hbar k_{\text{ex}}^{\rightarrow} - \hbar k_g^{\rightarrow} - \hbar\vec{k} = 0$, which reflect the conservation of energy and momentum in optical processes, respectively.

For the design of optoelectronic devices, energy and momentum conservation is a fundamental starting point. In solar cells, only photons with energy larger than the band gap could be absorbed. The smaller the band gap of a semiconductor, the more photons it can absorb. On the other hand, a smaller band gap results in lower energy for the generation of electron–hole pairs from absorbed photons. To effectively utilize the visible portion of the solar spectrum, the optimal band gap for a single-junction photovoltaic cell is theoretically estimated to be around 1.3 eV. Momentum conservation restricts the type of semiconductors, because direct band gap semiconductors have an absorption rate much higher than indirect band gap semiconductors.

Based on these fundamental principles, Hao et al. employed a series of strategies to improve the optoelectronic performance of perovskite $Cs_{1-x}FA_xPbI_3$ nanocrystals. An optimal band gap is achieved by tuning the composition for improved utilization of the solar spectrum, and the defect concentration is reduced to control the loss during transport. Meanwhile, the structural stability is enhanced by limiting phase separation. As a result, a certified efficiency of 16.6% for quantum dot solar cells is achieved (Fig. 1.7, right) (Hao et al., 2020).

1.7 Summary

This chapter introduces the theoretical foundation of low-dimensional materials. Starting from the Schrödinger equation, we briefly discuss the energy forms, dispersion relations, transport processes, and conversion mechanisms involving electrons, phonons, and photons as carriers.

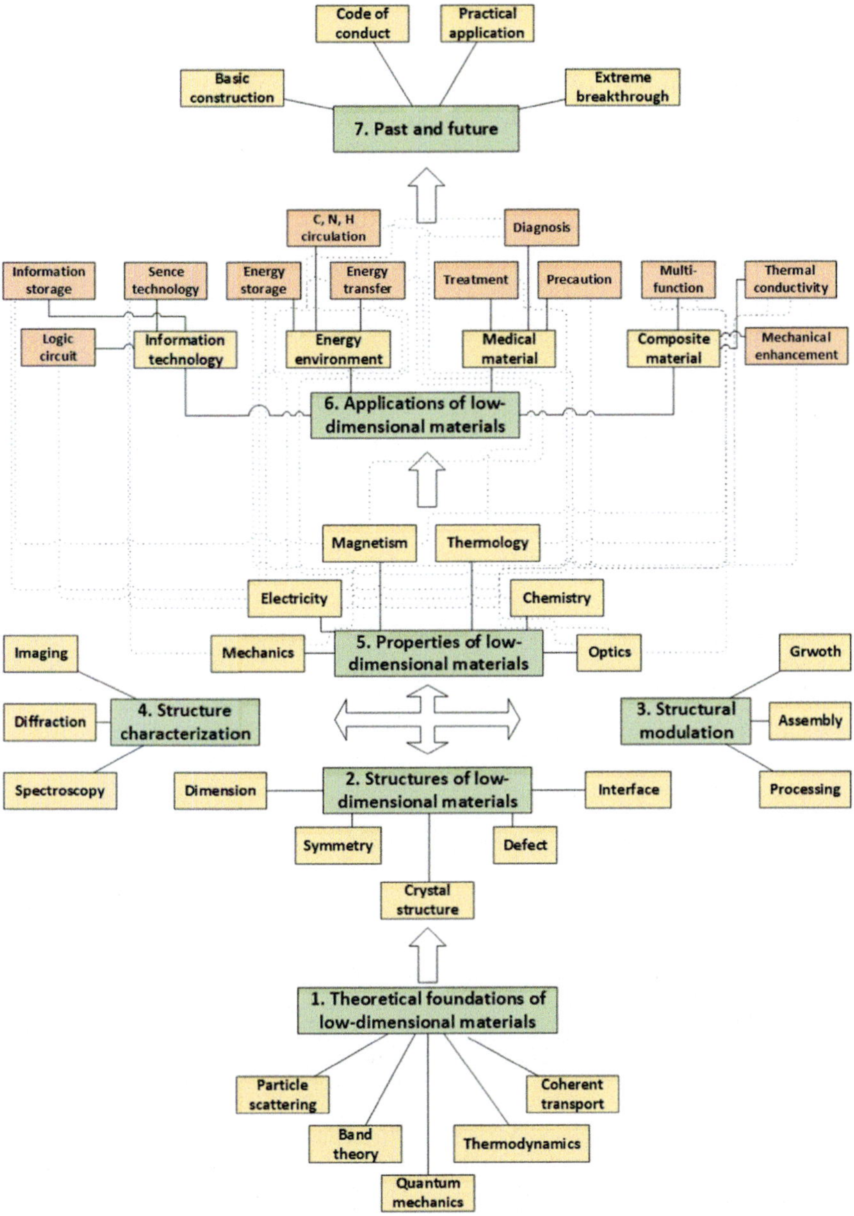

Figure 1.8 The structure of this book.

The quantum effects generated by quantum confinement in low-dimensional systems are demonstrated by discussing energy quantization and ballistic transport at small scales, as well as coherent propagation. The commonalities in energy distribution and transport processes among different types of particles, including electrons, phonons, and photons are emphasized. The quantum theory of photon–electron interaction naturally leads to the principles of energy and momentum conservation in electronic transitions, as well as the fundamental principles in the design of optoelectronic devices.

In the subsequent chapters, we will apply these foundational theories to elucidate how these principles determine the properties of low-dimensional materials, to guide their controlled fabrication, and to design novel principle-based high-performance functional devices (Fig. 1.8).

References

Anufriev, R., Ramiere, A., Maire, J., & Nomura, M. (2017). Heat guiding and focusing using ballistic phonon transport in phononic nanostructures. *Nature Communications*, *8*, 15505.

Chang, C. Z., Zhang, J. S., Feng, X., Shen, J., Zhang, Z. C., Guo, M. H., Li, K., Ou, Y. B., Wei, P., Wang, L. L., Ji, Z. Q., Feng, Y., Ji, S. H., Chen, X., Jia, J. F., Dai, X., Fang, Z., Zhang, S. C., He, K., & Xue, Q. K. (2013). Experimental observation of the quantum anomalous Hall effect in a magnetic topological insulator. *Science (New York, N.Y.)*, *340*(6129), 167–170.

Charlier, J. C., Blase, X., & Roche, S. (2007). Electronic and transport properties of nanotubes. *Reviews of Modern Physics*, *79*(2), 677–732.

Cheianov, V. V., Fal'ko, V., & Altshuler, B. L. (2007). The focusing of electron flow and a Veselago lens in graphene junctions. *Science (New York, N.Y.)*, *315*(5816), 1252–1255.

Chen, X., Gu, Z. C., Liu, Z. X., & Wen, X. G. (2013). Symmetry protected topological orders and the group cohomology of their symmetry group. *Physical Review B*, *87*(15), 155114.

Dresselhaus, M. S., Jorio, A., & Saito, R. (2010). Characterizing graphene, graphite, and carbon nanotubes by Raman spectroscopy. *Annual Review of Condensed Matter Physics*, *1*, 89–108.

Haldane, F. D. M. (1988). Model for a quantum Hall effect without Landau levels: Condensed-matter realization of the "parity anomaly". *Physical Review Letters*, *61*(18), 2015–2018.

Haldane, F. D. M., & Raghu, S. (2008). Possible realization of directional optical waveguides in photonic crystals with broken time-reversal symmetry. *Physical Review Letters*, *100*(1), 013904.

Hao, M. M., Bai, Y., Zeiske, S., Ren, L., Liu, J. X., Yuan, Y. B., Zarrabi, N., Cheng, N. Y., Ghasemi, M., Chen, P., Lyu, M. Q., He, D. X., Yun, J. H., Du, Y., Wang, Y., Ding, S. S., Armin, A., Meredith, P., Liu, G., & Wang, L. Z. (2020). Ligand-assisted cation-exchange engineering for high-efficiency colloidal $Cs_{1-X}FA_xPbI_3$ quantum dot solar cells with reduced phase segregation. *Nature Energy*, *5*(1), 79–88.

Hasan, M. Z., & Kane, C. L. (2010). Colloquium: Topological insulators. *Reviews of Modern Physics*, *82*(4), 3045–3067.

Hohenberg, P. C. (1967). Existence of long-range order in one and two dimensions. *Physical Review, 158*(2), 383.

Jorio, A., Saito, R., Dresselhaus, G., & Dresselhaus, M. S. (2011). *Raman spectroscopy: From graphite to* sp^2 *nanocarbons. Raman spectroscopy in graphene related systems* (pp. 73–101). Wiley-VCH.

Kong, J., Yenilmez, E., Tombler, T. W., Kim, W., Dai, H. J., Laughlin, R. B., Liu, L., Jayanthi, C. S., & Wu, S. Y. (2001). Quantum interference and ballistic transmission in nanotube electron waveguides. *Physical Review Letters, 87*(10), 106801.

Luckyanova, M. N., Garg, J., Esfarjani, K., Jandl, A., Bulsara, M. T., Schmidt, A. J., Minnich, A. J., Chen, S., Dresselhaus, M. S., Ren, Z. F., Fitzgerald, E. A., & Chen, G. (2012). Coherent phonon heat conduction in superlattices. *Science (New York, N.Y.), 338*(6109), 936–939.

Magnin, Y., Amara, H., Ducastelle, F., Loiseau, A., & Bichara, C. (2018). Entropy-driven stability of chiral single-walled carbon nanotubes. *Science (New York, N.Y.), 362*(6411), 212–215.

Mermin, N. D., & Wagner, H. (1966). Absence of ferromagnetism or antiferromagnetism in one- or two-dimensional isotropic Heisenberg models. *Physical Review Letters, 17*(22), 1133.

Prodan, E., & Prodan, C. (2009). Topological phonon modes and their role in dynamic instability of microtubules. *Physical Review Letters, 103*(24), 248101.

Schwab, K., Henriksen, E. A., Worlock, J. M., & Roukes, M. L. (2000). Measurement of the quantum of thermal conductance. *Nature, 404*(6781), 974–977.

Thouless, D. J., Kohmoto, M., Nightingale, M. P., & Dennijs, M. (1982). Quantized Hall conductance in a two-dimensional periodic potential. *Physical Review Letters, 49*(6), 405–408.

Tonomura, A., Endo, J., Matsuda, T., Kawasaki, T., & Ezawa, H. (1989). Demonstration of single-electron buildup of an interference pattern. *American Journal of Physics, 57*(2), 117–120.

Vonklitzing, K., Dorda, G., & Pepper, M. (1980). New method for high-accuracy determination of the fine-structure constant based on quantized Hall resistance. *Physical Review Letters, 45*(6), 494–497.

Wallace, P. R. (1947). The band theory of graphite. *Physical Review, 71*(9), 622–634.

Wang, P., Lu, L., & Bertoldi, K. (2015). Topological phononic crystals with one-way elastic edge waves. *Physical Review Letters, 115*(10), 104302.

Wang, Z., Chong, Y. D., Joannopoulos, J. D., & Soljacic, M. (2009). Observation of unidirectional backscattering-immune topological electromagnetic states. *Nature, 461*(7265), 772–775.

Yablonovitch, E. (1987). Inhibited spontaneous emission in solid-state physics and electronics. *Physical Review Letters, 58*(20), 2059–2062.

Yu, R., Zhang, W., Zhang, H. J., Zhang, S. C., Dai, X., & Fang, Z. (2010). Quantized anomalous Hall effect in magnetic topological insulators. *Science (New York, N.Y.), 329*(5987), 61–64.

Zhang, L. F., Ren, J., Wang, J. S., & Li, B. W. (2010). Topological nature of the phonon Hall effect. *Physical Review Letters, 105*(22), 225901.

CHAPTER TWO

Unique structures of low-dimensional materials

Materials structures could be defined by the features such as symmetry, periodic lattice, boundary conditions, and defects, on multiple length scales from the electronic structure, chemical bonds, molecules, clusters, grains, phases, surface, and macroscale assembly (Fig. 2.1). The structures are formed on the principles of thermodynamics and kinetics, with the influence from the environmental conditions.

Generally, the structure of materials at ground and stable state is determined by principle of lowest energy. Such a fundamental principle is effective at different length scales. Electrons in atoms are arranged according to Hund's rules to form the electronic structure of each element in the periodic table. Molecules are constructed with atoms following the hybrid theory of electron orbitals. The symmetry of molecules is described as the point group, reflecting the symmetry of atomic orbitals. For example, the shapes of methane and benzene molecules show the symmetry of

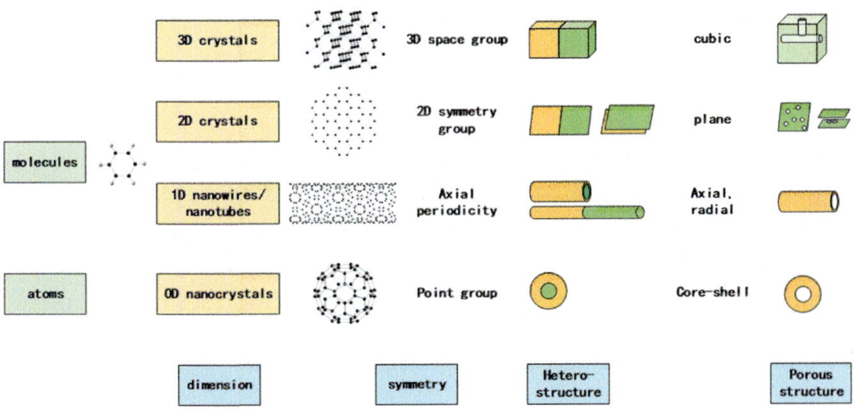

Figure 2.1 ***Structures of low-dimensional materials.*** Zero-, one-, two-dimensional materials are constructed with atoms and molecules as building blocks, under influence and restrictions of symmetry and dimensions. In addition, heterostructures and porous structures consist with low-dimensional structures and "voids" as the components.

Low-Dimensional Materials.
DOI: https://doi.org/10.1016/B978-0-443-29046-6.00008-0

the sp3 and sp2 hybridized orbitals, respectively. In a condensed state of materials, atoms are arranged into periodic lattice when crystallized. The symmetry of crystals is described by space group, under restriction of translational symmetry. The types of crystalline structures are closely related with the properties of chemical bonds. Metals tend to form densely packed structures such as face-centered cubic (FCC), hexagonal closest packed (HCP), and body-centered cubic (BCC) structures, due to the absence of directionality and saturation of metallic bonds. In contrast, covalent crystals inherit the basic shapes of molecular units, due to the directionality and saturation characteristics of covalent bonds. As a result, graphite and diamond have planar hexagonal and tetrahedral structures resembling the sp2 and sp3 hybridized orbitals, respectively. The length scale of low-dimensional materials is between that of molecules and bulk materials, with molecular features along certain directions, while keeping periodic conditions along other directions. Consequently, there are many unique structural features for low-dimensional materials.

Ideal crystals have infinite periodic structures, while real materials inevitably have surfaces that break the translational symmetry. In low-dimensional materials, surface energy plays an important role due to the high proportion of surface atoms. For a given volume, a spherical shape has the smallest surface area. In crystals, densely packed surfaces have lower specific surface energy. Surface energy drives low-dimensional materials to form near-spherical structures composed of densely packed surfaces; however, such structure is incompatible with the anisotropy of crystals, resulting in lattice strain. Consequently, elastic energy of lattice strain becomes another important factor determining the structure (Fig. 2.2). Surface energy is proportional to the area and has a quadratic relationship with the characteristic size, while elastic energy is proportional to the volume and has a cubic dependence on the characteristic size. The competition between surface energy and elastic energy differs for different length scales. The smaller the size, as such in low-dimensional materials, the higher the proportion of surface atoms, and the more dominant the surface energy becomes.

Structure of materials is also influenced by the environment. Depending on the conditions such as temperature and pressure, materials will form different phases that are governed by principles of thermodynamics. In classical thermodynamic models, one of the assumptions is that the system size is infinite; however, in real materials, especially low-dimensional materials, the system is finite, leading to much greater

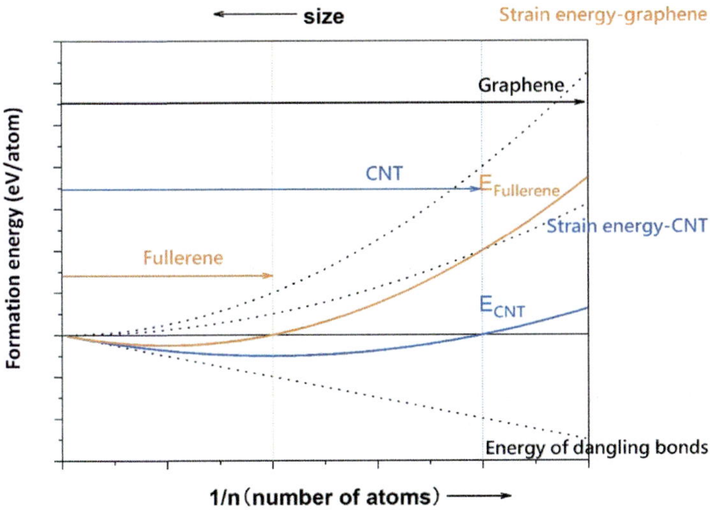

Figure 2.2 ***Size-dependent structural stability of low-dimensional carbon materials.*** When flat 2D graphene is wrapped into 1D carbon nanotube and 0D fullerene, the surface (edge) energy is lowered by reducing the dangling bonds, though elastic bending energy is introduced by bending the bonds. The size (atom number)-dependent competition between the surface and strain energies determines the critical sizes for the stability of fullerenes and carbon nanotubes. *0D*, zero-dimensional; *1D*, one-dimensional; *2D*, two-dimensional.

fluctuations and diverse structures. This is the structural origin of many unique physicochemical properties of low-dimensional materials, and at the same time poses a great challenge to structure-controlled synthesis.

2.1 Zero-dimensional materials

Zero-dimensional (0D) materials are confined in all three-dimensional directions. The number of atoms is estimated to be ranging from about 10^2 to 10^5, for the size ranging from 1 to 10 nm. The 0D materials are larger than a small molecule and much smaller than a macroscopic solid. Such size effects have an important influence on the structure of 0D materials. Compared with small molecules, 0D materials tend to form long-range ordered periodic structures. Compared with macroscopic solids, 0D materials have a high proportion of surface atoms and a dominant surface energy. The competition of energies such as cohesion energy, surface energy, and lattice elasticity energy on small scales determines the stable structure of 0D materials, including geometry, symmetry, and topology.

In the following sections, classical examples of 0D materials will be introduced, including metal clusters, and nanocrystals bonded by metallic bonds, semiconductor nanocrystals bonded by ionic bonds, and fullerenes bonded by covalent bonds. Unique structural features will be emphasized, such as magic number of the stability of metal clusters, crystallographic surfaces and shapes of semiconductor nanocrystals, and the icosahedral symmetry of C60.

2.1.1 Magic number of metal clusters

Research of metal nanoparticles can be traced back to Faraday's classic work on gold colloids in the 19th century, when he described the Tyndall phenomenon and hypothesized that there were very small particles which could not be resolved by microscopy at that time. More than 150 years later, this colloidal system remains surprisingly stable (Edwards & Thomas, 2007). Turkevich et al. (1951) confirmed that there are nanocrystals with the size ranging from 2 to 6 nm in the colloid by using a modern transmission electron microscope.

In the 1980s scientists discovered that for the metal clusters consisted of up to hundreds of atoms, there were certain atom numbers exhibiting extraordinarily high stability, as reflected in the prominent intensities of corresponding peaks in the mass spectra, for example, 8, 20, 40, and 58. Such numbers were called the "magic numbers." These distribution patterns of the magic numbers are similar to the atomic numbers in the periodic table, indicating an atom-like model to explain such a phenomenon. The electrons are delocalized in the clusters, and their distribution is governed by the quantum mechanics of a spherical geometry, similar to the case of atoms.

In this model, a single-electron approximation is considered in an average potential field: $U(r) = -\frac{U_0}{e^{(r-r_0)/\varepsilon}+1}$, where U_0 is the energy associated with the Fermi surface and work function, r_0 is the effective size, and ε is a parameter of the potential gradient. By solving the Schrödinger equation in this potential field, a multishell structure similar to the arrangement of electrons in atoms can be obtained to form shells with the number of electrons to be 2, 8, 18, 20, 34, 40, 58, etc. In simple metals, considering that each atom contributes one valence electron, the magic number associated with the stability of metal clusters could be understood.

As the number of atoms increases to hundreds and thousands, another magic number appeared. As mentioned earlier, surface energy has a strong influence on the structure of nanoparticles. The lower coordination number at the surface steps leads to higher surface energy. When crystalline planes could be closed to form polyhedrons, the surface energy could be minimized, corresponding to certain stable configurations (Martin, 1996). The magic numbers of the atoms (n) and the shells (K) show the following relations depending on the geometry of the polyhedrons: For cubes, $n = K^3$; for tetrahedrons, $n = \frac{1}{6}K^3 + \frac{1}{2}K^2 + \frac{1}{3}K$, for octahedrons, $n = \frac{2}{3}K^3 + \frac{1}{3}K$; for decahedrons, $n = \frac{5}{6}K^3 + \frac{1}{6}K$, and for icosahedrons, $n = \frac{10}{3}K^3 - 5K^2 + \frac{11}{3}K - 1$.

2.1.2 Facets and shapes of nanocrystals

The shape of nanocrystals is dominated by the surface energy, as reflected in Wulff's law $\frac{\gamma_{hkl}}{R_{hkl}} = \text{constant}$, where γ is the surface energy, $h\ k\ l$ is the crystallographic index, and R is the radius in the direction corresponding to ($h\ k\ l$) plane. In the case of FCC metals, the densely packed surface is (1 1 1). Based on the surface energy ratio between (1 1 1) and (0 0 1) planes, FCC nanocrystals should have a truncated octahedral shape. The surface energy could be further reduced if all surfaces consisted with (1 1 1) planes to form a decahedron or icosahedron. However, such a configuration will have to introduce a fivefold symmetry and a lattice strain of about 5% (Howie & Marks, 1984). Since lattice elastic energy is proportional to the volume, nanocrystals are able to accommodate higher stress and strain energy due to their small size. Under such conditions, the reduction in surface energy is able to overcome the rise in the strain energy. Indeed, Koga et al. (2004) and Li et al. (2008) found that gold nanocrystals tend to form an icosahedral structure at very small sizes and then transform into the icosahedral structure along with the increased size.

The shape of nanocrystals is influenced by the nature of chemical bonds. Semiconductor nanocrystals are bonded by the directional ionic and covalent bonds, instead of the nondirectional metallic bonds. In addition, ionic and covalent bonds are ionized to different extents. Crystalline planes of semiconducting nanocrystals also have different surface energy and chemical activities depending on the packing density and surface dangling bonds.

Under normal conditions, the surface of semiconductor nanocrystals consists of thermodynamically stable surfaces with low surface energy. In rutile phase titanium oxide nanocrystals, (1 0 1) plane constitutes 94% of the surface in Wulff-constructed crystal model. However, thermodynamically stable planes are typically not active for applications compared with other planes with higher surface energy, such as (0 0 1) plane. By systematically calculating the surface energies of titanium oxide crystals with various adsorbed atoms on different planes, Sun et al. found that fluorine atoms were able to invert the relative stability of (1 0 1) and (0 0 1) surfaces, leading to the experimental discovery of truncated octahedral-shaped titanium oxide single crystals with predominantly (0 0 1) planes (Fig. 2.3) (Liu et al., 2014; Yang et al., 2008). Menzel et al. (2013) further increased the ratio of (0 0 1) planes by increasing fluorine content and obtained titanium oxide nanosheets.

For CdSe quantum dots, Peng et al. (2000) successfully controlled the shape by utilizing the difference in anion and cation densities on different surfaces. In the CdSe nanocrystals, two end surfaces consist of (0 0 −1) and (0 0 1), terminated by Cd^{2+} and Se^{2-}, respectively. In contrast, on the side faces, anions and cations are uniformly distributed. Cd^{2+} ions on (0 0 1) plane have one dangling bond, which can easily be passivated by phosphate ions, while Cd^{2+} ions on (0 0 1) plane have three dangling bonds, which cannot be completely passivated. By utilizing the anisotropic surface properties, anisotropic growth was realized to control the shapes with different aspect ratios.

2.1.3 C_{60} the most beautiful molecule

Discovery of C_{60} stems from Kroto's (1997b) curiosity in the matters in the interstellar space, especially the long-chain carbon molecules. In 1985 he collaborated with Curl and Smalley at Rice University to simulate the

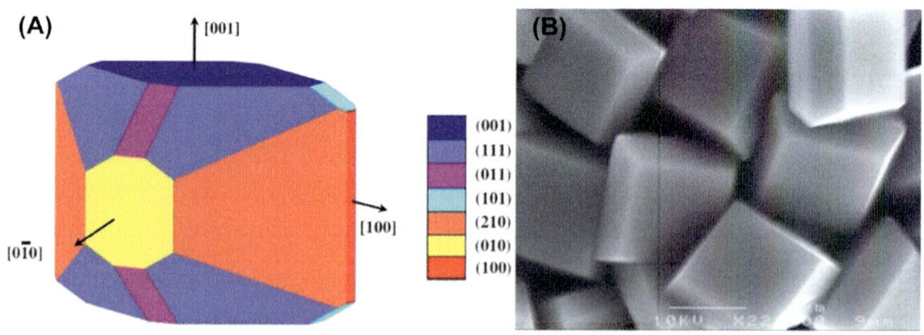

Figure 2.3 ***Facets and shapes of nanocrystals.*** (A) Wulff model of TiO_2 crystals (Gong & Selloni, 2007); (B) Scanning electron microscope (SEM) image of TiO_2 crystals with predominantly (0 0 1) plane (Yang et al., 2008).

high-energy plasma states in interstellar space and to study the formation of carbon-chain molecules, by using high-intensity laser ablation of graphite. Carbon clusters and molecules were indeed detected and then analyzed by using gas mass spectrometry. Unexpectedly, the intensity of the peak at 60 carbon atoms (C_{60}) was extraordinarily high along with other miner peaks at C_{70} and so on, suggesting the exceptional stability of C_{60} (Kroto et al., 1985). Assuming that C_{60} consists of known carbon isomers such as the sp3-hybridized diamond and the sp2-hybridized graphite, there must be dangling bonds at surface or edge. They proposed a new carbon isomer, by introducing five-member rings to fold the graphitic lattice into a closed polyhedron. Surprisingly, they found that such a polyhedron with 60 vertices is exactly a soccer ball, with a beautiful icosahedral symmetry. The newly discovered C_{60} was named Buckminsterfullerene, for the inspirations from the roof designed by architect Buckminster Fuller.

In 1990 Huffman et al. obtained a larger amount of fullerene by using arc discharge method and made single crystals large enough for fine structural characterizations. The single crystal was determined to be an FCC structure by X-ray diffraction. Characteristic molecular vibrational modes consisted of the theoretical calculations of C_{60} molecules were detected by infrared spectroscopy (Krätschmer et al., 1990; Taylor et al., 1990), confirming the model proposed by Kroto et al. For the discovery of fullerenes, Curl, Kroto, and Smalley were awarded the 1996 Nobel Prize in Chemistry (Kroto, 1997a, 1997b).

A C_{60} molecule consists of 60 carbon atoms at the vertices of a truncated icosahedron, forming 12 nonadjacent pentagons and 20 adjoining hexagons. The topology of a polyhedron follows Euler's theorem: $f + v = e + 2$, where f, v, and e are the number of facets, vertices, and edges, respectively. For a fullerene composed of any number of atoms, if considering only the hexagons (h) and the pentagons (p), the following relations are obtained: $f = h + p$; $2e = 5p + 6h$; $3v = 5p + 6h$. It can be deduced that p is always 12, meaning that a polyhedron requires 12 pentagons to form a closed structure.

The high stability of C_{60} is closely related to its atomic configuration and electronic arrangement. Adjacent five-membered rings will lead to localized stress concentrations, and the smallest molecule formed by 12 nonadjacent pentagons is C_{60}. In C_{60}, the environment is the same for all carbon atoms. Each atom is bonded to three neighboring atoms in sp2 hybridization with no dangling bonds, resulting in a stable structure.

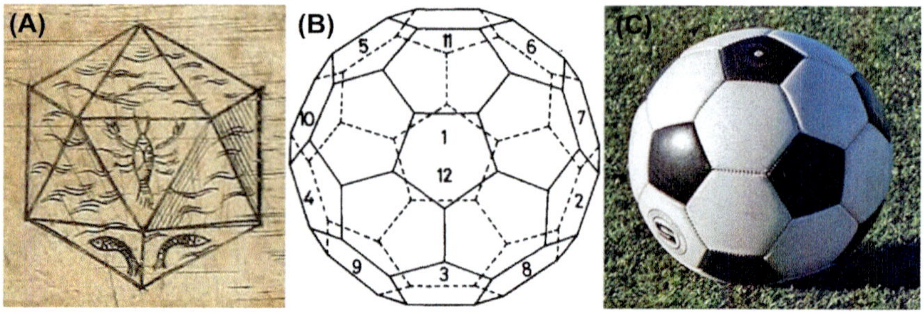

Figure 2.4 ***Icosahedral symmetry.*** (A) Platonic solids; (B) C_{60} molecular structure (Stone & Wales, 1986); and (C) modern soccer ball.

The C_{60} structure belongs to the Ih symmetry group, consisting of 6 fivefold symmetry axes through the centers of the 12 pentagons, 10 threefold symmetry axes through the centers of 20 hexagons, and 15 twofold symmetry axes of at the middle of the edges between two hexagons. In total, there are 120 symmetry operations, which is the highest symmetry of any known molecules. Therefore it is named the most perfect molecule (Kroto, 1997b).

Icosahedron is one of the five polyhedrons known as Platonic solids, dating back to ancient Greece. A volume of "Euclid's elements" (elements of geometry) was devoted to the properties of this class of geometries. During the Renaissance, icosahedron appeared in paintings of Leonardo da Vinci and his book "De Divina Proportione" (on the divine proportion). In modern times, icosahedron has become the classic shape of soccer ball. Maybe it is a coincidence, or maybe it is inevitable. It seems that appreciation and pursuit of harmony, symmetry, and beauty are the eternal theme of human beings (Fig. 2.4).

2.2 One-dimensional structures

One-dimensional (1D) crystals are strongly anisotropy. They are macroscopic in the longitudinal direction with translational symmetry, while they are nanoscale in the transverse direction with quantum confinement effects. In this section, representative 1D structures will be introduced, with the cross-sectional sizes ranging from polyacetylene at the atomic scale, to single-walled carbon nanotubes (CNTs) and ultrafine metal nanowires at a 1-nm scale, and to semiconductor nanowires at a 10-nm scale. Particularly, we will discuss the unique structures resulting from electronic structure, electronic interactions, geometrical configuration, and the competition between surface and strain energies.

2.2.1 Polyacetylene: the ultimate one-dimensional structure

The ultimate structure of a 1D material is an infinitely long chain of single atoms, which has a linear $C_{\infty v}$ symmetry, that is, infinite rotational symmetry in the axial direction and mirror symmetry in the perpendicular direction. However, this single-atom chain structure is highly unstable. The number of atoms in the longest single carbon chains is about 30, as revealed by mass spectrometry and electron microscopy (Jin, Lan, et al., 2009; Lou & Nordlander, 1996).

Polyacetylene is a long-chain hydrocarbon molecule consisting of a chain of carbon atoms as the backbone. Each of the carbon atoms is bonded to two neighboring carbon atoms and a hydrogen atom through sp2 hybridized σ-bonds, and the remaining one valence electron contributes to the π-bond. Since each carbon contributes one electron, normally, a half-filled conduction band will be formed. However, polyacetylene is a semiconductor with a band gap of about 1.5 eV, due to the Peierls distortion in the 1D structures. In the 1930s Mott and Peierls (1937) theoretically predicted that 1D structures spontaneously form a more stable dimeric structure, where the neighboring atoms form a pair, the period in real space becomes twice as long, and a corresponding band gap is formed at $ka = \pm \pi/2$ in k-space. The Peierls distortion of polyacetylene was verified by Fincher et al. (1982) with X-ray diffraction.

In 1977 Japanese scientist Hideki Shirakawa obtained crystalline polyacetylene. In collaboration with American scientists Alan Heeger and Alan MacDiarmid, they discovered that Br doping can dramatically increase the electrical conductivity by seven orders of magnitude, even approaching the conductivity of silver (Shirakawa et al., 1977). They shared the 2000 Nobel Prize in Chemistry "for the discovery and development of conductive polymers" (Heeger, 2001; Shirakawa, 2001).

2.2.2 Chirality of carbon nanotubes

CNTs are quasi1D tubular structures composed of sp2-bonded carbon atoms, which can be stable as free-standing individual molecules (Iijima, 1991). In 1991 Japanese scientist Iijima (1991) discovered CNTs and identified their unique helical structures using high-resolution transmission electron microscopy and electron diffraction. Mildred Dresselhaus et al. at Massachusetts Institute of Technology (MIT) applied group theory to analyze the symmetry of single-walled CNTs (SWCNTs) and made predictions about their electrical and vibrational properties, including the peculiar phenomenon that depending on the chirality, SWCNTs could be metallic or semiconducting (Dresselhaus et al., 1992; Saito et al., 1992).

The structure of CNTs is closely related to fullerenes. C_{70} can be viewed as formed by cutting through the equatorial plane of C_{60} to insert a ring of 10 carbon atoms. If this process is repeated 10,000 times, a CNT (C_{100060}) with a diameter of ~0.7 nm and a length up to micrometers could be formed. As the 0D C_{60} is extended into a 1D nanotube, three-dimensional quantum-confinement is replaced by an axial translational symmetry with radial periodic boundaries.

Generally, the stability of a material is determined by the formation energy. The structural stability of CNTs is related to the competition between dangling bonds-associated edge energy and curvature-dependent elastic energy. Robertson et al. (1992) used first principles and empirical potential calculations to show that strain energy is inversely proportional to the square of diameter. Thus CNTs with small diameters are subjected to large elastic strains. Compared with fullerenes, average strain energy of CNTs is lower due to the fact that bending of CNTs occurs mainly in the circumferential direction only. Compared with graphene nanoribbons with a similar width, stability of CNTs is higher due to the reduced edge energy (Lucas et al., 1993; Robertson et al., 1992; Sawada & Hamada, 1992).

As seen from the above construction process, CNTs have inherited some structural features of fullerenes, such as symmetry and molecular topology. The end caps can be regarded as half fullerenes, and the tube walls form a "donut" configuration when treated under the periodic boundary conditions. Different from the closed polyhedron topology of fullerenes, the "donut" configuration of CNTs conforms to Euler's law as follows: $v - e + f = 0$, where v is the number of vertices, e is the number of edges, and f is the number of planes. If only five- and six-membered rings are considered, the number of five-membered rings is calculated to be zero, and the walls are composed entirely of six-membered rings. And if we consider five-, six-, and seven-membered rings, it follows that five and seven-membered rings are equal in number (Monastyrsky, 2006).

These topological rules have profound implications in the defect structures and physical properties of CNTs.

It is worth mentioning that it is a common process to fold a two-dimensional (2D) space with periodic boundary conditions into a "donut" structure. In the reciprocal space, 2D Brillouin zone is folded into a donut, to understand the topology of the energy band structure.

As shown in Fig. 2.5, following construction of graphene rolled into a CNT, the direction of rolling or the chiral vector (C_h), as defined by the chiral index or chirality, completely determines the main structural features of CNTs. The chiral index (n, m) can be determined from the projections of a_1 and a_2 directions in the 2D graphene lattice, $C_h = na_1 + ma_2$. The length of C_h, which is the circumference of the CNT, determines the diameter: $d = \frac{L}{\pi} = \frac{a\sqrt{n^2 + nm + m^2}}{\pi}$.

Depending on the chiral angle, defined as the angle between C_h and the a_1 direction, CNTs could be categorized as nonchiral and chiral types. In chiral nanotubes, only helical symmetry is kept, while rotation and mirror symmetries are lost (Saito et al., 1998). If the chiral indices n and m have a common divisor d, there will be a d-fold rotational symmetry. In 2017 Zhang et al. (2017) at Peking University obtained (8,4) and (12,6) chiral-enriched CNTs by designing catalysts with four- and sixfold symmetry. The chiral index is the most fundamental structural characteristics of CNTs, which in turn determines their physical properties, making chirality-controlled growth and chirality-dependent properties to be one of the most active frontiers and important goals of CNT research.

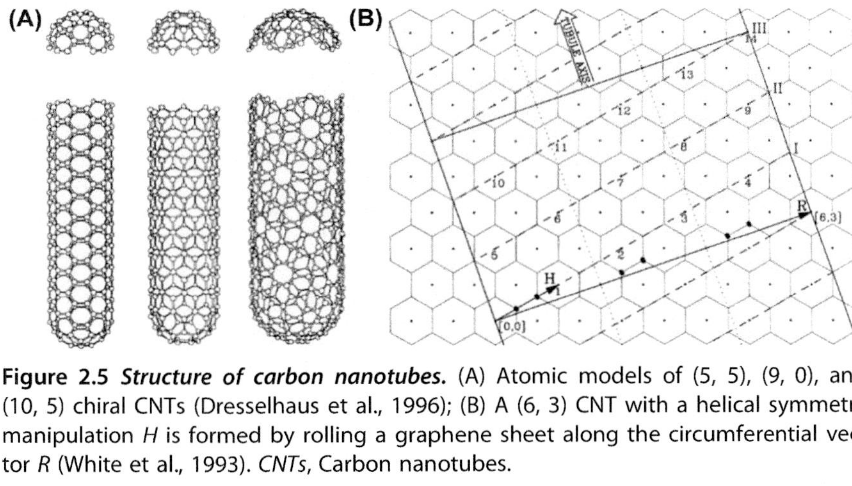

Figure 2.5 ***Structure of carbon nanotubes.*** (A) Atomic models of (5, 5), (9, 0), and (10, 5) chiral CNTs (Dresselhaus et al., 1996); (B) A (6, 3) CNT with a helical symmetry manipulation H is formed by rolling a graphene sheet along the circumferential vector R (White et al., 1993). *CNTs*, Carbon nanotubes.

2.2.3 Ultrafine metal nanowire and atomic chains

Due to the high chemical activity associated with the high specific surface area, ultrasmall particles and nanowires of pure metals are normally not stable in air. Under the protection of ultrahigh vacuum in a transmission electron microscope, Takayanagi group of Tokyo Institute of Technology successfully fabricated ultrafine nanowires and investigated the atomic formation process, structural characteristics, and electrical properties (Kondo & Takayanagi, 1997, 2000; Kondo et al., 1999; Ohnishi et al., 1998; Oshima et al., 2003). By using a scanning tunneling microscope probe to contact and pull against a gold thin edge, nanowires with a few atoms in width, and even metal atomic chains, were fabricated. Quantized conductance was observed during the step-by-step thinning processes (Ohnishi et al., 1998).

As the diameter of metal nanowires is reduced to $\sim$1 nm, peculiar structures, such as coaxial tubular helices, were observed (Kondo & Takayanagi, 2000). For the nanowires with the diameter to be $\sim$2 nm, surface reconstruction from the square to the hexagonal configurations was recorded (Kondo & Takayanagi, 1997). By using molecular dynamics simulations, Diao et al. (2003) found that the transition is driven by surface compressive stresses. The surface stress could be estimated $\sigma = 4fL/S$, where σ is surface compressive stress, f is a surface stress factor, and L and S are the length and cross-sectional areas of the nanowire, respectively. For gold nanowires with a width of 1.8 nm, the surface compressive stress can reach up to $\sim$6 GPa, driving the reconstruction from an FCC structure to a body-centered orthorhombic structure.

Utilizing the quantized conductance of ultrafine metal nanowires, Aono group at the National Institute for Materials Science in Japan has produced atomic switches that enabled memory and learning functions. The atomic switches have been commercialized through Nippon Electric Company (NEC) for application on satellites in the harsh electromagnetic environment of space (Hasegawa et al., 2010, 2012; Terabe et al., 2005; Waser & Aono, 2007).

In addition to noble metals in the previous works, we have fabricated iron atomic chains clamped by CNTs (Fig. 2.6), by applying electron beam irradiation, tensile elongation, and Joule heating to metal-filled CNTs. Chemically active metals are protected by chemically inert carbon shells during the growth by chemical vapor deposition. Quantized conductance was measured for the hybrid 1D nanotube-atomic chain system, showing potential applications of CNTs as interconnections at quantum limit (Tang, 2010). Furthermore, first-principles calculations show that the iron atomic chain retains its spin-polarization and can function as spin filters at atomic scale.

While it is an advantage for in situ electron microscopy to explore the extreme conditions and to investigate the atomic processes, it is difficult to apply the fabricated structures in "real" world. In 2008 Yu et al. (2016) at the University of California, Berkeley, reported synthesis of ultrafine metal nanowires with a diameter of less than two nanometers. By using oleylamine as solvent, reducing agent as well as surfactant, gold nanowires followed 1D growth during the reduction of chloroauric acid ($HAuCl_4$). Using spherical aberration-corrected high-resolution transmission electron microscopy, a large number of stacking faults were found in the ultrafine gold nanowires grown along the [1 1 1] direction (Fig. 2.6B). Formation of the stacking faults is closely related to the growth mechanism through oriented assembly of nanocrystals (Halder & Ravishankar, 2007). Along the growth direction of [1 1 1], the energy difference between ABABAB and ABCABC stacking is rather small for FCC metals such as gold. Since there is no effective constriction during the assembly, it is to form the stacking faults and to see the alternation of stacking orders (Yu et al., 2016). Because of the small diameter, the bending rigidity of ultrafine metal nanowires is very low. Roy et al. (2014) found that the (1 1 1) plane of Au nanowires could be distorted by anisotropic surface energy. Due to their excellent flexibility and high electrical conductivity, ultrafine metal nanowires have promising application prospects in wearable pressure

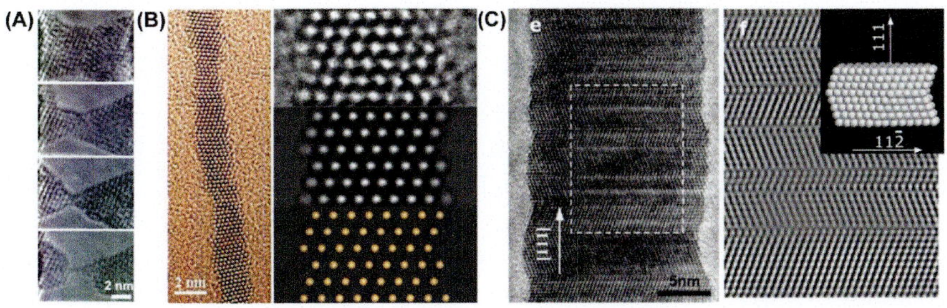

Figure 2.6 ***One-dimensional nanowire structures.*** (A) Formation processes of an iron atomic chain (Tang et al., 2010); (B) ultrafine gold nanowires and stacking faults (Yu et al., 2016); and (C) GaP nanowires and twin boundaries (Xiong et al., 2006).

sensors, transparent conductive films, etc. (Gong et al., 2014; Sánchez-Iglesias et al., 2012).

2.2.4 Growth direction of semiconductor nanowires

Crystalline structure and growth direction of semiconductor nanowires are closely related to the length scales. For nanowires with larger diameters, the crystalline structures are generally the same as those of bulk semiconductors, such as diamond-type structures of silicon and germanium, cubic zinc-blend structures of III–V semiconductors GaAs, GaP, GaN, etc., and hexagonal wurtzite structures of II–VI semiconductors CdS, ZnS, and ZnO. On the other hand, for nanowires with smaller diameters, surface energy plays an important role to determine the crystalline structure and growth directions.

Semiconductor nanowires are generally grown through a gas–liquid–solid (V–L–S) process guided by an eutectic liquid catalyst particle. The growth direction is determined by the interfaces between nanowire with the substrate and catalyst (Wei & Charles, 2006). Under steady-state growth conditions, the following equilibrium condition should be satisfied for the interfacial energies: $\sigma_l \cos(\beta) = -\sigma_{ls}$ along transverse direction, and $\sigma_l > \sigma_s/\sqrt{2}$ along longitudinal direction, where σ_l, σ_s, and σ_{ls} are the interfacial energies for the liquid–, solid–, and solid–liquid interfaces, respectively (Schmidt et al., 2009).

To lower the surface energy, semiconductor nanowires tend to grow along the densely packed crystalline planes, such as (1 1 1) face of cubic diamond-structured Si (Hochbaum et al., 2005), (1 1 1) face of cubic zinc-blend GaAs (Persson et al., 2004), and (0 0 0 2) face of hexagonal wurtzite structured ZnO (Yang et al., 2002). In the meantime, due to the large proportion of lateral surfaces, the surface energy is minimum when all facets are consisted with densely packed surfaces, which may not be compatible with the symmetry of the relevant crystalline structure. Typically, the growth direction of silicon nanowires is [1 1 1] direction, and the lateral sides consist of (1 1 0) planes.

When the nanowire diameter is smaller than 5 nm, the lateral surface energy becomes dominating, which may result in a change of growth direction. Lieber group at Harvard University reported silicon nanowires growing along [1 1 0] direction with (1 1 1) and (0 0 1) lateral planes (Wu et al., 2004).

The effect of lateral surface energy is also reflected in the defect structure of semiconductor nanowires. Stacking faults and twinning structures are frequently observed in semiconductor nanowires such as III–V GaP,

InP, and GaAs. Normally, the growth direction is along [1 1 1] direction with (1 1 2) lateral planes. With the formation of stacking faults and resulting twinning structures by slip in the [1 1 2] direction, the side surface consists of steps of densely packed (1 1 1) planes, to minimize the surface energy (Fig. 2.6C) (Caroff et al., 2009; Xiong et al., 2006).

2.3 Two-dimensional structures

2D materials are the ultimately thin structures with only one atom or one unit cell in thickness. Historically, 2D structures have only been considered idealized theoretical models, with thermodynamic theory demonstrating that no long-range ordering can be stable in infinite 2D systems. However, in 2004 Andre Geim's group at University of Manchester isolated the first 2D material, graphene, using a simple method of repeatedly peeling of graphite crystals by scotch tapes. Surprisingly, it was found that not only could 2D material be stable, but also graphene had an extraordinarily high electron mobility and exhibited unique Dirac fermion transport properties (Novoselov et al., 2004, 2005; Zhang et al., 2005). Andre Geim and Konstantin Novoselov were awarded the 2010 Nobel Prize in Physics for "for groundbreaking experiments regarding the 2D material graphene" (Geim, 2011).

A total of 20 years after the discovery, 2D materials have grown into a large family covering a wide range, from metals, semimetals, semiconductors, insulators, and topological insulators to superconductors (Miró et al., 2014), including oxidized (Kim et al., 2012; Zhu et al., 2010), hydrogenated (Elias et al., 2009), and fluorinated graphene (Nair et al., 2010) derived from graphene, graphdiyne formed from sp2-hybridized benzene rings connected with sp-hybridized acetylene bonds (Li et al., 2010; Li, Xu, et al., 2014), silicene (Feng et al., 2016; Takeda & Shiraishi, 1994; Vogt et al., 2012), and germanene (Li, Lu, et al., 2014; Liu et al., 2011) in the fourth main group; 2D black phosphorus (Li, Yu, et al., 2014; Ling et al., 2015) in the fifth main group; and hexagonal boron nitride (h-BN), an isoelectronic structure of graphene (Pakdel et al., 2014; Song et al., 2015), and so on.

In addition, transition-metal dichalcogenide (TMDC) monolayers include a group of 2D materials denoted by MX_2, where M stands for transition metals Mo, W, etc., and X stands for chalcogen elements S, Se, Te, etc. (Chhowalla et al., 2013; Manzeli et al., 2017; Wang, Kalantar-Zadeh, et al., 2012).

MXenes are another class of 2D materials of atomically thin transition metal carbides, nitrides, or carbonitrides, derived from the layered MAX phases (Naguib et al., 2014). Metal oxide and hydroxide nanosheets are charged 2D materials that could be stably dispersed in solutions (Ma & Sasaki, 2010; Osada & Sasaki, 2012). There are also organic materials forming 2D metal-organic framework structures and covalent organic framework structures (Chandra et al., 2013).

2.3.1 Stability of two-dimensional materials

As a classical model for sp2-hybridized graphitic materials, fundamental physical properties of graphene, such as its crystal structure, symmetry, and electronic structure, were intensively studied by Wallace et al. since the 1940s. Through calculations of the electron band structure, Semenoff predicted a unique linear dispersion relation near the Fermi level (Semenoff, 1984; Slonczewski & Weiss, 1958; Wallace, 1947). However, according to the Mermin–Wagner theorem, because the energy of long-wavelength phonons does not converge, resulting undulations will destroy the long-range order of 2D structures (Hohenberg, 1967; Mermin & Wagner, 1966; Mermin, 1968); therefore they cannot be stable.

In addition to the stability of 2D crystals, the stability of long-range ordered phases in two and lower dimensions is a universal question. The 2016 Nobel Prize in Physics was awarded to Kosterlitz, Haldane, and Thouless for their theoretical discoveries about topological phases and topological phase transitions. One of the most fundamental and important theoretical models is the vortex topological phase in 2D system (Kosterlitz & Thouless, 1972, 1973).

The relative stability of 2D materials can be considered in terms of thermodynamics of crystallization. The nucleation Gibbs free energy can be described by the following equation: $\Delta G = V\Delta\mu + A\gamma$, where $\Delta\mu$ is volumetric free energy, and γ is specific surface energy associated with the newly formed interface. The higher the volume free energy $\Delta\mu$ and the lower the specific surface energy γ, the lower the potential energy and the more stable the structure of the corresponding crystals. For isotropic crystals, both terms are related to their interatomic potential, the stronger the interatomic forces, the higher the cohesion and surface energies. Therefore stable 2D materials are typically from anisotropic layered materials, with a strong in-plane cohesion and a weak interplanner interaction and a low surface energy. Taking graphene as an example, its enthalpy of formation is 716 kJ/mol, higher than that of silicon

(450 kJ/mol), while the surface energy of graphite is 46.7 mJ/m^2, much lower than that of silico (1240 mJ/m^2) (Gilman, 1960; Wang et al., 2009). Therefore 2D crystals of graphite should be much more stable than that of silicon. A rule of thumb of necessary conditions of a stable 2D material is the strong chemical bonding within the layers, mainly covalent bonds, and weak interactions between the layers, typically van der Waals forces. Therefore 2D materials are also often named van der Waals crystals.

Stability of 2D materials can be theoretically considered in terms of lattice vibrations (Wirtz & Rubio, 2004). In the case of graphene, for example, there are two atoms in its unit cell and therefore two acoustic and two optical vibrational modes in the 2D lattice plane. Since graphene actually exists in three-dimensional space, it also has two modes of vibration perpendicular to the plane. An out-of-plane acoustic mode corresponds to an overall translation. The other out-of-plane optical mode corresponds to alternating oscillations of neighboring atoms. The dispersion of the frequency ω with respect to the wave number k is given by $\omega_{flex}(k) = \left(\frac{\kappa}{\sigma}\right)^{1/2} k^2 \propto k^2$, where κ and σ are the bending and tensile elasticity moduli, respectively. The number of phonon modes per unit area can be calculated by $N_{ph} = \frac{1}{2\pi}\int_0^\infty dk \frac{k}{e^{\beta\sqrt{\kappa/\sigma k^2}-1}}$, where $\beta = 1/k_B T$. For any temperature $T > 0$, as the wave number approaches to 0, this integral will tend to diverge, indicating thermodynamic instability.

Despite theoretical predictions, actual 2D materials exist and are stable. Their stability is attributed to the spontaneous formation of ripple structures or stabilization effects from substrates. Naturally occurring corrugated structures in suspended 2D materials have been visualized by using molecular dynamics simulations (Fasolino et al., 2007) and high-resolution electron microscopy (Meyer et al., 2007). 2D materials are generally attached to a substrate where, due to tensile stresses, the out-of-plane phonon dispersion relations become $\omega(k) = k\left(\frac{\kappa}{\sigma} k^2 + \frac{\gamma}{\sigma}\right)^{1/2}$, where γ is the interlayer coupling coefficient. In such case, as the wave number approaches 0, the dispersion relation will tend to be linear rather than square, and thus no divergent integral results will occur, avoiding thermodynamic instability (Castro Neto et al., 2009).

2.3.2 Symmetry of two-dimensional materials

2D crystals are subject to different symmetry restrictions from zero- and 1D crystals. Symmetry of 0D C_{60} belongs to point group. 1D CNTs have translational symmetry in the axial direction and nonrestricted rotational symmetries in the circumferential direction. 2D crystals, on the other hand, are constrained by long-range translational symmetry in the plane. 2D symmetry space group has 17 symmetry elements belonging to the monoclinic, orthorhombic, rhombic, tetragonal, and hexagonal systems (Hahn, 2006). Ideal 2D materials can be viewed as the extension of the unit cells indefinitely in the 2D directions, and its basic unit should also have a planar structure. Consequently, most of the 2D materials are planar hexagonal honeycomb-shaped structures, such as graphene, silicene, germanene, h-BN, and TMDCs. Such a group with hexagons as the basic building blocks is categorized as "six-membered-ring inorganic materials" (Fig. 2.7) (Liu et al., 2021).

A carbon atom has four valence electrons. Three of them form σ-bonds in the plane at an angle of 120 degrees, and the other valence electron forms π-bond. Graphene lattice is formed by extending in the 2D plane with six-membered benzene ring as the basic unit. Graphene unit cell includes two carbon atoms. The Hamiltonian can be represented by a 2×2 matrix. The two carbon atoms have an inversion symmetry which, together with the time-reversal symmetry, is the fundamental reason for the appearance of Dirac cones in the electronic structure of graphene, whereas the threefold rotational

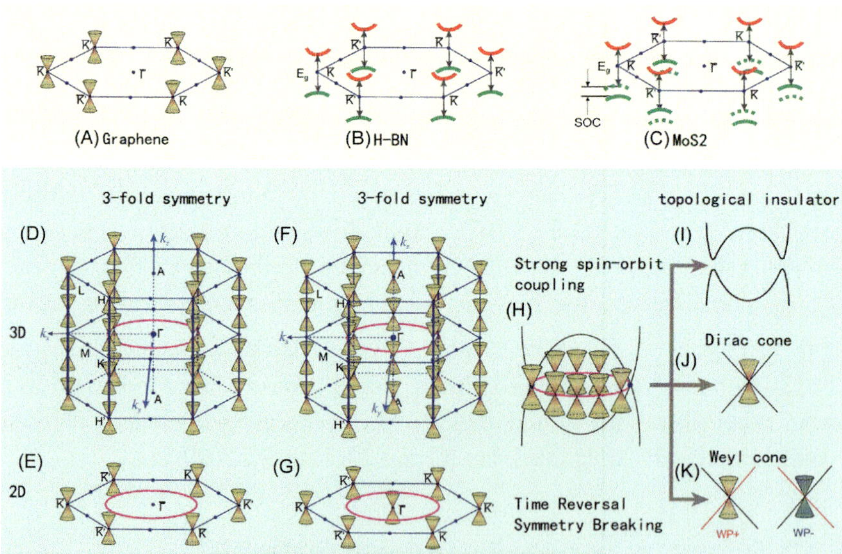

Figure 2.7 ***Two-dimensional material systems with six-member rings as basic building blocks.*** (A) Graphene; (B) h-BN; (C) MoS_2; (D–G) Dirac cones protected by symmetry; (H) Dirac nodes; and (I–K) formation of topological insulators and Weyl cones from Dirac cones due to strong spin–orbit coupling and breaking of time-reversal symmetry (Liu et al., 2021). *h-BN*, Hexagonal boron nitride.

symmetry determines the positions of the Dirac cones (at K and K′) (Castro Neto et al., 2009).

Silicene, germanene, and stanene all belong to the fourth main group as graphene. Since there are four valence electrons, if they can be hybridized by sp2 to form a graphene-like planar honeycomb structure, they will also have similar electronic structures as graphene. Vogt et al. (2012) found that silicene deposited on a silver surface can be stabilized in a honeycomb planar configuration and confirmed the electronic linear energy dispersion relation, that is, the existence of Dirac cone. Under normal conditions, since silicon and germanium atoms are larger than carbon atoms, the overlap of p_z electrons for sp2 hybridization is not sufficient to form graphene-like π-bonds. Takeda and Shiraishi (1994) found that silicon and germanium can form distorted honeycomb structures by hybrid sp2 and sp3 bonding. Cahangirov proved that this distorted honeycomb structure still has spatial inversion symmetry and threefold rotation symmetry, guaranteeing the existence of Dirac cones (Cahangirov et al., 2009). Liu et al. (2011) predicted the appearence of quantum spin Hall effect by considering the spin−orbit coupling and associated gap opening at the Dirac point.

h-BN has three and five valence electrons for boron and nitrogen atoms, respectively. Boron atoms form σ-bonds with nitrogen atoms through sp2 hybridization, constituting a honeycomb-type structure that has almost the same lattice parameter as that of graphene. Therefore h-BN is known as "white graphene." Since the boron and nitrogen atoms do not possess the space inversion symmetry, there are no symmetry-protected K points. h-BN is an insulator with a band gap of 5.9 eV. On the other hand, owing to its identical crystal structure to graphene, highly insulating properties, and atomic-level flatness, h-BN is an ideal substrate for fabricating and studying the electron transport properties of electronic devices based on graphene and other 2D materials (Dean et al., 2010).

TMDC monolayers have a sandwich hexagonal structure composed of two layers of chalcogen atoms and a middle layer of metal atoms. According to the stacking order, TMDC can be divided into ABA arrangement of 2H phase and ABC arrangement of 1 T phase. Under certain conditions, the two phases can be transformed into each other.

2.3.3 Defects in two-dimensional materials

Defects are inevitable in materials. On the one hand, there are thermally activated defects such as vacancies, which have a thermodynamic equilibrium concentration at a certain temperature. On the other hand, actual materials with finite size inevitably have edges, interfaces, and surfaces. For 2D materials, according to the Mermin–Wagner theorem (Mermin & Wagner, 1966; Mermin, 1968), long-range ordered 2D crystals are unstable and can be stabilized by ripples that deviate from perfect structures.

In 2D lattices, there are 0D and 1D defects. 0D point defects include rotation of chemical bonds, vacancies, substitution atoms, and impurity atoms. Point defects usually form localized electronic states, with defect energy levels in the energy bands. Therefore they can be applied as high-quality single-photon sources for quantum communication (Aharonovich & Toth, 2017; Tran et al., 2016).

An important class of point defects in graphene is the Stone–Thrower–Wales (S–T–W) defect formed by rotating a carbon–carbon bond by 90 degrees, which have relatively a low formation energy (4.5–5.3 eV) (Stone & Wales, 1986; Thrower, 1969). The 5/7 pair in the S–T–W defect is also the core of edge dislocations in graphene lattice. The dislocation is not mobile at room temperature, due to the high activation energy (~10 eV) (Banhart et al., 2011). Under high temperature, these dislocations are then the main mechanisms for plastic deformation of graphene and CNTs consisted of rolled-up graphene sheets (Warner et al., 2012). Another important point defect in graphene is the vacancy. Due to the dangling bonds, the formation energy of vacancies is quite high (~7.5 eV). Depending on the number of missing atoms, they can be categorized as mono-, diatomic-, and polyatomic vacancies (Banhart et al., 2011). Vacancies and impurity atoms can form stable complex defect structures, incorporating platinum, copper, iron, silicon, etc. (Krasheninnikov et al., 2009; Ramasse et al., 2013; Robertson et al., 2013). Defects are important for the functions of materials. In graphene, chemical bonding and charge transfer between transition group metal atoms and carbon atoms could stabilize the single-atom catalysts (Yang et al., 2013). Merging of vacancies with S–T–W defects and continued rotation of chemical bonds will result in more complex defects and can ultimately be transformed into 2D amorphous structures (Kotakoski et al., 2011; Meyer et al., 2008).

Crystalline structure of h-BN is similar to that of graphene and can be viewed as a six-membered honeycomb network formed by the substitution of neighboring carbon atoms by boron and nitrogen atoms. Boron and nitrogen vacancies have close formation energies according to the first principles calculations (Azevedo et al., 2007). Under electron beam irradiations, due to the mass difference between boron and nitrogen atoms, it is more likely to generate boron vacancies. Under high temperature, vacancies will merge into triangular voids with edges terminated by nitrogen atom, as revealed by in situ transmission electron microscopy (TEM) observations (Jin, Lin, et al., 2009).

Point defects of 2D TMDCs are more complex, due to the more complicated chemical structures. In case of MoS_2, for example, vacancy and substitution atoms can be categorized into monoatomic sulfur (V_S), diatomic sulfur (V_{S2}), MoS_3 (V_{MoS3}), MoS_6 vacancies (V_{MoS6}), as well as exchange of positions between Mo and S_2 (Mo_{S2}, $S2_{Mo}$) (Zhou et al., 2013).

After generation of vacancies, due to the high activity of the edge atoms, vacancies will merge and grow into larger holes. The edge configuration is determined by both the surface energy (edge energy) and the kinetics during growth and etching processes. The growth and etching rates of the armchair-type edges are higher than those of the zigzag edges. By repeated etching growth, Ma et al. obtained large single crystals of graphene with armchair-type edges (Fig. 2.8) (Ma et al., 2013).

The edges of MoS_2 can be divided into Mo- and S-atom-terminated edges. The ratio of S-atoms can be 0%, 50%, 75%, and 100%, resulting in

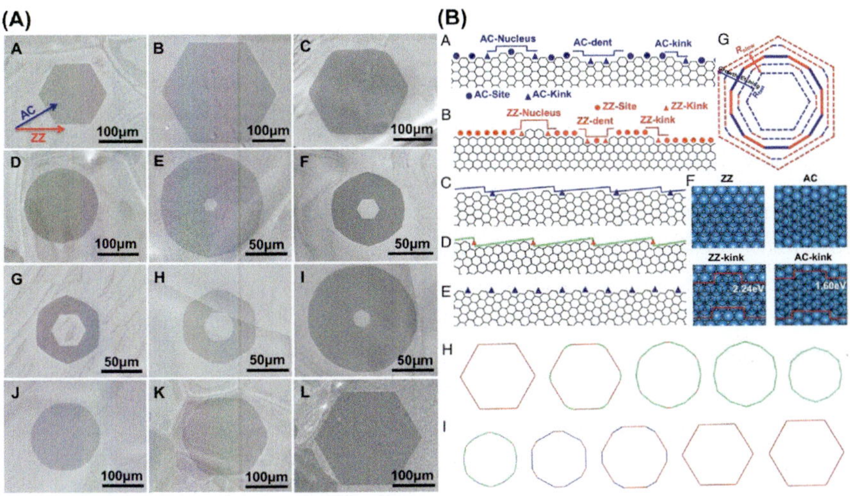

Figure 2.8 ***Edge structure of graphene.*** (A) Experimental and (B) theoretical simulation of growing and etching rates associated with edge configuration. Millimeter-sized hexagonal graphene single crystals were obtained by repeated growing-etching processes (Ma et al., 2013).

distinct electronic states and catalytic properties (Lauritsen et al., 2007). Interestingly, as two neighboring holes grow, a narrow nanoribbon could be formed between them. Under extreme conditions, MoS_2 nanoribbons with an atomic scale width become 1D metallic wires (Lin et al., 2014), though MoS_2 is a semiconductor.

In 2D crystals, grain boundaries are important line defects. For polycrystalline graphene with neighboring grains having different orientations, the grain boundaries consist of 5–7 rings and edge-type dislocations (Huang et al., 2011). The formation energy of grain boundary depends on the misorientation angle following a parabolic relationship. According to theoretical calculations by Liu et al., grain boundaries with 15 degrees misorientation have the highest strain energy (Liu & Yakobson, 2010). In contrast, grain boundaries in h-BN mainly consist of 4–8 rings to avoid electrostatic repulsion effects of the B–B and N–N bonds, as predicted by theory (Liu et al., 2012) and validated by in situ TEM observations (Cretu et al., 2014). The complex crystalline structure of 2D TMDCs is also manifested in their rich grain boundary structures. Zhou et al. (2013) observed a variety of MoS_2 grain boundary structures, such as the combinations of 4–4, 4–6, 4–8, 5–7, and 6–8 rings, by using atomically resolved high-angle annular dark-field scanning transmission electron microscopy (HAADF-STEM) images.

2.4 Low-dimensional heterostructures

Interfaces and junctions are the basic and crucial components of semiconductor devices. In addition, interface structures also have important influence to the mechanical, thermal, and catalysis properties. In this section, we will introduce the interfaces and heterojunctions in low-dimensional materials and discuss how the energies associated with the cohesion, interface, surface, and elasticity determine the interface structure.

Heterointerfaces are the interfaces between two different phases, which could be treated as the matrix and a second phase according to the ratio. When mixing different atoms, homogeneous solid solutions or separated phases could be formed. In classical theory, it is the Hume–Rothery rule that describes the tendency to form a solid solution, depending on the parameters such as atomic size, crystalline structure, valence electrons, and electronegativity. Elements with similar features tend to form solutions. Among the four parameters, the first two are about the lattice strain energy, and the latter two are about the chemical bonding.

Heterointerfaces could be categorized as coherent and noncoherent types. For coherent interfaces, the lattice is continuous by interfacial chemical bonding. However, elastic strain energy is introduced due to the difference of lattice constants. Competition of interfacial energy and elastic energy depends on the thickness and volume of the system. When the thickness reaches the threshold value, mismatch dislocations will be generated to release strain energy. In traditional thin film heterojunctions and superlattices, lattice mismatch could not be larger than 1% for coherent interfaces. However, for low-dimensional materials, due to reduced size, higher strain energy could be tolerated; therefore coherent heterointerfaces could be formed in a wider range of the compositions (Raychaudhuri & Yu, 2006; Spencer & Tersoff, 2000).

2.4.1 Zero-dimensional core–shell structures

In 0D systems, multiple elements could form homogeneous solid solutions, ordered intermetallic phases, separated phase, core–shell structures, and so on (Ferrando et al., 2008). Due to the small size of 0D systems, hetero structures that are distinct from those of the bulk systems could be formed. A classic example is the Au–Ag system, which forms a homogeneous solution for the whole composition range for the bulk system, because of the same crystalline structure and similar chemical properties. However, at nanometer scale, due to the difference in surface energy, Au–Ag core–shell nanoparticles could be formed (Guisbiers et al., 2016).

Chen et al. (2016) developed the dip-pen lithography method based on Atomic Force Microscope (AFM) and systematically investigated the phase structures of multielemental nanoparticles with up to five elements. Guisbiers et al. (2014, 2016) summarized two rules for the phase structures of core–shell nanoparticles, based on binary systems of Ag, Cu, Pt, Pd, Ni, and Co: (1) for elements with different melting points, elements with higher melting points normally form the shells; (2) for elements with similar melting points, elements with lower surface energy are normally located at the outer shells. Under far-from-equilibrium conditions, metastable structures distinct from the thermodynamically stable phases could be formed. Yao et al. (2018) found that high-entropy nanoparticles of up to eight elements could be synthesized under a high-speed impact.

Among the heterostructures, core–shell nanoparticles are special for 0D structures and have important applications in catalysis and optics. In catalysis, there is a compromise of catalytic activity and thermal stability. For the core–shell nanoparticles, expensive metals with high chemical activities are located at the surface, and the structure could be stabilized by the heterointerface with the core. In addition, the electronic structures could be tuned by strain at the interface. Currently, multiple-element metal nanoparticles are under intensive investigations for development of high-activity and high-stability catalysts (Bu et al., 2016; Tao et al., 2008; Wang, Xin, et al., 2012).

Since, lattice strain energy is proportional to the volume, due to the small size, coherent interface could be formed for the 0D nanoparticles that can tolerate a much larger lattice mismatch (Fig. 2.9A and B). For nanoparticles with a diameter smaller than 4 nm, Andrew Smith et al. reported the coherent core–shell CdTe–ZnSe quantum dots of which

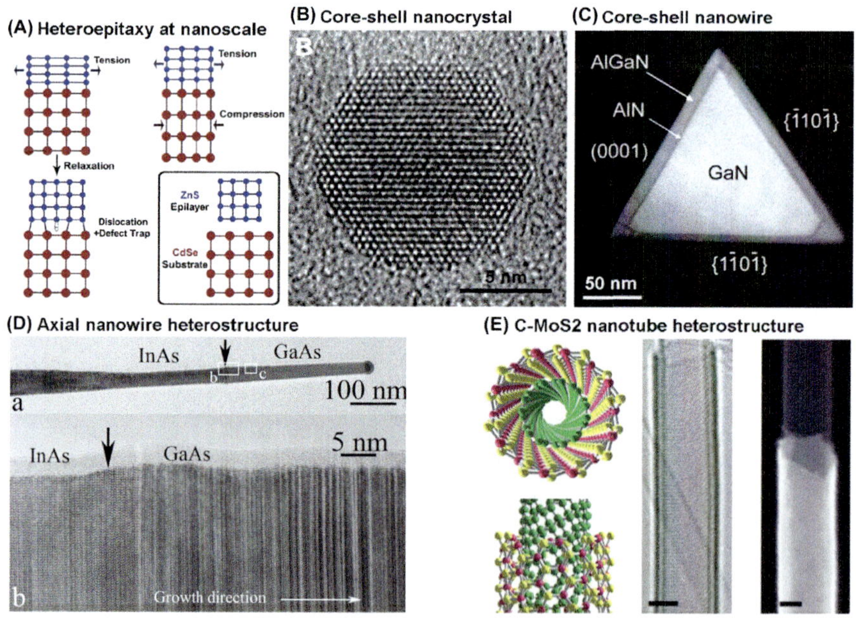

Figure 2.9 ***Heterostructures of zero- and one-dimensional materials.*** (A) Low-dimensional coherent heterointerface with high tolerance of lattice mismatch; (B) CdSe–CdTe core–shell nanocrystals (Smith & Nie, 2010); (C) cross-sectional TEM image of a GaN–AlN–AlGaN nanowire (Li et al., 2006); (D) InAs–GaAs nanowires with a heterointerface marked by an arrow (Paladugu et al., 2008); (E) van der Waals heterostructure of carbon and MoS_2 nanotubes (Xiang et al., 2020).

the lattice mismatch is as high as 11.4%. The high lattice strain becomes an effective approach to modulate the band structure and the wavelength of light emission (Smith & Nie, 2010; Smith et al., 2009).

For optoelectronic applications, the high density of defect sites at surface could become traps of electron–hole pairs to annihilate the luminescence and to reduce the quantum efficiency. Peng et al. (1997) designed CdSe/CdS coherent core–shell nanoparticles, where excited holes are confined at the core region to avoid defect sites at surface, and effectively enhanced the quantum yield. Because of reduced blinking, CdSe/CdS core–shell nanocrystals showed high potential for the bioimaging and diagnosis applications (Chen et al., 2013).

2.4.2 One-dimensional heterostructures

Since the 1D structures are anisotropic, so are the 1D heterostructures. There are heterointerfaces along transverse and longitudinal directions in nanowires and along radial and axial directions in nanotubes. By introducing heterointerfaces in transverse direction, 0D quantum dots are confined in segments of the 1D systems. On the other hand, heterointerfaces along longitudinal direction could confine charge carriers along the interface, so that scattering could be reduced to increase the mobility.

There has been a lot of research in the heterojunctions in semiconductor nanowires of the IV, III–V, and II–VI groups, as shown in Fig. 2.9C and D (Nilsson et al., 2008; Paladugu et al., 2008; Qian et al., 2004; Wang et al., 2010; Xiang et al., 2006). High-mobility transistors were fabricated using Si–Ge core–shell nanowires (Xiang et al., 2006). High intensity light-emitting diodes (LEDs) were realized by using high-mobility III–V group heterostructure semiconductor nanowires (Li et al., 2006; Qian et al., 2004). Because of the small diameters, they could tolerate high lattice match (Chuang et al., 2007; Kästner & Gösele, 2004), as shown by GaAs/InSb epitaxial nanowires with the lattice mismatch as high as 14.6% (Philippe et al., 2009).

For CNTs, distinct from the dopant-controlled electrical properties in conventional semiconductors, it is the chirality that fundamentally determines the electronic structure and transport properties. Therefore, by connecting CNTs with different chiralities, it is possible to fabricate junction devices with the same composition, but distinct electrical properties and functions (Ouyang et al., 2001; Zhen et al., 1999). Yakobson and colleagues studied chirality transformations theoretically and proposed dislocation-based mechanisms to predict chirality changes during plastic deformations (Yakobson, 1998). In our recent experiments, local chirality of an individual CNT was altered by applying tension and Joule heating, leading to controlled metal-to-semiconductor transition. By using the developed thermomechanical method, CNT junction transistors with the channel as short as 2.8 nm have been fabricated (Tang et al., 2018). In addition to junctions in axial direction, Maruyama group in University of Tokyo realized van der Waals heterostructure in the radial direction, by growing BN and MoS_2 layers on SWCNTs (Fig. 2.9E) (Zheng et al., 2021). While no epitaxial relation was identified for the nanotubes heterostructures, statistical correlation has been found between the handedness of the inner and outer shells.

2.4.3 Two-dimensional heterostructures

Traditional semiconductor devices are based on thin films and heterointerfaces. Especially, heterojunctions with coherent interfaces are important for high-mobility optoelectronic devices. 2D materials could be seen as the ultimate form of thin films, with the thickness of a single atom or a crystalline unit cell. However, 2D materials are not simple extension; instead, there are fundamental difference of the interface in 2D materials as compared with bulk semiconductors, because of the different chemical bonding within and in-between layers.

In 2013, Andre Geim et al. in the University of Manchester proposed the concept of 2D van der Waals heterostructures (Fig. 2.10) because of the van der Waals bonding between layers in 2D materials (Geim & Grigorieva, 2013). In contrast to the strict requirement for defect-free heterointerfaces in bulk semiconductors, 2D heterostructures could be fabricated by simply stacking 2D materials as Lego blocks with various compositions and to realize various functions.

From the history of semiconductors, we could learn that a "clean" system of high-mobility interfaces could bring novel electrical properties. In 2D materials, H-BN was found to be an ideal substrate because of the atomically flat surface and absence of dangling bonds, effectively enhancing the carrier mobility by a magnitude of order (Zomer et al., 2011). Moiré pattern formed with h-BN and graphene imposes a weak yet effective periodic modulation to the electronic structure, turning into a complex fractal band structure known as the "Hofstadter Butterfly" (Dean et al., 2013; Hunt et al., 2013; Yu et al., 2014). In 2018 Pablo Jarillo-Herrero group at MIT published two papers back-to-back in Nature on the electronic

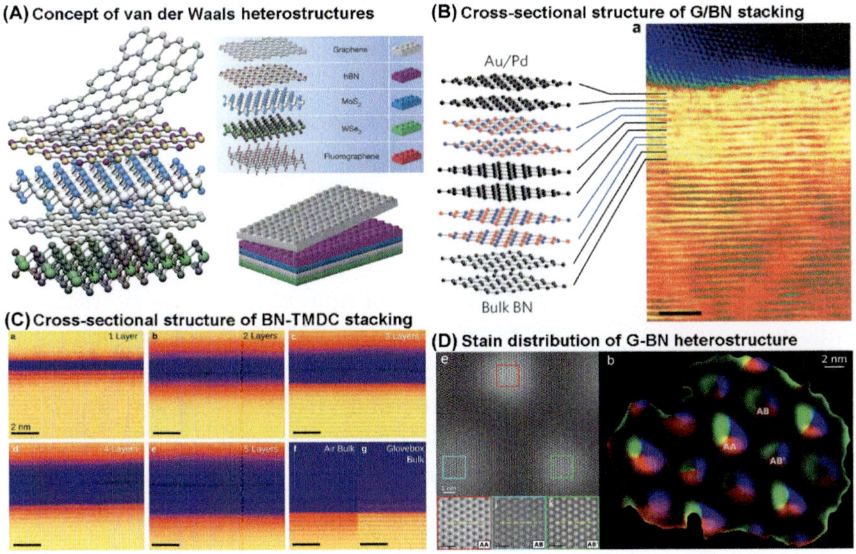

Figure 2.10 ***Two-dimensional van der Waals heterostructures.*** (A) Concept of constructing heterostructures by stacking 2D materials as Lego blocks (Geim & Grigorieva, 2013); (B) cross-sectional TEM image of graphene/h-BN heterostructures (Haigh et al., 2012); (C) TEM images of h-BN/TMDC heterostructures with different numbers of layers showing atomic-level residual impurity layers (Rooney et al., 2017); and (D) strain distribution of graphene/h-BN heterostructures (Argentero et al., 2017). *2D*, Two-dimensional; *h-BN*, hexagonal boron nitride; *TMDC*, transition-metal dichalcogenide.

structure and transport properties of "magic angle" bilayer graphene, in which Moiré patterns formed due to interlayer interactions lead to localization of electrons, forming Mott insulators with quantum phase transitions similar to those of high-temperature oxide superconductors (Cao, Fatemi, Demir, et al., 2018; Cao, Fatemi, Fang, et al., 2018).

2.5 Summary

From the magic number of the metal clusters' stability, the most beautiful molecule C_{60}, polyhedral nanocrystals, and CNTs' chirality, to theoretically predicted unstable but actually existing 2D materials, this chapter describes the diverse and unique structures of low-dimensional materials. We have seen how fundamental principles such as thermodynamics determine the basic structure of materials in low-dimensional systems, including symmetry, defects, interfaces, and surface structure.

Due to the small size and therefore large thermodynamic fluctuations of low-dimensional structures, it is possible to have multiple configurations coexisting, which is a great challenge for controlled growth and fabrication. For example, because of the small difference in the formation energy of CNTs with different chiral structures, it is still a bottleneck to grow CNTs with controlled chirality and uniform electrical property, for large-scale application in electronics such as high-performance transistors. In future research, it is necessary to explore new methods of controlling the structure, combining in situ structural characteristics and theoretical simulations. On the other hand, it is a great opportunity to explore new applications by utilizing the diverse structures and properties of low-dimensional materials.

References

Aharonovich, I., & Toth, M. (2017). Quantum emitters in two dimensions, American Association for the Advancement of Science, Australia*Science (New York, N.Y.), 358* (6360), 170–171. Available from https://doi.org/10.1126/science.aao6951, http://science.sciencemag.org/content/sci/358/6360/170.full.pdf.

Argentero, G., Mittelberger, A., Reza Ahmadpour Monazam, M., Cao, Y., Pennycook, T. J., Mangler, C., Kramberger, C., Kotakoski, J., Geim, A. K., & Meyer, J. C. (2017). Unraveling the 3D atomic structure of a suspended graphene/hBN van der Waals heterostructure. *Nano Letters, 17*(3), 1409–1416. Available from https://doi.org/10.1021/acs.nanolett.6b04360, http://pubs.acs.org/journal/nalefd.

Azevedo, S., Kaschny, J. R., de Castilho, C. M. C., & de Brito Mota, F. (2007). A theoretical investigation of defects in a boron nitride monolayer. *Nanotechnology, 18*(49). Available from https://doi.org/10.1088/0957-4484/18/49/495707.

Banhart, F., Kotakoski, J., & Krasheninnikov, A. V. (2011). Structural defects in graphene. *ACS Nano, 5*(1), 26–41. Available from https://doi.org/10.1021/nn102598m.

Bu, L., Zhang, N., Guo, S., Zhang, X., Li, J., Yao, J., Wu, T., Lu, G., Ma, J.-Y., Su, D., & Huang, X. (2016). Biaxially strained PtPb/Pt core/shell nanoplate boosts oxygen reduction catalysis. *Science (New York, N.Y.), 354*(6318), 1410–1414. Available from https://doi.org/10.1126/science.aah6133.

Cahangirov, S., Topsakal, M., Aktürk, E., Şahin, H., & Ciraci, S. (2009). Two- and one-dimensional honeycomb structures of silicon and germanium. *Physical Review Letters, 102*(23). Available from https://doi.org/10.1103/physrevlett.102.236804.

Cao, Y., Fatemi, V., Demir, A., Fang, S., Tomarken, S. L., Luo, J. Y., Sanchez-Yamagishi, J. D., Watanabe, K., Taniguchi, T., Kaxiras, E., Ashoori, R. C., & Jarillo-Herrero, P. (2018). Correlated insulator behaviour at half-filling in magic-angle graphene superlattices. *Nature, 556*(7699), 80–84. Available from https://doi.org/10.1038/nature26154, http://www.nature.com/nature/index.html.

Cao, Y., Fatemi, V., Fang, S., Watanabe, K., Taniguchi, T., Kaxiras, E., & Jarillo-Herrero, P. (2018). Unconventional superconductivity in magic-angle graphene superlattices. *Nature*, *556*(7699), 43–50. Available from https://doi.org/10.1038/nature26160, http://www.nature.com/nature/index.html.

Caroff, P., Dick, K. A., Johansson, J., Messing, M. E., Deppert, K., & Samuelson, L. (2009). Controlled polytypic and twin-plane superlattices in iii–v nanowires. *Nature Nanotechnology*, *4*(1), 50–55. Available from https://doi.org/10.1038/nnano.2008.359.

Castro Neto, A. H., Guinea, F., Peres, N. M. R., Novoselov, K. S., & Geim, A. K. (2009). The electronic properties of graphene. *Reviews of Modern Physics*, *81*(1), 109–162. Available from https://doi.org/10.1103/revmodphys.81.109.

Chandra, S., Kandambeth, S., Biswal, B. P., Lukose, B., Kunjir, S. M., Chaudhary, M., Babarao, R., Heine, T., & Banerjee, R. (2013). Chemically stable multilayered covalent organic nanosheets from covalent organic frameworks via mechanical delamination. *Journal of the American Chemical Society*, *135*(47), 17853–17861. Available from https://doi.org/10.1021/ja408121p.

Chen, O., Zhao, J., Chauhan, V. P., Cui, J., Wong, C., Harris, D. K., Wei, H., Han, H. S., Fukumura, D., Jain, R. K., & Bawendi, M. G. (2013). Compact high-quality CdSe–CdS core-shell nanocrystals with narrow emission linewidths and suppressed blinking. *Nature Materials*, *12*(5), 445–451. Available from https://doi.org/10.1038/nmat3539.

Chen, P. C., Liu, X., Hedrick, J. L., Xie, Z., Wang, S., Lin, Q. Y., Hersam, M. C., Dravid, V. P., & Mirkin, C. A. (2016). Polyelemental nanoparticle libraries. *Science (New York, N.Y.)*, *352*(6293), 1565–1569. Available from https://doi.org/10.1126/science.aaf8402, http://science.sciencemag.org/content/sci/352/6293/1565.full.pdf.

Chhowalla, M., Shin, H. S., Eda, G., Li, L. J., Loh, K. P., & Zhang, H. (2013). The chemistry of two-dimensional layered transition metal dichalcogenide nanosheets. *Nature Chemistry*, *5*(4), 263–275. Available from https://doi.org/10.1038/nchem.1589.

Chuang, L. C., Moewe, M., Chase, C., Kobayashi, N. P., Chang-Hasnain, C., & Crankshaw, S. (2007). Critical diameter for III–V nanowires grown on lattice-mismatched substrates. *Applied Physics Letters*, *90*(4). Available from https://doi.org/10.1063/1.2436655.

Cretu, O., Lin, Y. C., & Suenaga, K. (2014). Evidence for active atomic defects in monolayer hexagonal boron nitride: A new mechanism of plasticity in two-dimensional materials. *Nano Letters*, *14*(2), 1064–1068. Available from https://doi.org/10.1021/nl404735w.

Dean, C. R., Wang, L., Maher, P., Forsythe, C., Ghahari, F., Gao, Y., Katoch, J., Ishigami, M., Moon, P., Koshino, M., Taniguchi, T., Watanabe, K., Shepard, K. L., Hone, J., & Kim, P. (2013). Hofstadter's butterfly and the fractal quantum Hall effect in moiré superlattices. *Nature*, *497*(7451), 598–602. Available from https://doi.org/10.1038/nature12186.

Dean, C. R., Young, A. F., Meric, I., Lee, C., Wang, L., Sorgenfrei, S., Watanabe, K., Taniguchi, T., Kim, P., Shepard, K. L., & Hone, J. (2010). Boron nitride substrates for high-quality graphene electronics. *Nature Nanotechnology*, *5*(10), 722–726. Available from https://doi.org/10.1038/nnano.2010.172.

Diao, J., Gall, K., & Dunn, M. L. (2003). Surface-stress-induced phase transformation in metal nanowires. *Nature Materials*, *2*(10), 656–660. Available from https://doi.org/10.1038/nmat977, http://www.nature.com/nmat/.

Dresselhaus, M.S., Dresselhaus, G., & Eklund, P.C. (1996). *Science of fullerenes and carbon nanotubes. Their properties and applications*. Elsevier.

Dresselhaus, M. S., Dresselhaus, G., & Saito, R. (1992). Carbon fibers based on C_{60} and their symmetry. *Physical Review B*, *45*(11), 6234–6242. Available from https://doi.org/10.1103/physrevb.45.6234.

Edwards, P. P., & Thomas, J. M. (2007). Gold in a metallic divided state—From Faraday to present-day nanoscience. *Angewandte Chemie—International Edition*, *46*(29), 5480–5486. Available from https://doi.org/10.1002/anie.200700428.

Elias, D. C., Nair, R. R., Mohiuddin, T. M. G., Morozov, S. V., Blake, P., Halsall, M. P., Ferrari, A. C., Boukhvalov, D. W., Katsnelson, M. I., Geim, A. K., & Novoselov, K. S. (2009). Control of graphene's properties by reversible hydrogenation: Evidence for graphane. *Science (New York, N.Y.)*, *323*(5914), 610–613. Available from https://doi.org/10.1126/science.1167130.

Fasolino, A., Los, J. H., & Katsnelson, M. I. (2007). Intrinsic ripples in graphene. *Nature Materials*, *6*(11), 858–861. Available from https://doi.org/10.1038/nmat2011.

Feng, Y., Liu, D., Feng, B., Liu, X., Zhao, L., Xie, Z., Liu, Y., Liang, A., Hu, C., Hu, Y., He, S., Liu, G., Zhang, J., Chen, C., Xu, Z., Chen, L., Wu, K., Liu, Y.-T., Lin, H., ... Zhou, X. J. (2016). Direct evidence of interaction-induced Dirac cones in a monolayer silicene/Ag(1 1 1) system. *Proceedings of the National Academy of Sciences*, *113* (51), 14656–14661. Available from https://doi.org/10.1073/pnas.1613434114.

Ferrando, R., Jellinek, J., & Johnston, R. L. (2008). Nanoalloys: From theory to applications of alloy clusters and nanoparticles. *Chemical Reviews*, *108*(3), 845–910. Available from https://doi.org/10.1021/cr040090g, http://pubs.acs.org/journal/chreay.

Fincher, C. R., Chen, C.-E., Heeger, A. J., MacDiarmid, A. G., & Hastings, J. B. (1982). Structural determination of the symmetry-breaking parameter in trans—(CH)*x*. *Physical Review Letters*, *48*(2), 100–104. Available from https://doi.org/10.1103/physrevlett.48.100.

Geim, A. K., & Grigorieva, I. V. (2013). Van der Waals heterostructures. *Nature*, *499* (7459), 419–425. Available from https://doi.org/10.1038/nature12385.

Geim, A. K. (2011). Nobel lecture: Random walk to graphene. *Reviews of Modern Physics*, *83* (3), 851–862. Available from https://doi.org/10.1103/RevModPhys.83.851undefined, http://oai.aps.org/filefetch?identifier = 10.1103/RevModPhys.83.851&component = fulltext&description = markup&format = xml.

Gilman, J. J. (1960). Direct measurements of the surface energies of crystals. *Journal of Applied Physics*, *31*(12), 2208–2218. Available from https://doi.org/10.1063/1.1735524.

Gong, S., Schwalb, W., Wang, Y., Chen, Y., Tang, Y., Si, J., Shirinzadeh, B., & Cheng, W. (2014). A wearable and highly sensitive pressure sensor with ultrathin gold nanowires. *Nature Communications*, *5*. Available from https://doi.org/10.1038/ncomms4132.

Gong, X.-Q., & Selloni, A. (2007). First-principles study of the structures and energetics of stoichiometric brookite TiO_2 surfaces. *Physical Review B*, *76*(23). Available from https://doi.org/10.1103/physrevb.76.235307.

Guisbiers, G., Mejia-Rosales, S., Khanal, S., Ruiz-Zepeda, F., Whetten, R. L., & José-Yacaman, M. (2014). Gold-copper nano-alloy, "tumbaga," in the era of nano: Phase diagram and segregation. *Nano Letters*, *14*(11), 6718–6726. Available from https://doi.org/10.1021/nl503584q, http://pubs.acs.org/journal/nalefd.

Guisbiers, G., Mendoza-Cruz, R., Bazán-Díaz, L., Velázquez-Salazar, J. J., Mendoza-Perez, R., Robledo-Torres, J. A., Rodriguez-Lopez, J. L., Montejano-Carrizales, J. M., Whetten, R. L., & José-Yacamán, M. (2016). Electrum, the gold-silver alloy, from the bulk scale to the nanoscale: Synthesis, properties, and segregation rules. *ACS Nano*, *10*(1), 188–198. Available from https://doi.org/10.1021/acsnano.5b05755, http://pubs.acs.org/journal/ancac3.

Hahn, T. (2006). *The 17 plane groups (two-dimensional space groups)*. International Union of Crystallography (IUCr), A. Available from https://doi.org/10.1107/97809553602060000512.

Haigh, S. J., Gholinia, A., Jalil, R., Romani, S., Britnell, L., Elias, D. C., Novoselov, K. S., Ponomarenko, L. A., Geim, A. K., & Gorbachev, R. (2012). Cross-sectional imaging of individual layers and buried interfaces of graphene-based heterostructures and superlattices. *Nature Materials*, *11*(9), 764–767. Available from https://doi.org/10.1038/nmat3386.

Halder, A., & Ravishankar, N. (2007). Ultrafine single-crystalline gold nanowire arrays by oriented attachment. *Advanced Materials*, *19*(14), 1854–1858. Available from https://doi.org/10.1002/adma.200602325.

Hasegawa, T., Ohno, T., Terabe, K., Tsuruoka, T., Nakayama, T., Gimzewski, J. K., & Aono, M. (2010). Learning abilities achieved by a single solid-state atomic switch. *Advanced Materials*, *22*(16), 1831–1834. Available from https://doi.org/10.1002/adma.200903680.

Hasegawa, T., Terabe, K., Tsuruoka, T., & Aono, M. (2012). Atomic switch: Atom/ion movement controlled devices for beyond von-Neumann computers. *Advanced Materials*, *24*(2), 252–267. Available from https://doi.org/10.1002/adma.201102597.

Heeger, A. J. (2001). Semiconducting and metallic polymers: The fourth generation of polymeric materials (nobel lecture). *Angewandte Chemie—International Edition*, *40*(14), 2591–2611. Available from https://doi.org/10.1002/1521–3773(20010716)40:14 < 2591::AID-ANIE2591 > 3.0.CO;2-0.

Hochbaum, A. I., Fan, R., He, R., & Yang, P. (2005). Controlled growth of Si nanowire arrays for device integration. *Nano Letters*, *5*(3), 457–460. Available from https://doi.org/10.1021/nl047990x.

Hohenberg, P. C. (1967). Existence of long-range order in one and two dimensions. *Physical Review*, *158*(2), 383–386. Available from https://doi.org/10.1103/physrev.158.383.

Howie, A., & Marks, L. D. (1984). Elastic strains and the energy balance for multiply twinned particles. *Philosophical Magazine A: Physics of Condensed Matter, Structure, Defects and Mechanical Properties*, *49*(1), 95–109. Available from https://doi.org/10.1080/01418618408233432.

Huang, P. Y., Ruiz-Vargas, C. S., Van Der Zande, A. M., Whitney, W. S., Levendorf, M. P., Kevek, J. W., Garg, S., Alden, J. S., Hustedt, C. J., Zhu, Y., Park, J., McEuen, P. L., & Muller, D. A. (2011). Grains and grain boundaries in single-layer graphene atomic patchwork quilts. *Nature*, *469*(7330), 389–392. Available from https://doi.org/10.1038/nature09718.

Hunt, B., Sanchez-Yamagishi, J. D., Young, A. F., Yankowitz, M., LeRoy, B. J., Watanabe, K., Taniguchi, T., Moon, P., Koshino, M., Jarillo-Herrero, P., & Ashoori, R. C. (2013). Massive Dirac fermions and Hofstadter butterfly in a van der Waals heterostructure. *Science (New York, N.Y.)*, *340*(6139), 1427–1430. Available from https://doi.org/10.1126/science.1237240.

Iijima, S. (1991). Helical microtubules of graphitic carbon. *Nature*, *354*(6348), 56–58. Available from https://doi.org/10.1038/354056a0.

Jin, C., Lan, H., Peng, L., Suenaga, K., & Iijima, S. (2009). Deriving carbon atomic chains from graphene. *Physical Review Letters*, *102*(20). Available from https://doi.org/10.1103/PhysRevLett.102.205501, http://oai.aps.org/oai?verb = GetRecord&Identifier = oai:aps.org:PhysRevLett.102.205501&metadataPrefix = oai_apsmeta_2, Japan.

Jin, C., Lin, F., Suenaga, K., & Iijima, S. (2009). Fabrication of a freestanding boron nitride single layer and its defect assignments. *Physical Review Letters*, *102*(19). Available from https://doi.org/10.1103/PhysRevLett.102.195505, http://oai.aps.org/oai?verb = GetRecord&Identifier = oai:aps.org:PhysRevLett.102.195505&metadataPrefix = oai_apsmeta_2, Japan.

Kästner, G., & Gösele, U. (2004). Stress and dislocations at cross-sectional heterojunctions in a cylindrical nanowire. *Philosophical Magazine*, *84*(35), 3803–3824. Available from https://doi.org/10.1080/1478643042000281389.

Kim, J., Cote, L. J., & Huang, J. (2012). Two dimensional soft material: New faces of graphene oxide. *Accounts of Chemical Research*, *45*(8), 1356–1364. Available from https://doi.org/10.1021/ar300047s.

Koga, K., Ikeshoji, T., & Sugawara, K. I. (2004). Size- and temperature-dependent structural transitions in gold nanoparticles. *Physical Review Letters*, *92*(11), 1–115507. Available from https://doi.org/10.1103/PhysRevLett.92.115507, https://journals.aps.org/prl/issues.

Kondo, Y., Ru, Q., & Takayanagi, K. (1999). Thickness induced structural phase transition of gold nanofilm. *Physical Review Letters*, *82*(4), 751–754. Available from https://doi.org/10.1103/physrevlett.82.751.

Kondo, Y., & Takayanagi, K. (1997). Gold nanobridge stabilized by surface structure. *Physical Review Letters*, *79*(18), 3455–3458. Available from https://doi.org/10.1103/PhysRevLett.79.3455.

Kondo, Y., & Takayanagi, K. (2000). Synthesis and characterization of helical multi-shell gold nanowires. *Science (New York, N.Y.)*, *289*(5479), 606–608. Available from https://doi.org/10.1126/science.289.5479.606.

Kosterlitz, J. M., & Thouless, D. J. (1972). Long range order and metastability in two dimensional solids and superfluids. (Application of dislocation theory). *Journal of Physics C: Solid State Physics*, *5*(11), L124–L126. Available from https://doi.org/10.1088/0022-3719/5/11/002.

Kosterlitz, J. M., & Thouless, D. J. (1973). Ordering, metastability and phase transitions in two-dimensional systems. *Journal of Physics C: Solid State Physics*, *6*(7), 1181–1203. Available from https://doi.org/10.1088/0022-3719/6/7/010.

Kotakoski, J., Krasheninnikov, A. V., Kaiser, U., & Meyer, J. C. (2011). From point defects in graphene to two-dimensional amorphous carbon. *Physical Review Letters*, *106*(10). Available from https://doi.org/10.1103/physrevlett.106.105505.

Krasheninnikov, A. V., Lehtinen, P. O., Foster, A. S., Pyykkö, P., & Nieminen, R. M. (2009). Embedding transition-metal atoms in graphene: Structure, bonding, and magnetism. *Physical Review Letters*, *102*(12). Available from https://doi.org/10.1103/physrevlett.102.126807.

Krätschmer, W., Lamb, L. D., Fostiropoulos, K., & Huffman, D. R. (1990). Solid C60: A new form of carbon. *Nature*, *347*(6291), 354–358. Available from https://doi.org/10.1038/347354a0.

Kroto, H. (1997a). Symmetry, space, stars and C_{60}. *Reviews of Modern Physics*, *69*(3), 703–722. Available from https://doi.org/10.1103/revmodphys.69.703, https://journals.aps.org/rmp.

Kroto, H. (1997b). Symmetry, space, stars, and C_{60} (Nobel lecture). *Angewandte Chemie (International Edition in English)*, *36*(15), 1578–1593. Available from https://doi.org/10.1002/anie.199715781, http://onlinelibrary.wiley.com/journal/10.1002/(ISSN)1521-3773.

Kroto, H. W., Heath, J. R., O'Brien, S. C., Curl, R. F., & Smalley, R. E. (1985). C_{60}: Buckminsterfullerene. *Nature*, *318*(6042), 162–163. Available from https://doi.org/10.1038/318162a0.

Lauritsen, J. V., Kibsgaard, J., Helveg, S., Topsøe, H., Clausen, B. S., Lægsgaard, E., & Besenbacher, F. (2007). Size-dependent structure of MoS_2 nanocrystals. *Nature Nanotechnology*, *2*(1), 53–58. Available from https://doi.org/10.1038/nnano.2006.171, http://www.nature.com/nnano/index.html.

Li, G., Li, Y., Liu, H., Guo, Y., Li, Y., & Zhu, D. (2010). Architecture of graphdiyne nanoscale films. *Chemical Communications*, *46*(19), 3256–3258. Available from https://doi.org/10.1039/b922733d, http://pubs.rsc.org/en/journals/journal/cc.

Li, L., Lu, S. Z., Pan, J., Qin, Z., Wang, Y. Q., Wang, Y., Cao, G. Y., Du, S., & Gao, H. J. (2014). Buckled germanene formation on Pt(1 1 1). *Advanced Materials*, *26*(28), 4820–4824. Available from https://doi.org/10.1002/adma.201400909, http://www3.interscience.wiley.com/journal/119030556/issue.

Li, L., Yu, Y., Ye, G. J., Ge, Q., Ou, X., Wu, H., Feng, D., Chen, X. H., & Zhang, Y. (2014). Black phosphorus field-effect transistors. *Nature Nanotechnology*, *9*(5), 372–377. Available from https://doi.org/10.1038/nnano.2014.35, http://www.nature.com/nnano/index.html.

Li, Y., Xiang, J., Qian, F., Gradečak, S., Wu, Y., Yan, H., Blom, D. A., & Lieber, C. M. (2006). Dopant-free GaN/AlN/AlGaN radial nanowire heterostructures as high electron mobility transistors. *Nano Letters*, *6*(7), 1468–1473. Available from https://doi.org/10.1021/nl060849z.

Li, Y., Xu, L., Liu, H., & Li, Y. (2014). Graphdiyne and graphyne: From theoretical predictions to practical construction. *Chemical Society Reviews*, *43*(8), 2572–2586. Available from https://doi.org/10.1039/c3cs60388a, http://pubs.rsc.org/en/journals/journal/cs.

Li, Z. Y., Young, N. P., Di Vece, M., Palomba, S., Palmer, R. E., Bleloch, A. L., Curley, B. C., Johnston, R. L., Jiang, J., & Yuan, J. (2008). Three-dimensional atomic-scale structure of size-selected gold nanoclusters. *Nature*, *451*(7174), 46–48. Available from https://doi.org/10.1038/nature06470.

Lin, J., Cretu, O., Zhou, W., Suenaga, K., Prasai, D., Bolotin, K. I., Cuong, N. T., Otani, M., Okada, S., Lupini, A. R., Idrobo, J. C., Caudel, D., Burger, A., Ghimire, N. J., Yan, J., Mandrus, D. G., Pennycook, S. J., & Pantelides, S. T. (2014). Flexible metallic nanowires with self-adaptive contacts to semiconducting transition-metal dichalcogenide monolayers. *Nature Nanotechnology*, *9*(6), 436–442. Available from https://doi.org/10.1038/nnano.2014.81, http://www.nature.com/nnano/index.html.

Ling, X., Wang, H., Huang, S., Xia, F., & Dresselhaus, M. S. (2015). The renaissance of black phosphorus. *Proceedings of the National Academy of Sciences*, *112*(15), 4523–4530. Available from https://doi.org/10.1073/pnas.1416581112.

Liu, C. C., Feng, W., & Yao, Y. (2011). Quantum spin Hall effect in silicene and two-dimensional germanium. *Physical Review Letters*, *107*(7). Available from https://doi.org/10.1103/PhysRevLett.107.076802, http://oai.aps.org/filefetch?identifier = 10.1103/PhysRevLett.107.076802&component = fulltext&description = markup&format = xml, China.

Liu, C.-C., Jiang, H., & Yao, Y. (2011). Low-energy effective Hamiltonian involving spin-orbit coupling in silicene and two-dimensional germanium and tin. *Physical Review B*, *84*(19). Available from https://doi.org/10.1103/physrevb.84.195430.

Liu, G., Chen, X. Q., Liu, B., Ren, W., & Cheng, H. M. (2021). Six-membered-ring inorganic materials: Definition and prospects. *National Science Review*, *8*(1). Available from https://doi.org/10.1093/nsr/nwaa248, http://nsr.oxfordjournals.org/.

Liu, G., Yang, H. G., Pan, J., Yang, Y. Q., Lu, G. Q. M., & Cheng, H. M. (2014). Titanium dioxide crystals with tailored facets. *Chemical Reviews*, *114*(19), 9559–9612. Available from https://doi.org/10.1021/cr400621z, http://pubs.acs.org/journal/chreay.

Liu, Y., & Yakobson, B. I. (2010). Cones, Pringles, and grain boundary landscapes in graphene topology. *Nano Letters*, *10*(6), 2178–2183. Available from https://doi.org/10.1021/nl100988r.

Liu, Y., Zou, X., & Yakobson, B. I. (2012). Dislocations and grain boundaries in two-dimensional boron nitride. *ACS Nano*, *6*(8), 7053–7058. Available from https://doi.org/10.1021/nn302099q.

Lou, L., & Nordlander, P. (1996). Carbon atomic chains in strong electric fields. *Physical Review B*, *54*(23), 16659–16662. Available from https://doi.org/10.1103/physrevb.54.16659.

Lucas, A. A., Lambin, P. H., & Smalley, R. E. (1993). On the energetics of tubular fullerenes. *Journal of Physics and Chemistry of Solids*, *54*(5), 587–593. Available from https://doi.org/10.1016/0022-3697(93)90237-l.

Ma, R., & Sasaki, T. (2010). Nanosheets of oxides and hydroxides: Ultimate 2D charge-bearing functional crystallites. *Advanced Materials*, *22*(45), 5082–5104. Available from https://doi.org/10.1002/adma.201001722.

Ma, T., Ren, W., Zhang, X., Liu, Z., Gao, Y., Yin, L.-C., Ma, X.-L., Ding, F., & Cheng, H.-M. (2013). Edge-controlled growth and kinetics of single-crystal graphene domains by chemical vapor deposition. *Proceedings of the National Academy of Sciences*, *110*(51), 20386–20391. Available from https://doi.org/10.1073/pnas.1312802110.

Manzeli, S., Ovchinnikov, D., Pasquier, D., Yazyev, O. V., & Kis, A. (2017). 2D transition metal dichalcogenides. *Nature Reviews Materials, 2*. Available from https://doi.org/10.1038/natrevmats.2017.33, http://www.nature.com/natrevmats/.

Martin, T. P. (1996). Shells of atoms. *Physics Reports, 273*(4), 199–241. Available from https://doi.org/10.1016/0370-1573(95)00083-6.

Menzel, R., Duerrbeck, A., Liberti, E., Yau, H. C., McComb, D., & Shaffer, M. S. P. (2013). Determining the morphology and photocatalytic activity of two-dimensional anatase nanoplatelets using reagent stoichiometry. *Chemistry of Materials, 25*(10), 2137–2145. Available from https://doi.org/10.1021/cm400785z.

Mermin, N. D. (1968). Crystalline order in two dimensions. *Physical Review, 176*(1), 250–254. Available from https://doi.org/10.1103/physrev.176.250.

Mermin, N. D., & Wagner, H. (1966). Absence of ferromagnetism or antiferromagnetism in one- or two-dimensional isotropic Heisenberg models. *Physical Review Letters, 17*(22), 1133–1136. Available from https://doi.org/10.1103/physrevlett.17.1133.

Meyer, J. C., Geim, A. K., Katsnelson, M. I., Novoselov, K. S., Booth, T. J., & Roth, S. (2007). The structure of suspended graphene sheets. *Nature, 446*(7131), 60–63. Available from https://doi.org/10.1038/nature05545.

Meyer, J. C., Kisielowski, C., Erni, R., Rossell, M. D., Crommie, M. F., & Zettl, A. (2008). Direct imaging of lattice atoms and topological defects in graphene membranes. *Nano Letters, 8*(11), 3582–3586. Available from https://doi.org/10.1021/nl801386m.

Miró, P., Audiffred, M., & Heine, T. (2014). An atlas of two-dimensional materials. *Chemical Society Reviews, 43*(18), 6537–6554. Available from https://doi.org/10.1039/c4cs00102h, http://pubs.rsc.org/en/journals/journal/cs.

Monastyrsky, M. I. (2006). *Topology in condensed matter*. Springer-Verlag.

Mott, N. F., & Peierls, R. (1937). Discussion of the paper by de Boer and Verwey. *Proceedings of the Physical Society, 49*(4S), 72–73. Available from https://doi.org/10.1088/0959-5309/49/4s/308.

Naguib, M., Mochalin, V. N., Barsoum, M. W., & Gogotsi, Y. (2014). 25th anniversary article: MXenes: A new family of two-dimensional materials. *Advanced Materials, 26*(7), 992–1005. Available from https://doi.org/10.1002/adma.201304138.

Nair, R. R., Ren, W., Jalil, R., Riaz, I., Kravets, V. G., Britnell, L., Blake, P., Schedin, F., Mayorov, A. S., Yuan, S., Katsnelson, M. I., Cheng, H. M., Strupinski, W., Bulusheva, L. G., Okotrub, A. V., Grigorieva, I. V., Grigorenko, A. N., Novoselov, K. S., & Geim, A. K. (2010). Fluorographene: A two-dimensional counterpart of Teflon. *Small (Weinheim an der Bergstrasse, Germany), 6*(24), 2877–2884. Available from https://doi.org/10.1002/smll.201001555.

Nilsson, H. A., Duty, T., Abay, S., Wilson, C., Wagner, J. B., Thelander, C., Delsing, P., & Samuelson, L. (2008). A radio frequency single-electron transistor based on an InAs/InP heterostructure nanowire. *Nano Letters, 8*(3), 872–875. Available from https://doi.org/10.1021/nl0731062.

Novoselov, K. S., Geim, A. K., Morozov, S. V., Jiang, D., Katsnelson, M. I., Grigorieva, I. V., Dubonos, S. V., & Firsov, A. A. (2005). Two-dimensional gas of massless Dirac fermions in graphene. *Nature, 438*(7065), 197–200. Available from https://doi.org/10.1038/nature04233.

Novoselov, K. S., Geim, A. K., Morozov, S. V., Jiang, D., Zhang, Y., Dubonos, S. V., Grigorieva, I. V., & Firsov, A. A. (2004). Electric field effect in atomically thin carbon films. *Science (New York, N.Y.), 306*(5696), 666–669. Available from https://doi.org/10.1126/science.1102896.

Ohnishi, H., Kondo, Y., & Takayanagi, K. (1998). Quantized conductance through individual rows of suspended gold atoms. *Nature, 395*(6704), 780–783. Available from https://doi.org/10.1038/27399.

Osada, M., & Sasaki, T. (2012). Two-dimensional dielectric nanosheets: Novel nanoelectronics from nanocrystal building blocks. *Advanced Materials*, *24*(2), 210–228. Available from https://doi.org/10.1002/adma.201103241.

Oshima, Y., Onga, A., & Takayanagi, K. (2003). Helical gold nanotube synthesized at 150 k. *Physical Review Letters*, *91*(20). Available from https://doi.org/10.1103/PhysRevLett.91.205503.

Ouyang, M., Huang, J. L., Cheung, C. L., & Lieber, C. M. (2001). Atomically resolved single-walled carbon nanotube intramolecular junctions. *Science (New York, N.Y.)*, *291* (5501), 97–100. Available from https://doi.org/10.1126/science.291.5501.97.

Pakdel, A., Bando, Y., & Golberg, D. (2014). Nano boron nitride flatland. *Chemical Society Reviews*, *43*(3), 934–959. Available from https://doi.org/10.1039/c3cs60260e, http://pubs.rsc.org/en/journals/journal/cs.

Paladugu, M., Zou, J., Guo, Y. N., Zhang, X., Kim, Y., Joyce, H. J., Gao, Q., Tan, H. H., & Jagadish, C. (2008). Nature of heterointerfaces in GaAs/InAs and InAs/GaAs axial nanowire heterostructures. *Applied Physics Letters*, *93*(10). Available from https://doi.org/10.1063/1.2978959.

Peng, X., Manna, L., Yang, W., Wickham, J., Scher, E., Kadavanich, A., & Alivisatos, A. P. (2000). Shape control of CdSe nanocrystals. *Nature*, *404*(6773), 59–61. Available from https://doi.org/10.1038/35003535.

Peng, X., Schlamp, M. C., Kadavanich, A. V., & Alivisatos, A. P. (1997). Epitaxial growth of highly luminescent CdSe/CdS core/shell nanocrystals with photostability and electronic accessibility. *Journal of the American Chemical Society*, *119*(30), 7019–7029. Available from https://doi.org/10.1021/ja970754m.

Persson, A. I., Larsson, M. W., Stenström, S., Ohlsson, B. J., Samuelson, L., & Wallenberg, L. R. (2004). Solid-phase diffusion mechanism for GaAs nanowire growth. *Nature Materials*, *3*(10), 677–681. Available from https://doi.org/10.1038/nmat1220, http://www.nature.com/nmat/.

Philippe, C., Maria, E. M., Borg, B. M., Kimberly, A. D., Knut, D., & Lars-Erik, W. (2009). InSb heterostructure nanowires: MOVPE growth under extreme lattice mismatch. *Nanotechnology*, *20*(49).

Qian, F., Li, Y., Gradečak, S., Wang, D., Barrelet, C. J., & Lieber, C. M. (2004). Gallium nitride-based nanowire radial heterostructures for nanophotonics. *Nano Letters*, *4*(10), 1975–1979. Available from https://doi.org/10.1021/nl0487774.

Ramasse, Q. M., Seabourne, C. R., Kepaptsoglou, D.-M., Zan, R., Bangert, U., & Scott, A. J. (2013). Probing the bonding and electronic structure of single atom dopants in graphene with electron energy loss spectroscopy. *Nano Letters*, *13*(10), 4989–4995. Available from https://doi.org/10.1021/nl304187e.

Raychaudhuri, S., & Yu, E. T. (2006). Critical dimensions in coherently strained coaxial nanowire heterostructures. *Journal of Applied Physics*, *99*(11). Available from https://doi.org/10.1063/1.2202697.

Robertson, A. W., Montanari, B., He, K., Kim, J., Allen, C. S., Wu, Y. A., Olivier, J., Neethling, J., Harrison, N., Kirkland, A. I., & Warner, J. H. (2013). Dynamics of single Fe atoms in graphene vacancies. *Nano Letters*, *13*(4), 1468–1475. Available from https://doi.org/10.1021/nl304495v.

Robertson, D. H., Brenner, D. W., & Mintmire, J. W. (1992). Energetics of nanoscale graphitic tubules. *Physical Review B*, *45*(21), 12592–12595. Available from https://doi.org/10.1103/physrevb.45.12592.

Rooney, A. P., Kozikov, A., Rudenko, A. N., Prestat, E., Hamer, M. J., Withers, F., Cao, Y., Novoselov, K. S., Katsnelson, M. I., Gorbachev, R., & Haigh, S. J. (2017). Observing imperfection in atomic interfaces for van der Waals heterostructures. *Nano Letters*, *17*(9), 5222–5228. Available from https://doi.org/10.1021/acs.nanolett.7b01248, http://pubs.acs.org/journal/nalefd.

Roy, A., Kundu, S., Müller, K., Rosenauer, A., Singh, S., Pant, P., Gururajan, M. P., Kumar, P., Weissmüller, J., Singh, A. K., & Ravishankar, N. (2014). Wrinkling of atomic planes in ultrathin Au nanowires. *Nano Letters*, *14*(8), 4859–4866. Available from https://doi.org/10.1021/nl502259w, http://pubs.acs.org/journal/nalefd.

Saito, R., Dresselhaus, G., & Dresselhaus, M. S. (1998). *Physical properties of carbon nanotubes*. Imperial College Press.

Saito, R., Fujita, M., Dresselhaus, G., & Dresselhaus, M. S. (1992). Electronic structure of chiral graphene tubules. *Applied Physics Letters*, *60*(18), 2204–2206. Available from https://doi.org/10.1063/1.107080.

Sánchez-Iglesias, A., Rivas-Murias, B., Grzelczak, M., Pérez-Juste, J., Liz-Marzán, L. M., Rivadulla, F., & Correa-Duarte, M. A. (2012). Highly transparent and conductive films of densely aligned ultrathin Au nanowire monolayers. *Nano Letters*, *12*(12), 6066–6070. Available from https://doi.org/10.1021/nl3021522.

Sawada, S. I., & Hamada, N. (1992). Energetics of carbon nano-tubes. *Solid State Communications*, *83*(11), 917–919. Available from https://doi.org/10.1016/0038-1098(92)90911-R.

Schmidt, V., Wittemann, J. V., Senz, S., & Gösele, U. (2009). Silicon nanowires: A review on aspects of their growth and their electrical properties. *Advanced Materials*, *21* (25–26), 2681–2702. Available from https://doi.org/10.1002/adma.200803754.

Semenoff, G. W. (1984). Condensed-matter simulation of a three-dimensional anomaly. *Physical Review Letters*, *53*(26), 2449–2452. Available from https://doi.org/10.1103/PhysRevLett.53.2449.

Shirakawa, H., Louis, E. J., MacDiarmid, A. G., Chiang, C. K., & Heeger, A. J. (1977). Synthesis of electrically conducting organic polymers: Halogen derivatives of polyacetylene, (CH)*x*. *Journal of the Chemical Society, Chemical Communications* (16), 578–580. Available from https://doi.org/10.1039/C39770000578.

Shirakawa, H. (2001). The discovery of polyacetylene film: The dawning of an era of conducting polymers (Nobel lecture). *Angewandte Chemie—International Edition*, *40*(14), 2574–2580. Available from http://onlinelibrary.wiley.com/journal/10.1002/(ISSN)1521-3773, https://doi.org/10.1002/1521-3773(20010716)40:14 < 2574::aid-anie2574 > 3.0.co;2-n.

Slonczewski, J. C., & Weiss, P. R. (1958). Band structure of graphite. *Physical Review*, *109* (2), 272–279. Available from https://doi.org/10.1103/physrev.109.272.

Smith, A. M., & Nie, S. (2010). Semiconductor nanocrystals: Structure, properties, and band gap engineering. *Accounts of Chemical Research*, *43*(2), 190–200. Available from https://doi.org/10.1021/ar9001069.

Smith, A. M., Mohs, A. M., & Nie, S. (2009). Tuning the optical and electronic properties of colloidal nanocrystals by lattice strain. *Nature Nanotechnology*, *4*(1), 56–63. Available from https://doi.org/10.1038/nnano.2008.360.

Song, X., Gao, J., Nie, Y., Gao, T., Sun, J., Ma, D., Li, Q., Chen, Y., Jin, C., Bachmatiuk, A., Rümmeli, M. H., Ding, F., Zhang, Y., & Liu, Z. (2015). Chemical vapor deposition growth of large-scale hexagonal boron nitride with controllable orientation. *Nano Research*, *8*(10), 3164–3176. Available from https://doi.org/10.1007/s12274-015-0816-9, http://www.springer.com/materials/nanotechnology/journal/12274.

Spencer, B. J., & Tersoff, J. (2000). Dislocation energetics in epitaxial strained islands. *Applied Physics Letters*, 77(16), 2533–2535. Available from https://doi.org/10.1063/1.1319186.

Stone, A. J., & Wales, D. J. (1986). Theoretical studies of icosahedral C_{60} and some related species. *Chemical Physics Letters*, *128*(5–6), 501–503. Available from https://doi.org/10.1016/0009-2614(86)80661-3.

Takeda, K., & Shiraishi, K. (1994). Theoretical possibility of stage corrugation in Si and Ge analogs of graphite. *Physical Review B*, *50*(20), 14916–14922. Available from https://doi.org/10.1103/PhysRevB.50.14916.

Tang, D. M., Kvashnin, D. G., Cretu, O., Nemoto, Y., Uesugi, F., Takeguchi, M., Zhou, X., Hsia, F. C., Liu, C., Sorokin, P. B., Kawamoto, N., Mitome, M., Cheng, H. M., Golberg, D., & Bando, Y. (2018). Chirality transitions and transport properties of individual few-walled carbon nanotubes as revealed by in situ TEM probing. *Ultramicroscopy*, *194*, 108–116. Available from https://doi.org/10.1016/j.ultramic.2018.07.012, http://www.elsevier.com/locate/ultramic.

Tang, D.-M. (2010). *Situ transmission electron microscopy studies of carbon nanotube nucleation mechanism and carbon nanotube-clamped metal atomic chains*. Chinese Academy of Sciences.

Tang, D.-M., Yin, L.-C., Li, F., Liu, C., Yu, W.-J., Hou, P.-X., Wu, B., Lee, Y.-H., Ma, X.-L., & Cheng, H.-M. (2010). Carbon nanotube-clamped metal atomic chain. *Proceedings of the National Academy of Sciences*, *107*(20), 9055–9059. Available from https://doi.org/10.1073/pnas.0914970107.

Tao, F., Grass, M. E., Zhang, Y., Butcher, D. R., Renzas, J. R., Liu, Z., Chung, J. Y., Mun, B. S., Salmeron, M., & Somorjai, G. A. (2008). Reaction-driven restructuring of Rh–Pd and Pt–Pd core–shell nanoparticles. *Science (New York, N.Y.)*, *322*(5903), 932–934. Available from https://doi.org/10.1126/science.1164170.

Taylor, R., Hare, J. P., Abdul-Sada, A. K., & Kroto, H. W. (1990). Isolation, separation and characterisation of the fullerenes C_{60} and C_{70}: The third form of carbon. *Journal of the Chemical Society, Chemical Communications* (20), 1423–1425. Available from https://doi.org/10.1039/C39900001423.

Terabe, K., Hasegawa, T., Nakayama, T., & Aono, M. (2005). Quantized conductance atomic switch. *Nature*, *433*(7021), 47–50. Available from https://doi.org/10.1038/nature03190.

Thrower, P. A. (1969). The study of defects in graphite by transmission electron microscopy. *Chemistry And Physics of Carbon*, *5*, 217–320.

Tran, T. T., Bray, K., Ford, M. J., Toth, M., & Aharonovich, I. (2016). Quantum emission from hexagonal boron nitride monolayers. *Nature Nanotechnology*, *11*(1), 37–41. Available from https://doi.org/10.1038/nnano.2015.242, http://www.nature.com/nnano/index.html.

Turkevich, J., Stevenson, P. C., & Hillier, J. (1951). A study of the nucleation and growth processes in the synthesis of colloidal gold. *Discussions of the Faraday Society*, *11*, 55–75. Available from https://doi.org/10.1039/DF9511100055.

Vogt, P., De Padova, P., Quaresima, C., Avila, J., Frantzeskakis, E., Asensio, M. C., Resta, A., Ealet, B., & Le Lay, G. (2012). Silicene: Compelling experimental evidence for graphenelike two-dimensional silicon. *Physical Review Letters*, *108*(15). Available from https://doi.org/10.1103/PhysRevLett.108.155501, http://oai.aps.org/filefetch?identifier = 10.1103/PhysRevLett.108.155501&component = fulltext&escription = markup&format = xml, France.

Wallace, P. R. (1947). The band theory of graphite. *Physical Review*, *71*(9), 622–634. Available from https://doi.org/10.1103/physrev.71.622.

Wang, D., Xin, H. L., Hovden, R., Wang, H., Yu, Y., & Muller, D. A. (2012). Structurally ordered intermetallic platinum–cobalt core–shell nanoparticles with enhanced activity and stability as oxygen reduction electrocatalysts. *Nature Materials*, *12*, 81–87.

Wang, K., Chen, J. J., Zeng, Z. M., Tarr, J., Zhou, W. L., Zhang, Y., Yan, Y. F., Jiang, C. S., Pern, J., & Mascarenhas, A. (2010). Synthesis and photovoltaic effect of vertically aligned ZnO/ZnS core/shell nanowire arrays. *Applied Physics Letters*, *96*(12). Available from https://doi.org/10.1063/1.3367706.

Wang, Q. H., Kalantar-Zadeh, K., Kis, A., Coleman, J. N., & Strano, M. S. (2012). Electronics and optoelectronics of two-dimensional transition metal dichalcogenides. *Nature Nanotechnology*, 7(11), 699–712. Available from https://doi.org/10.1038/nnano.2012.193, http://www.nature.com/nnano/index.html.

Wang, S., Zhang, Y., Abidi, N., & Cabrales, L. (2009). Wettability and surface free energy of graphene films. *Langmuir: The ACS Journal of Surfaces and Colloids*, *25*(18), 11078–11081. Available from https://doi.org/10.1021/la901402f, http://pubs.acs.org/doi/pdfplus/10.1021/la901402f, United States.

Warner, J. H., Margine, E. R., Mukai, M., Robertson, A. W., Giustino, F., & Kirkland, A. I. (2012). Dislocation-driven deformations in graphene. *Science (New York, N.Y.)*, *337*(6091), 209–212. Available from https://doi.org/10.1126/science.1217529, http://www.sciencemag.org/content/337/6091/209.full.pdf.

Waser, R., & Aono, M. (2007). Nanoionics-based resistive switching memories. *Nature Materials*, *6*(11), 833–840. Available from https://doi.org/10.1038/nmat2023.

Wei, L., & Charles, M. L. (2006). Semiconductor nanowires. *Journal of Physics D: Applied Physics*, *39*(21), R387–R406.

White, C. T., Robertson, D. H., & Mintmire, J. W. (1993). Helical and rotational symmetries of nanoscale graphitic tubules. *Physical Review B*, *47*(9), 5485–5488. Available from https://doi.org/10.1103/physrevb.47.5485.

Wirtz, L., & Rubio, A. (2004). The phonon dispersion of graphite revisited. *Solid State Communications*, *131*(3–4), 141–152. Available from https://doi.org/10.1016/j.ssc.2004.04.042.

Wu, Y., Cui, Y., Huynh, L., Barrelet, C. J., Bell, D. C., & Lieber, C. M. (2004). Controlled growth and structures of molecular-scale silicon nanowires. *Nano Letters*, *4*(3), 433–436. Available from https://doi.org/10.1021/nl035162i.

Xiang, J., Lu, W., Hu, Y., Wu, Y., Yan, H., & Lieber, C. M. (2006). Ge/Si nanowire heterostructures as high-performance field-effect transistors. *Nature*, *441*(7092), 489–493. Available from https://doi.org/10.1038/nature04796, http://www.nature.com/nature/index.html.

Xiang, R., Inoue, T., Zheng, Y., Kumamoto, A., Qian, Y., Sato, Y., Liu, M., Tang, D., Gokhale, D., Guo, J., Hisama, K., Yotsumoto, S., Ogamoto, T., Arai, H., Kobayashi, Y., Zhang, H., Hou, B., Anisimov, A., Maruyama, M., ... Maruyama, S. (2020). One-dimensional van der Waals heterostructures. *Science (New York, N.Y.)*, *367*(6477), 537–542. Available from https://doi.org/10.1126/science.aaz2570, https://science.sciencemag.org/content/367/6477/537/tab-pdf.

Xiong, Q., Wang, J., & Eklund, P. C. (2006). Coherent twinning phenomena: Towards twinning superlattices in III-V semiconducting nanowires. *Nano Letters*, *6*(12), 2736–2742. Available from https://doi.org/10.1021/nl0616983.

Yakobson, B. I. (1998). Mechanical relaxation and "intramolecular plasticity" in carbon nanotubes. *Applied Physics Letters*, *72*(8), 918–920. Available from https://doi.org/10.1063/1.120873.

Yang, H. G., Sun, C. H., Qiao, S. Z., Zou, J., Liu, G., Smith, S. C., Cheng, H. M., & Lu, G. Q. (2008). Anatase TiO_2 single crystals with a large percentage of reactive facets. *Nature*, *453*(7195), 638–641. Available from https://doi.org/10.1038/nature06964.

Yang, P., Yan, H., Mao, S., Russo, R., Johnson, J., Saykally, R., Morris, N., Pham, J., He, R., & Choi, H. J. (2002). Controlled growth of ZnO nanowires and their optical properties. *Advanced Functional Materials*, *12*(5), 323–331. Available from https://doi.org/10.1002/1616-3028(20020517)12:5 < 323::AID-ADFM323 > 3.0.CO;2-G.

Yang, X. F., Wang, A., Qiao, B., Li, J., Liu, J., & Zhang, T. (2013). Single-atom catalysts: A new frontier in heterogeneous catalysis. *Accounts of Chemical Research*, *46*(8), 1740–1748. Available from https://doi.org/10.1021/ar300361m.

Yao, Y., Huang, Z., Xie, P., Lacey, S. D., Jacob, R. J., Xie, H., Chen, F., Nie, A., Pu, T., Rehwoldt, M., Yu, D., Zachariah, M. R., Wang, C., Shahbazian-Yassar, R., Li, J., & Hu, L. (2018). Carbothermal shock synthesis of high-entropy-alloy nanoparticles. *Science (New York, N.Y.)*, *359*(6383), 1489–1494. Available from https://doi.org/10.1126/science.aan5412, http://science.sciencemag.org/content/359/6383/1489/tab-pdf.

Yu, G. L., Gorbachev, R. V., Tu, J. S., Kretinin, A. V., Cao, Y., Jalil, R., Withers, F., Ponomarenko, L. A., Piot, B. A., Potemski, M., Elias, D. C., Chen, X., Watanabe, K., Taniguchi, T., Grigorieva, I. V., Novoselov, K. S., Fal'ko, V. I., Geim, A. K., & Mishchenko, A. (2014). Hierarchy of Hofstadter states and replica quantum Hall ferromagnetism in graphene superlattices. *Nature Physics*, *10*(7), 525–529. Available from https://doi.org/10.1038/nphys2979.

Yu, Y., Cui, F., Sun, J., & Yang, P. (2016). Atomic structure of ultrathin gold nanowires. *Nano Letters*, *16*(5), 3078–3084. Available from https://doi.org/10.1021/acs.nanolett.6b00233, http://pubs.acs.org/journal/nalefd.

Zhang, S., Kang, L., Wang, X., Tong, L., Yang, L., Wang, Z., Qi, K., Deng, S., Li, Q., Bai, X., Ding, F., & Zhang, J. (2017). Arrays of horizontal carbon nanotubes of controlled chirality grown using designed catalysts. *Nature*, *543*(7644), 234–238. Available from https://doi.org/10.1038/nature21051, http://www.nature.com/nature/index.html.

Zhang, Y., Tan, Y. W., Stormer, H. L., & Kim, P. (2005). Experimental observation of the quantum Hall effect and Berry's phase in graphene. *Nature*, *438*(7065), 201–204. Available from https://doi.org/10.1038/nature04235.

Zhen, Y., Postma, H. W. C., Balents, L., & Dekker, C. (1999). Carbon nanotube intramolecular junctions. *Nature*, *402*(6759), 273–276. Available from https://doi.org/10.1038/46241.

Zheng, Y., Kumamoto, A., Hisama, K., Otsuka, K., Wickerson, G., Sato, Y., Liu, M., Inoue, T., Chiashi, S., Tang, D.-M., Zhang, Q., Anisimov, A., Kauppinen, E. I., Li, Y., Suenaga, K., Ikuhara, Y., Maruyama, S., & Xiang, R. (2021). One-dimensional van der Waals heterostructures: Growth mechanism and handedness correlation revealed by nondestructive TEM. *Proceedings of the National Academy of Sciences*, *118*(37). Available from https://doi.org/10.1073/pnas.2107295118.

Zhou, W., Zou, X., Najmaei, S., Liu, Z., Shi, Y., Kong, J., Lou, J., Ajayan, P. M., Yakobson, B. I., & Idrobo, J. C. (2013). Intrinsic structural defects in monolayer molybdenum disulfide. *Nano Letters*, *13*(6), 2615–2622. Available from https://doi.org/10.1021/nl4007479.

Zhu, Y., Murali, S., Cai, W., Li, X., Suk, J. W., Potts, J. R., & Ruoff, R. S. (2010). Graphene and graphene oxide: Synthesis, properties, and applications. *Advanced Materials*, *22*(35), 3906–3924. Available from https://doi.org/10.1002/adma.201001068, http://onlinelibrary.wiley.com/doi/10.1002/adma.201001068/pdf, United States.

Zomer, P. J., Dash, S. P., Tombros, N., & van Wees, B. J. (2011). A transfer technique for high mobility graphene devices on commercially available hexagonal boron nitride. *Applied Physics Letters*, *99*(23). Available from https://doi.org/10.1063/1.3665405.

CHAPTER THREE

Controlled fabrication of low-dimensional materials

Essentially, materials science is about the structure–property relationship, and structural control is a prerequisite of realizing the function and applications of materials. Synthesis and processing of traditional metal materials includes controlling the microstructure, surface state and morphology of components, as well as the connection among components. It can be divided into three types: material growth, assembly and processing, more specifically smelting, casting, forging, heat treatment, machining, surface modification, welding, riveting, etc. (Fig. 3.1). For low-dimensional materials, the preparation and structural control is the material basis for exploring and testing low-dimensional theories, and a practical way to realize the excellent properties and unique functions.

As shown in Fig. 3.1, as is the case with conventional materials, the fabrication of low-dimensional materials involves growth, processing, and assembly. In the following, Section 3.1 describes the basic thermodynamics and kinetics of low-dimensional material growth and assembly; Section 3.2 introduces the bottom-up approaches to grow low-dimensional material; Section 3.3 discusses about the assembly of

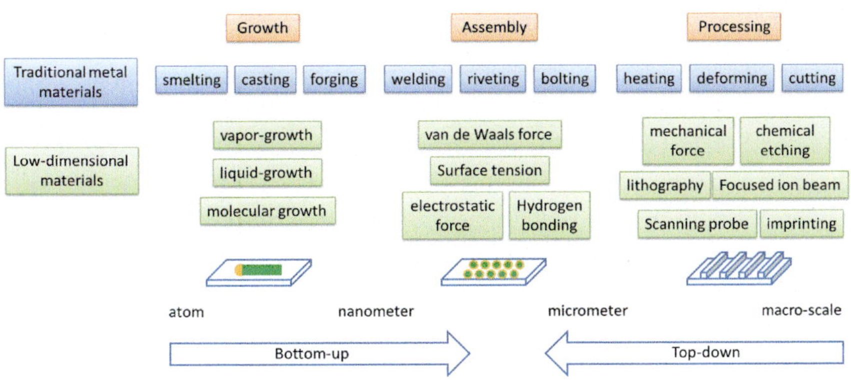

Figure 3.1 Preparation technologies for growth, assembly and processing of traditional metal materials and low-dimensional materials. Both can be divided into bottom-up and top-down strategies.

Low-Dimensional Materials.
DOI: https://doi.org/10.1016/B978-0-443-29046-6.00009-2

low-dimensional materials for the construction of hierarchical structures; Section 3.4 introduces top-down methods for processing low-dimensional materials; and Section 3.5 summarizes achievements, challenges, and future directions in controlled fabrication of low-dimensional materials.

3.1 Fundamentals of low-dimensional materials fabrication

3.1.1 Bottom-up and top-down approaches

Material fabrication can be categorized into two routes: "bottom-up" growth and assembly, and "top-down" processing. Chemical synthesis is a typical bottom-up approach, which could produce a large number of identical molecules with molecular uniformity. Traditional material processing, on the other hand, is a typical top-down approach, which removes, deforms, and reorganizes bulk materials. Low-dimensional materials are in-between molecules and bulk materials, with the size being close to the molecules in certain directions and approaching macroscale in the unconfined directions. Therefore, it is necessary to control the structures during synthesis at multiple levels in terms of composition, scale, dimensions, shapes, surfaces and interfaces, etc.

Bottom-up approaches include the growth and assembly of low-dimensional materials. They both follow the same basic principles of thermodynamics and dynamics, but the specific roles, energies, and processes involved are far different. The growth of low-dimensional structures is mainly the formation of strong chemical bonds (e.g. covalent bonds between atoms). Whereas, assembly is the connection and stacking of building blocks by weak interactions, such as van der Waals (vdW) forces. A top-down approach is mainly a process in which materials with smaller size are formed by size reduction, morphology control, and reconnection of materials under external forces. For a practical fabrication process, it is often a combination of both bottom-up and top-down approaches. Fig. 3.2 shows the typical template methods for a mass preparation of low-dimensional materials. By templating, regularly ordered pore or surface structures can be effectively prepared, and micro- and macroscopically ordered structures can also be controlled by confinement effects.

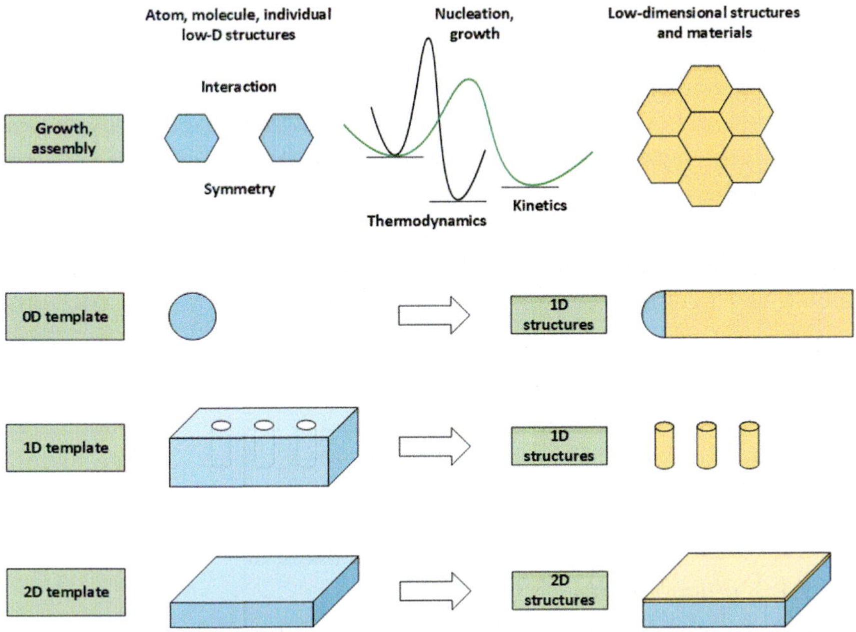

Figure 3.2 Growth and assembly of low-dimensional materials. Both processes undergo nucleation and growth processes to synthesize low-dimensional and assembled structures under the influence of monomer interactions and symmetry, which are controlled by thermodynamics and kinetics. While the growth takes atoms and molecules as building blocks, the assembly takes low-dimensional structures as building blocks. Template method is a commonly used strategy for controlling dimensionality and scale of materials, such as the growth of nanowires or nanotubes from zero-dimensional nanoparticles, the growth of nanowires from one-dimensional channels, and the growth of two-dimensional materials templated by two-dimensional surface.

3.1.2 Growth and assembly thermodynamics

Thermodynamics and kinetics of material growth and assembly are the basis for microstructural control. The former answers questions of the most likely stable structure under equilibrium conditions. The latter answers questions of the competitive formation process of different configurations under certain environment. Thermodynamic statistics are based on the basic assumption of a large system, which contains enough individual building blocks or traverses all possible configurations for thermodynamic equilibriums over a long enough period of time. Growth of low-dimensional materials, however, is a process that occurs in a finite system at a finite time, which requires the consideration of both spatial and temporal scales.

Taking isothermal and isobaric thermodynamic process as an example, $\Delta G = \Delta H - T\Delta S$, where G is the Gibbs free energy, H is the enthalpy related to the system's internal energies and interactions, T is the temperature, S is the entropy. Growth process of materials involves a multiphase process under the action of external fields such as gravity, electric field, and magnetic field; hence the chemical potentials in the external fields should be taken into account in the growth of materials. For an isothermal and isobaric process, $dG = \sum^{n} \mu_i dN_i$, where the action of external fields is included in the chemical potential μ. For the growth and assembly of low-dimensional materials, it is necessary to understand the interactions, the sources of energy, and entropy in the system under certain reaction conditions.

3.1.2.1 Binding energy

Material growth is primarily the formation of strong chemical bonds, so that the binding energy is the primary form of thermodynamic energy to be considered. The types of chemical bonding determine the material properties and to some extent the routes of synthesis and processing. Metallic low-dimensional materials are bonded by metallic bonds, while semiconductors and dielectric materials are bonded by covalent or ionic bonds. Free electrons in metallic materials become a shared sea of electrons, while ionic bonds form anions and cations through charge transfer. Metallic and ionic bonded materials can be synthesized through ionic reactions in solution. Covalently bonded materials, in which the electrons form localized shared electron pairs, have higher binding energies and usually react at higher activation energies and higher temperature.

Assembly of low-dimensional materials, on the other hand, is an ordering process under weak chemical bonds and their interactions, such as electrostatic forces, hydrogen bonds, vdW forces, and host–guest interactions. The activation energy is relatively low due to the weak bonding, so that the assembly is usually carried out under mild conditions. For example, Langmuir–Blodgett assembly process occurs at the surface of a liquid, where the interaction is kept at the level of $k_B T$ and progressively approaches the equilibrium structure with the assistance of thermal perturbations.

3.1.2.2 Surface and interface energy

The most striking feature of low-dimensional structures compared to bulk materials is the large proportion of surface atoms, and thus the surface or interface energy is an important energy term. For instance, the reason why fullerene and carbon nanotube (CNT) are thermodynamic stable is that all of the dangling bonds of graphene layers are closed to reduce interface energy. According to Wulff's theorem, the total surface energy of a system tends to be minimized at a given volume. Taking the face-centered cubic (FCC) structure as an example, the surface energy is ranked as (1 1 1) < (1 0 0) < (1 1 0) and thus tends to be a truncated octahedron surrounded by (1 1 1) and (1 0 0) facets. In general, the surface or interfacial energy of a material determines the equilibrium shape of a crystal.

3.1.2.3 Effect of entropy

Low-dimensional structures can be viewed as a macromolecule. As the number of atoms increases, several molecular configurations with comparable binding energies can be formed. The difference in configurational entropies is thus an important factor in determining their relative stability. In the case of fullerenes and CNTs, it is required to have 12 nonadjacent five-membered rings to form a closed cage-like graphite structure. Under topological and geometrical restrictions, the larger the number of constituent carbon atoms, the more isomers with similar binding energies are available. For instance, Bichara et al. proposed that single-walled CNTs can grow chiral because of the configurational entropy of tube edges, which can explain their temperature evolutions of chiral distributions (Magnin et al., 2018).

Due to the inner weak interaction, entropy plays a more important role for the assembly process of low-dimensional materials compared to the growth process (Manoharan, 2015). Entropy is usually thought of as a measure of disorder and is more accurately defined as the number of possible configurations with the number of variables (occupied volumes) in the phase space of a system. For solid-state materials, the entropy mainly comes from the vibrational and rotational degrees of freedoms of the atoms, when the degrees of freedom of translational motion have been frozen. For a simple hard-sphere stacking model, the configuration entropy is embodied in the free volume. Because close-packed structures make a full use of three-dimensional space with a large effective volume, their degrees of freedom and configuration entropy are higher. Therefore the ideal hard-sphere model under the entropy maximum principle is tightly stacked FCC structure (Bolhuis et al., 1997).

3.1.2.4 Effects of external field

Chemical bonding is usually limited to the interaction between neighboring atoms or assembly units and is classified as a near-neighbor interaction. The bonding interaction rapidly reduces as the increasing of the distance r between them. For instance, the coulomb forces are proportional to r^{-2}, and vdW forces conform to the relationship of r^{-6}. The external field effect, on the other hand, can create a relatively homogeneous field in a large space, such as gravity field, electric field, and magnetic field, which plays an important role in the formation of a large-scale-ordered assembly process.

3.1.3 Kinetics of growth and assembly

Growth process of materials involves a series of kinetic processes such as reaction, nucleation, growth, and diffusion. Understanding the kinetic features, such as activation energies corresponding to the different configurations and processes at each step, is the foundation for the structural control. Growth kinetics can be viewed as a process, in which the state points of a system take a walk on an energy landscape and then achieve the control over structures. It is usually hard and complex to obtain the information on this potential energy surface, leading to a great challenge in the study of growth kinetics of low-dimensional materials.

Nucleation is the key step in determining the final structures of materials. Generally, symmetry will be reduced during subsequent growth process, compared with the symmetry of nucleus. Nucleation is driven by a decrease in the Gibbs free energy of the system: $\Delta G = -V\rho\Delta\mu + A\gamma + E_{\text{strain}} - E_{\text{defect}}$, where V is volume, ρ is density of particles, $\Delta\mu$ is the difference between chemical potentials of two phases, A is area, γ is surface energy, the last two terms are strain energy and defect energy, respectively. There is a critical size for the nucleation, corresponding to the point at which the continued growth of the crystal embryo causes the Gibbs free energy to increase or decrease. Particles smaller than the critical size will be dissolved, while particles larger than the critical size can continue to grow.

Nucleation can be categorized into homogenous and heterogeneous ones, depending on the absence or presence of a nucleation center. Due to the existence of heterogeneous interfaces, the critical size corresponding to heterogeneous nucleation is much smaller than that of homogeneous nucleation. Thereby heterogenous nucleation is more likely to occur, and the nuclei structure can be controlled by interfacial interactions. Typically, the liquid-phase growth of nanocrystals is a homogeneous nucleation process, whereas the growth of one- and two-dimensional materials is a heterogeneous nucleation process that occurs at the interface of the catalyst or substrate.

The assembly process of low-dimensional materials shows different kinetic characteristics compared to the growth process, due to the larger assembly unit and weaker interactions. When it reaches thermal equilibrium at a certain temperature, the average kinetic energy of the particles is $k_B T$, which can also be expressed as $\frac{1}{2}mv^2$. For a certain kinetic energy, the larger the particle, the smaller the average reaction velocity and the longer it takes to reach equilibrium. Weidman et al. characterized the formation process of PbS nanocrystalline superlattices by means of in situ grazing-incidence X-ray scattering. The structures start from monodisperse colloidal particles and undergo an intermediate state from FCC to body-centered orthogonal and then approach final body-centered cubic superlattice structures, over the course of ~30 minutes (Weidman et al., 2016).

3.1.4 Strategies to control the structures of low-dimensional materials

Based on the above principles of growth and assembly, the structural control strategies of low-dimensional materials are categorized into thermodynamic and kinetic approaches. Under a certain condition, the longer the reaction time, the higher the probability of obtaining a thermodynamically stable configuration. The kinetic control, on the other hand, requires a consideration of the relative competition rate among different configurations, hence the optimal conditions are near the critical conditions under which the desired configurations can grow.

Structural control can be specifically categorized into three main strategies, which are determined by intrinsic properties, template induction, and external forces, respectively. The extreme low-dimensional materials are limited by the intrinsic structures, for example, two-dimensional materials are usually derived from the bulk-layered crystals. In the small-scale conditions of low-dimensional materials, intrinsic properties and stability will change if the size-dependent energy terms change. A classic example is the specific configuration of C_{60} molecules, whose particular stability is the result of edge closure and symmetry effects, although they are highly curved and subjected to large strains. The template induction strategy includes catalysts and substrates in chemical vapor deposition (CVD) methods, as well as micelles in liquid phase growth, etc. For two-dimensional materials that are predicted to be thermodynamically unstable, the key to improving the stability and enabling the controlled growth is the bonding to a substrate. Understanding the growth mechanism of low-dimensional materials, including interfacial reactions involving catalysts and mass transport processes, is fundamental to modulating the structure of one-dimensional materials such as CNTs and nanowires. External forces strategy plays an important role on a large-scale-ordered assembly. Recently, it has been found that the transport properties (either metallic or semiconducting) of CNTs can be strongly affected by the electrostatic fields during growth (Wang et al., 2018).

3.2 Growth of low-dimensional materials

Bottom-up growth of low-dimensional materials involves strong chemical bond connections, such as atomic-scale covalent bonds, which is controlled by the thermodynamics and kinetics of chemical reactions and crystal growth. The influencing factors consist of the intrinsic phase stability of crystals, symmetry, surface energy, chemical environment, template, etc. This section will be divided into three parts: molecular synthesis, liquid-phase growth, and vapor-phase growth. We will mainly introduce controllable strategies of low-dimensional materials in terms of intrinsic properties, template induction, external field regulation, along with the understanding of growth mechanisms such as the chemical reactions, crystal growth and mass transportation, etc. Finally, we will briefly introduce the application of new technologies such as machine learning in simulating and optimizing growth.

3.2.1 Molecular synthesis

Richard Feynman delivered a famous lecture titled "There is plenty of room at the bottom" (Feynman, 1960). In the past decades, with the invention and development of scanning probe electron microscopy, it becomes a reality to manipulate individual atoms and construct structures with quantum properties (Crommie et al., 1993; Eigler & Schweizer,

1990). Besides, Sauvage, Stoddart and Feringa won the 2016 Nobel Prize in Chemistry for "designing and fabricating molecular machines by means of molecular synthesis," demonstrating the promise of bottom-up design and fabrication of functional devices from atoms and molecules. In principle, molecules can be precisely controlled at the atomic-level through self-assembly or covalent bond coupling, thus precisely controlling their physicochemical properties. This is a particularly attractive fabrication route.

Molecular synthesis of low-dimensional materials is usually carried out using scanning probe microscopy (SPM) techniques, which form self-assembled nanostructures by increasing the opportunities for molecules to dwell, collide, and react on the surface of a metallic single-crystal substrate. The catalytic activity of metals is utilized to facilitate the coupling reaction between molecules to form functional macromolecules or low-dimensional structures by covalent bonding. Additionally, by means of scanning tunneling microscopy and atomic force microscopy (AFM), the molecular configurations before and after reactions as well as intermediate configurations are in situ characterized, and the properties such as electronic structures are also investigated.

As shown in Fig. 3.3, zero-dimensional graphene quantum dots, one-dimensional graphene nanoribbons, and two-dimensional graphene crystals can be constructed using benzene ring as building blocks (Chen et al.,

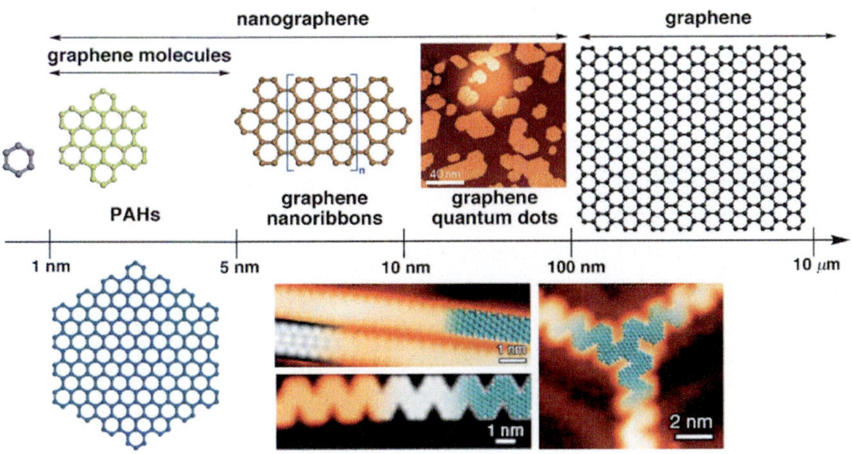

Figure 3.3 Molecular synthesis of graphene. From benzene ring to zero-dimensional graphene quantum dots, one-dimensional graphene nanoribbons, and two-dimensional graphene crystals (Chen et al., 2012).

2012). In 2007, Grill et al. prepared covalently bound nanostructures with controlled shape and size such as dimers, linear chains, and two-dimensional arrays by designing and synthesizing molecules with a different number of predefined connection points (Grill et al., 2007). In 2010, Fasel and Müllen et al. synthesized atomically precise graphene nanoribbons of a width only 1 nm, which used a surface-assisted coupling of molecular precursors into linear polyphenylenes and their subsequent cyclodehydrogenation (Cai et al., 2010).

The key to the molecular synthesis is the design of molecular precursors, which can be evolved into isotropic or anisotropic low-dimensional structures with different dimensions depending on their symmetry. Fasel et al. utilized polyphenylene molecules as precursors to synthesize triangular nanographenes on Cu (1 1 1) by intramolecular aryl–aryl coupling reactions (Treier et al., 2011). By designing specific molecular precursors as building blocks, low-dimensional porous structures can also be synthesized. With a surface-promoted aryl–aryl coupling of hexaiodo-substituted macrocycle cyclohexa-*m*-phenylene, Fasel et al. fabricated regular covalently linked two-dimensional polyphenylene networks with subnanometer periodicity (Bieri et al., 2009).

Although the structure and properties of C_{18} rings have been theoretically investigated as early as 1966, the experimental evidence remains a great challenge. In 2019, Kaiser et al. successfully manipulated the molecular reactions by using AFM and synthesized a pure carbon ring containing 18 carbon atoms (Kaiser et al., 2019). They first synthesized the cyclic carbon oxide $C_{24}O_6$ as precursors and then removed CO molecules one by one with the tip of a needle, leaving a C_{18} ring. In situ characterization showed that this is a structure with alternating triple and single bonds through sp hybridization. This work perfectly incorporates both bottom-up synthesis and top-down processing at the molecular level.

As the number of atoms increases, it becomes increasingly difficult to synthesize molecules in specific configurations. Therefore it is desirable to synthesize a seed by molecular synthesis methods and then maintain the seed structure by epitaxial growth, which can take the advantage of molecularly designed synthesis while overcoming the limitation of the magnitude of systems. For instance, Fasel et al. designed a polycyclic hydrocarbon $C_{96}H_{54}$ as a precursor molecule, synthesized a carbon cap structure of (6,6) single-walled CNTs by C–C coupling dehydrogenation reaction, and then obtained single-chiral nanotubes by epitaxial growth (Sanchez-Valencia et al., 2014).

3.2.2 Liquid-phase growth

The history of liquid-phase growth of low-dimensional materials dates back to the 19th century, when Faraday synthesized Au nanoparticles using the solution method. Using this method, a number of systems, including metals, oxides, sulfides, II-VI, and III-V semiconductors, can be effectively controlled in terms of their composition, size, shape, defects, and interfaces. As a result, a large quantity of structurally homogeneous liquid or sol−gel stabilized low-dimensional materials can be obtained. The mild reaction conditions facilitate large-scale production and real applications.

For the growth of nanocrystals, it is necessary to consider the interactions between precursors, surface ligands and solvent environments, including precursor chemistry and nanocrystal nucleation. The former mainly involves the reduction and decomposition of metal salts. With the increasing concentration, seeds nucleate homogeneously after reaching a supersaturation. The stability of nanocrystals in solution is closely related to the surface ligands and their interactions. The binding strength of the surface ligands to the different crystalline facets is an important means of regulating the surface energy of crystal facets, the growth rate, as well as the size and shape of the products.

3.2.2.1 Size control

Liquid-phase growth of nanocrystals is usually a homogeneous nucleation-growth process, and its size control depends on the understanding of the nucleation and growth mechanisms. One of the major challenges is Ostwald ripening, where small particles are subsumed or dissolved as the reaction proceeds and the concentration of reactants decreases, while large particles continue to grow due to surface energy, resulting in an uneven size distribution of crystals. To break through the ripening phenomenon, one route is to control nucleation to avoid the appearance of crystal embryos of inhomogeneous size, and the other route is to control the degree of supersaturation.

LaMer and Dinegar from Columbia university pointed out that the requirements for obtaining monodispersed colloid particles are single nucleation rather than sequential nucleation, and slow growth at a small supersaturation (Lamer & Dinegar, 1950). The Brus group at Bell laboratory and Bawendi group at Massachusetts Institute of Technology (MIT) proposed the method of "Organometallic Synthesis in Thermal

Coordination Solvents," in which reactant precursors are rapidly injected into solutions, resulting in instantaneous massive nucleation followed by a decrease in concentration and supersaturation. Nearly monodisperse CdSe and other II–VI nanocrystals were then achieved (Murray et al., 1993; Steigerwald & Brus, 1990). Alivisatos and Peng from Lawrence Berkeley National Laboratory (United States) proposed a method of distribution "focusing" to manipulate average size and size distribution of nanocrystals kinetically, which occurs when the nanocrystals in solutions are all slightly larger than critical size at a given monomer concentration and grow under near-equilibrium conditions. The smaller crystals grow faster than the larger ones, and thus the size distribution can be focused (Peng & Peng, 2001; Peng et al., 2000), which has been generally applied in the synthesis of II-VI and III-V semiconducting low-dimensional materials (Peng et al., 1998).

3.2.2.2 Shape control

The catalytic and other properties of nanocrystals vary significantly from one surface to another, so controlling the facets and shapes of nanocrystals is an important issue in nanocrystal growth. So far, the main regulating parameter has been the difference in surface energy between different crystal facets, which is realized by the selective absorption of surface ligands.

Xia et al. at the University of Washington summarized the principles for generating shape-controlled metal nanocrystals (Fig. 3.4) (Xiong & Xia, 2007). They emphasized the influence of defects including stacking faults and different numbers of twin defects, as well as the order of surface energies for different facets in determining the shape evolution during crystal growth. A variety of nanocrystal shapes such as spheres, cubes, octahedrons, decahedrons, icosahedrons, nanosheets, and nanorods have been achieved for a large number of different metals, including Pd, Ag, Au, Pt, Rh, Bi, Fe, and Co (Xia et al., 2009).

Nanocrystal surface at the atomic level is important for structure control, but is also quite complex. For instances, PbS nanocrystals can have either nonpolar (1 1 1) facets with one half of the terminal Pb atoms and equal number of S atoms, or full Pb-terminated polar (1 1 1) facets. These different arrangements of atoms determine the adsorption of molecules or ions on crystal surface. Wang et al. at Lawrence Berkeley National Laboratory, United States, successfully modulated the ratio of (1 1 1) and (0 0 1) surfaces to predict and prepare PbS nanocrystals of different shapes,

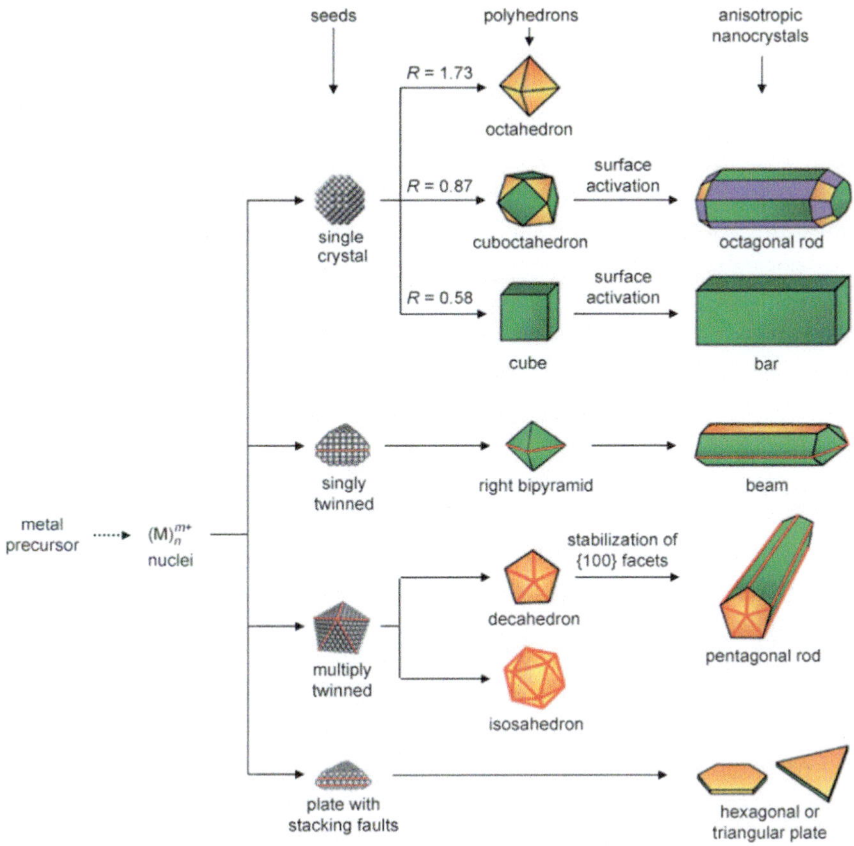

Figure 3.4 Different shapes of palladium nanocrystals under the control of growth thermodynamics and kinetics. The *R* is the ratio between the crystal growth rates along the <1 0 0> to <1 1 1> directions (Xiong & Xia, 2007).

by taking advantages of the differences in the binding energies between surface OH^- groups and different surfaces (Zherebetskyy et al., 2014).

In general, crystal surfaces tend to consist of facets with low surface energy and high stability; however, highly stable facets are usually less chemically active. For anatase TiO_2 nanocrystals, which is a kind of important photocatalysts, $\{1\ 0\ 1\}$ facets are the most stable surfaces under clean and hydrogenated conditions, while the more active ones are $\{0\ 0\ 1\}$ facets. The University of Queensland, Lu et al. found that surrounded fluorine atoms can change the relative stabilities of surfaces of anatase TiO_2, $\{0\ 0\ 1\}$ facets are energetically favorable to $\{1\ 0\ 1\}$ facets, and thus the TiO_2 crystals with a high percentage of the $\{0\ 0\ 1\}$ facets have been prepared (Yang et al., 2008). On this basis, Liu et al. from Institute of Metal Research, Chinese Academy of Sciences, tailored the surfaces of TiO_2 crystals to specific morphologies with different dominant facets (Liu et al., 2014). Recently, Domen's group at the University of Tokyo demonstrated overall photocatalytic water splitting at an external quantum efficiency of up to 96%, using aluminum-doped $SrTiO_3$ crystals consisting of {1 1 0} and {1 0 0} facets. By selectively photodepositing the cocatalysts Rh/Cr_2O_3 and CoOOH for the hydrogen and oxygen evolution reactions, respectively, on different facets of the crystals using anisotropic charge transport, either the hydrogen or oxygen evolution reactions have been promoted (Takata et al., 2020).

3.2.2.3 Composition control

Heterostructures are important in catalytic, optical, electrical, and other functional materials and devices. By rationally controlling the compositions of catalysts, highly active noble metals can segregate on surfaces and highly stable inexpensive metals can form core structures. This strategy can reduce the cost and improve catalytic activities and stabilities. In semiconductor devices, confining electrons, involved in photoelectron transfer process at heterogeneous interfaces rather than on surfaces, can avoid surface effects.

Alloy and heterogeneous nanocrystals can be obtained by multistep epitaxial growth. Nanocrystals are usually generated first, which then can be taken as a template for the controlled nucleation and growth of core—shell nanocrystals at heterogeneous interface. The key to this regulation is to control supersaturation of precursors during nanocrystal growth, so that homogeneous nucleation does not occur, while nucleation and growth of interior nanocrystals are prioritized.

Guyot-Sionnest et al. from University of Chicago used a two-step process to grow ZnS-capped CdSe nanocrystals. The ZnS capping has a higher bandgap than CdSe, increasing the luminescence quantum yield by 50% at room temperature (Hines & Guyot-Sionnest, 1996). Peng et al. synthesized a series of CdSe/CdS core—shell quantum dots with different shell thicknesses (4—16 monolayers) and found that the shell structure can hold on the blinking behaviors in the luminescence process, thus improving the overall photoluminescence quantum yield (Qin et al., 2014).

The combination of nanocrystal growth and selective etching can be used to obtain various heterogeneous structures such as cages, hollows, and multilayers. Wang et al. from Argonne National Laboratory, United

States, first synthesized PtNi alloy nanocrystals, then selectively removed the near-surface Ni by acid treatment, and then obtained a heterogeneous catalyst structure with multilayered Pt-skin surface upon high-temperature annealing (Wang, Chi, et al., 2011). In the oxygen reduction reaction, this heterogeneous catalyst shows an improvement factor of more than one order of magnitude in activity versus conventional Pt catalysts.

3.2.2.4 Pore structure control

In most cases, what we see is the entity of a material, but in many applications, "space" is also the indispensable part of material functionality. For the uses in mass transport and energy storage, "space" materials are required for transportation channels and storage locations. Nanospace materials, traditionally referred to as porous materials, include zeolites, metal–organic frameworks (MOFs), covalent organic-frameworks, and porous molecular solids.

The production of pores is crucial in synthesizing nanospace materials. The dominant method is the template method, of which the micelle template method is commonly used. For example, a self-assembly of amphiphilic molecules (e.g., surfactants) in aqueous phase is utilized to form zero-, one-, and two-dimensional interconnected polymers. These structures are then used as templates to generate ordered porous structures by reacting in a limited space. Due to space limitation, the preparation of nanospace materials is not included in this chapter, but interested readers are referred to recent reviews (Kitagawa, 2017; Li et al., 2016; Slater & Cooper, 2015; Zhu et al., 2015).

3.2.3 Vapor-phase growth

The previous section introduced methods for growing low-dimensional materials by liquid-phase growth route, which involves finding a suitable liquid-phase reaction and a stable ionized precursor in liquid. However, this is not applicable to covalently bonded low-dimensional materials.

This section describes the growth of low-dimensional materials in the vapor/gas phase, that is, the generation and decomposition of precursors by sputtering and heating, then nucleation and growth in the vapor phase by molecular or atomic reactions. Compared to liquid-phase synthesis route, vapor-phase growth provides higher energies and a wider range of fluctuations in the thermodynamic potential energy surface. As a result, more extreme structures can be produced. Since the 1980s, Gletiter from the University of Saarland, Germany, a pioneer in nanomaterials, has been working on the preparation of nanocrystalline powders by vapor-phase growth route and the formation of nanostructured solid materials by sintering (Gleiter, 1989).

3.2.3.1 Arc-discharge and laser ablation

Arc-discharge and laser ablation are two methods that can generate extreme high temperature up to about 4000 K and transform reactant gas into plasma far from equilibrium, which later condenses into solid low-dimensional materials. Typically, in 1985, Kroto et al accidentally discovered football-shaped C_{60} molecules when they intended to investigate the formation mechanism of long-chained carbon molecules in interstellar space, by ablating a graphite electrode with laser (Kroto et al., 1985). They proposed a truncated icosahedron closure structure model (football) to explain the special stability of C_{60}, which contains 60 vertices forming 12 pentagons and 20 hexagons. Later, Krätschmer et al. produced macroscopic quantities of C_{60} molecules by evaporating graphite electrodes in an inert atmosphere (Krätschmer et al., 1990).

In 1991, Iijima accidentally observed CNTs while characterizing fullerene products prepared by arc-discharge method using a high-resolution transmission electron microscope (TEM) (Iijima, 1991), which is another surprise for this extreme synthesis method. Iijima and Ichihashi (1993) and Bethune et al. (1993) independently synthesized single-walled CNTs using the arc-discharge method. The key to this control is the introduction of Fe and Co transition group metal particles as catalysts.

It is worth mentioning that low-dimensional materials produced by arc-discharge and laser ablation methods have very high crystallinity, but the growth conditions are hard to control precisely. Besides, the growth processes are difficult to be quantitatively studied and well understood. Thereby the large-scale preparation and application are somewhat limited.

3.2.3.2 Chemical vapor deposition

CVD is currently the dominant method for vapor-phase growth of low-dimensional materials. Compared to arc-discharge and laser ablation, CVD conditions are relatively mild, with temperatures typically within 2000 K. The precursors are usually gas molecules and the main energy source is high-temperature heating, known as thermal CVD. If CVD is combined with plasma activation, known as plasma-enhanced CVD, lower growth temperatures and higher growth rates are available.

Regardless of the specific method taken, catalysts and substrates can be used to guide and regulate the structure of low-dimensional materials during growth.

By CVD method, a wide variety of semiconductor nanowires can be fabricated, such as silicon, germanium, gallium nitride, gallium phosphide, and zinc oxide. The method allows for the control of their diameter, crystal structure, growth position, growth direction, and heterogeneous structure. Complex structures, such as spiral-structured nanowires, can be grown by inducing defects such as screw dislocation (Dasgupta et al., 2014; Ek & Filler, 2018). Numerous theoretical and in situ studies have provided insights into the growth mechanism of nanowires by CVD method (Fig. 3.5). As early as 1964, the vapor–liquid–solid (VLS) growth mechanism of Si nanowire was proposed by Wagner et al. It involves catalytic decomposition of gas-phase reactants, dissolution of Si into catalysts to form eutectic liquid-phase particles, as well as precipitation and continuous growth at the solid-liquid interface. The binary eutectic phase diagram can help us to understand the underlying processes (Magnin et al., 2015; Wagner & Ellis, 1964).

CVD is currently the primary method for synthesizing CNTs. For instance, the position of catalysts determines the growth location of CNTs (Kong et al., 1998). With the assistance of gas flow or fast heating, horizontally aligned arrays can be obtained, while vertically aligned CNT

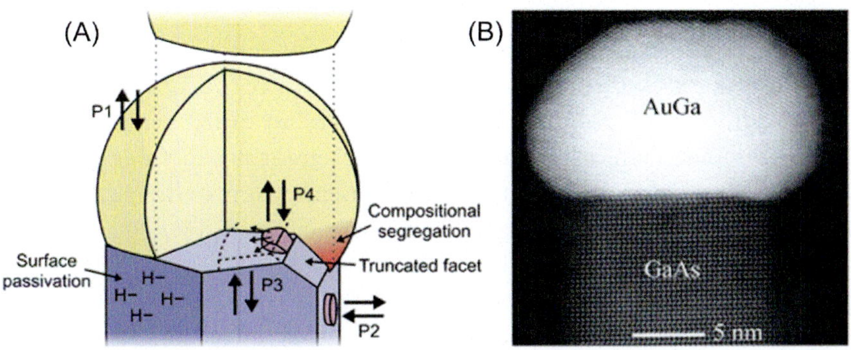

Figure 3.5 Growth mechanism of nanowires. (A) Overview of the transport processes occurring between the different phases during VLS growth: (P1) vapor-seed particle, (P2) vapor-nanowire sidewall, (P3) nanowire sidewall-seed particle, and (P4) seed particle-solid nanowire. (B) High-resolution STEM image of gallium arsenide nanowires with wurtzite structure and its solidified gold-gallium seed particle (Ek & Filler, 2018; Krogstrup et al., 2013). *STEM*, Scanning transmission electron microscopy; *VLS*, vapor–liquid–solid.

arrays can also be prepared owing to the crowding effect of neighboring CNTs (Hata et al., 2004; Jiang et al., 2002). Floating catalyst CVD method allows for large-scale and continuous growth of CNTs (Cheng et al., 1998), while regulating atmospheres can control the ratio of metallic and semiconducting CNTs (Li et al., 2005). The direct growth of CNTs with a specific chirality can be controlled to a certain extent by using solid alloy catalysts (Yang et al., 2014). Additionally, one-dimensional vdW heterostructures can be obtained by using CNTs as a template (Xiang et al., 2020). However, it still suffers from a lack of structural control over structures of CNTs, and the growth mechanisms involving catalyst particles and catalyst-CNT interactions are poorly understood (Jourdain and Bichara, 2013).

With the development of characterization techniques and in-depth theoretical studies, more knowledge has been gained about CVD growth mechanisms of CNTs. The authors discovered that single-walled CNTs can be grown from solid oxide catalysts. In situ TEM observation revealed that the catalyst particles maintain a solid state during CNT growth process, suggesting a vapor—solid—solid mechanism (Liu et al., 2011). Another in situ TEM observation of cobalt-catalyzed growth of CNTs disclosed that cobalt catalyst was in a solid state and reacted with carbon into cobalt carbide phase (Co_3C). In more detail, carbon atoms were supplied by diffusion through the surface of the catalyst and its interface with CNTs, thus explaining why multi-walled CNTs can grow rapidly in this system (Wang et al., 2020).

CVD is also widely applied to grow two-dimensional materials. It differs from the growth of one-dimensional nanowires and nanotubes, here the catalysts are not zero-dimensional particles but planar substrates to stabilize thermodynamically unstable two-dimensional structures (Li et al., 2009). Currently, CVD growth of two-dimensional materials has achieved effective control in terms of grain size (Yan et al., 2014), edge structure (Ma et al., 2013), and cleanliness (Lin et al., 2019). Liu and Peng et al. from Peking University were devoted to the development of "ultraclean" graphene. They systematically analyzed the sources of contamination on graphene surface. With various methods such as gas-phase reaction regulation and selective etching, they have obtained clean graphene surface (Zhang et al., 2020). Ren et al. at Institute of Metal Research, Chinese Academy of Sciences, carried out a series of experiments on the growth kinetics of graphene and proposed a strategy of "growth-etching-regrowth." By reducing the nucleation density and repairing structural defects, high-quality large monocrystalline graphene has been obtained (Ma et al., 2014). Moreover, by controlling reaction atmospheres and cooling rates, graphene polycrystals from micrometer to nanometer scale were obtained. The electrical, mechanical, and thermal properties of these polycrystals can be tailored by tuning densities of grain boundaries (Ma et al., 2017; Zhao et al., 2019).

Monolayer transition metal sulfides (TMDs) are a vital family of two-dimensional materials with similar structures but diverse properties. They can be utilized to construct optical and electronic devices with numerous functions. Monolayer crystals with high crystallinity can be grown on substrates by CVD, with sulfur group elements as precursors and transition metal oxides (TMOs) as metal precursors (Najmaei et al., 2013; Wang et al., 2014). Ajayan and Duan et al. fabricated interlayer and intralayered heterostructures with different TMDs by epitaxial growth and two-step growth (Duan et al., 2014; Gong et al., 2014, 2015). For metal oxide precursors with high melting point and finite vapor pressure, the growth temperature of TMD structures is relatively high. Li et al. found that the addition of NaCl can form liquid-phase intermediate compounds with low melting points and lower the growth temperature, resulting in the growth of two-dimensional materials based on VLS mechanism. The configurations include nanoribbons, wafer-scale monolayers, and doped heterostructures (Fig. 3.6) (Li et al., 2015, 2018, 2021). Liu et al. from Nanyang Technological University in Singapore developed this method and prepared a series of two-dimensional transition-metal chalcogenides with up to 47 different components (Zhou et al., 2018).

Two-dimensional layered nitrides and carbides, commonly referred to MXene (Vahid Mohammadi et al., 2021), are another family of two-dimensional materials that are typically delaminated from layered MAX phase, where M denotes transition metal, A stands for Al, Si, etc., and X refers to C and/or N. By selectively removing the A-layer, the exfoliated two-dimensional MXene can be obtained, meanwhile defects are inevitably introduced (Naguib et al., 2011). Ren's group discovered that the introduction of Si during the vapor-phase growth process can passivate the surface dangling bonds of two-dimensional molybdenum nitride, forming an $SiN-MoN_2-SiN$ sandwich structure. This growth behavior led to the transition of island-like to planar growth mode, and continuous monolayer films with centimeter-scale were obtained, which exhibited excellent mechanical strength and chemical stability (Hong et al., 2020).

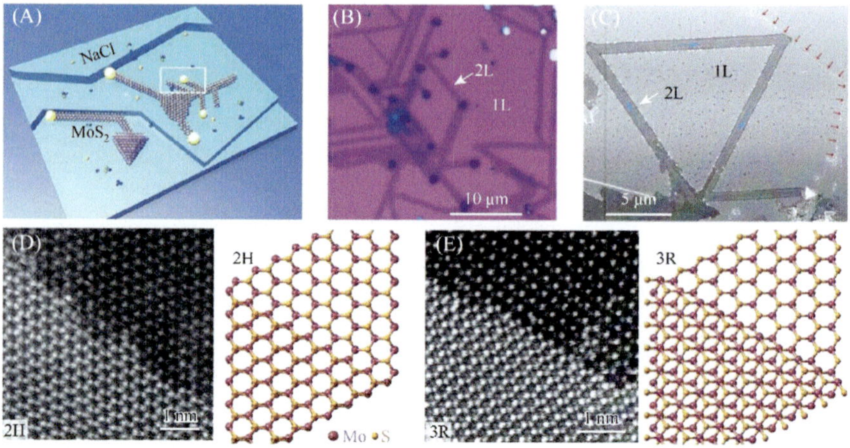

Figure 3.6 "Salt-assisted" VLS mechanism of growing 2D materials. (A) Schematic of VLS growth of MoS_2 ribbons on an NaCl crystal surface. (B, C) Optical and scanning electron microscope images of MoS_2 nanoribbons growing on monolayers of MoS_2. (D, E) STEM images and the corresponding atomic configurations of an edge of a ribbon showing the boundary between the MoS_2 supporting monolayer and the respective 2H-stacked bilayer and 3R-type bilayer (Li et al., 2018). *VLS*, Vapor–liquid–solid; *STEM*, Scanning transmission electron microscopy.

3.2.4 Machine learning-assisted growth

The structural controlled growth of low-dimensional materials is essentially an optimization process in a growth parameter space. The challenges lie in the complexity of the parameter space: (1) The information is not complete, resulting in an unclear nucleation and growth mechanism. (2) The information is not adequate where the parameter space is too large to traverse. (3) The information is not perfect due to the accompanied experiment errors.

Traditionally, the growth of low-dimensional materials is often empirically optimized in a small parameter space by trial-and-error, examining the independent effects of a single variable in a stepwise manner, which is less efficient. In the optimization of complex and huge parameter space with incomplete and imperfect information, machine learning has unique advantages and plays a role in three aspects: (1) data mining, including historical literature data, database data, high-throughput experiment, and calculation data. (2) Supervised or unsupervised learning methods to extract valuable information and establish relationship between synthesis parameter and structure. (3) Algorithms, for example, deep learning to predict and optimize the growth conditions for target properties. In recent years, machine learning has already been applied to the optimization, simulation, and design of different materials, such as quantum dots (Voznyy et al., 2019), CNTs (Nikolaev et al., 2014), MOFs (Fanourgakis et al., 2020), nanoporous zeolite (Moliner et al., 2019), and so forth.

The key in the field of materials science is the processing—structure—performance relationships. Machine learning algorithms can extract patterns hidden in the data. Ceder and Jain et al. from Lawrence Berkeley National Laboratory collected more than 3 million scientific abstracts and used an unsupervised machine learning approach to classify phrases through principal component analysis. They then built a model to simulate the associations between phrases and predicted the relationships between material compositions and their thermoelectric properties in a collective manner. (Tshitoyan et al., 2019).

The size uniformity of semiconductor quantum dots determines the uniformity of their optical properties and applications in the field of optoelectronics. The Sargent Group from the University of Toronto used Bayesian optimization algorithm to analyze 2300 groups of preparation conditions, accumulated over the past 6 years, for PbS quantum dots. It was found that the main factors affecting the size uniformity were Pb/S ratio, injection temperature of solution, concentration of oleylamine as a solvent, and surface ligand. The optimal synthesis conditions predicted by the model have been experimentally verified with an excitation luminescence peak line width of 55 meV at 950 nm wavelength (1.3 eV), which is close to the luminescence line width limit of a single quantum dot (50 meV) (Peterson & Krauss, 2006; Voznyy et al., 2019).

High-throughput experiment is an important method to achieve rapid screening, optimize material design, and establish a database for material synthesis. The author of this book used catalyst patterns for high-throughput growth of single-walled CNTs, and established a database of 1280 groups of growth conditions by combining with high-throughput Raman spectroscopy. We used machine learning to predict and verify the optimal growth conditions of high-quality CNTs, and obtained CNTs with a G- to D-mode intensity ratio of up to 138 in Raman spectra (Fig. 3.7) (Ji et al., 2021). Mirkin's group at Northwestern University in the United States used the polymer pen lithography (Huo et al., 2008), combined with an ink spray-coating method, to obtain the database of multicomponent gradient and size particle gradient. They successfully screened an unreported catalyst, Au_3Cu, for the growth of single-walled CNTs (Kluender et al., 2019).

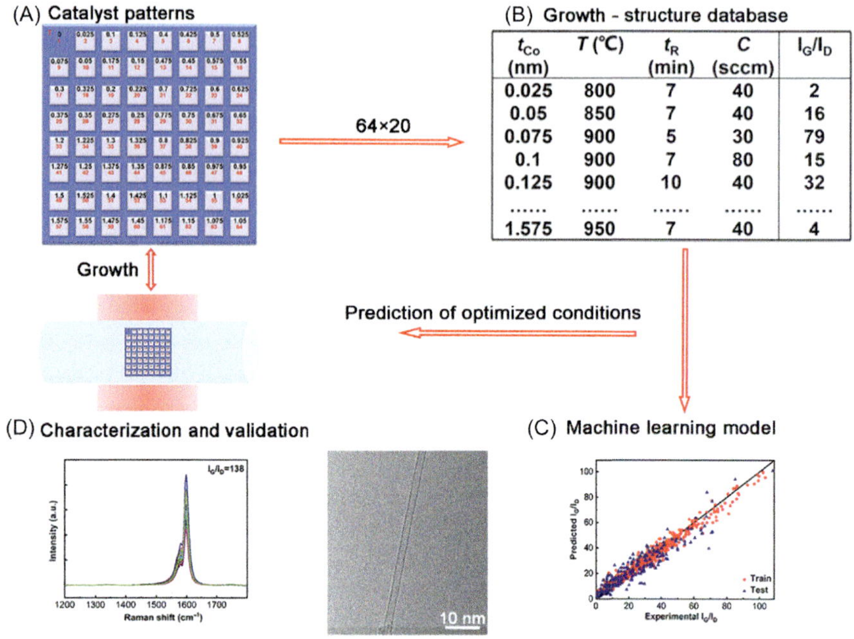

Figure 3.7 High-throughput-machine learning methods for optimizing the growth of highly crystalline CNTs. (A) Preparation of catalyst patterns. (B) Establishment of the growth parameters-material structure database, where t_{Co} is the nominal thickness of Co catalyst, T is the reaction temperature, t_R is the reduction time, C is the flow rate of carbon source, I_G/I_D is the G-mode and D-mode intensity ratio in Raman spectra. (C) Machine learning models for predicting optimal conditions. (D) Structural characterization and validation of nanotube growth (Ji et al., 2021).

Intelligent automated systems for material growth are also developed. For instance, Maruyama's group at the Air Force Research Laboratory (United States) reported an automated optimized growth system for CNTs. The adjustment parameters are mainly growth temperature and partial pressure of precursors. In situ Raman spectroscopy was used to characterize the ratio of multi- and single-walled CNTs, and a logistic regression algorithm was used to select parameters for the next round of growth. Both techniques effectively improved the optimization efficiency of material synthesis (Nikolaev et al., 2014). In 2020 researchers from the University of Liverpool used industrial robots and Bayesian optimization search algorithm to autonomously carry out experiments on the synthesis of photocatalytic materials and to optimize the conditions. They carried out 688 sets of experiments in 8 days and improved the catalytic performance by 6 times after optimization (Burger et al., 2020).

It is foreseeable that in the near future, repetitive tasks will be replaced by machines, and optimization problems will be handed over to machine learning algorithms. The creativity of human beings will be used to formulate conjectures, make judgments, and become partners with robots. "Automated scientific discovery," that is, the combination of high-throughput, online machine learning, feedback automation, and researcher expertise, will play a greater role in the discovery and optimization of new materials.

3.3 Assembly of low-dimensional structures

Self-assembly is a process by which components spontaneously form highly levels of ordered structures as a result of their interactions and external forces. Self-assembly occurs at all spatial and temporal scales. Generally, processes ranging from the growth of nanocrystals and the assembly of nanostructures, from biomolecules to cells and life patterns, all involve bottom-up assembly process. In 1912, Scottish biologist and mathematician Thomason published his famous book "On Growth and Form" (Thompson, 1992), which revealed the growth patterns of organisms from single cells to living organisms, from inorganic compounds to biomineral structures by using physical, mathematical, and mechanical principles. It is surprising that they all follow the similar patterns and mathematical laws.

The assembly of low-dimensional materials is a necessary process from individual low-dimensional structures to the real applications at large scale, which is related to many technical and scientific questions. For instance, how to maintain unique properties at the macroscale associated with the nanoscale and low-dimensionality of nanostructures? How to ensure that the assembled macrostructure has sufficient structural stiffness, mechanical strength, and thermal stability? How to realize the novel properties and functions through the assembly of low-dimensional structures, which the building blocks do not have? In this section, we will briefly introduce the principles and methods of assembly of low-dimensional materials.

3.3.1 Principles of self-assembly

The assembly of low-dimensional structures is a bonding and growth process of components under their interactions and external fields, similar to the crystal growth process. The thermodynamic phase diagram, which consists of the relationship between the stability of the assembled structure and the external conditions, guides the design of the assembly. Thermodynamics is about the most stable and the lowest energy state, while kinetics considers assembly time and specific processes, so that it is possible to assembly structures in a metastable state. For an open energy-dissipation system, the assembly configuration forms dynamic and complex assemblies in response to environmental changes, energy exchanges, and mass transports.

Unlike the growth/synthetic methods, the assembly of low-dimensional structures is a process of ordering through weaker chemical bonds at a larger molecular scale. The three key elements influencing the assembly of low-dimensional materials are components, interactions, and external environments (Fig. 3.8). The components (building blocks) for assembly are individual low-dimensional structures, including quantum dots, nanocrystals, nanowires, nanotubes, and nanoplatelets. As aforementioned, the growth of materials can be seen as the assembly process from atoms and molecules, but essentially, the atoms and molecules are nondistinguished isotropic particles. The building blocks for assembly are

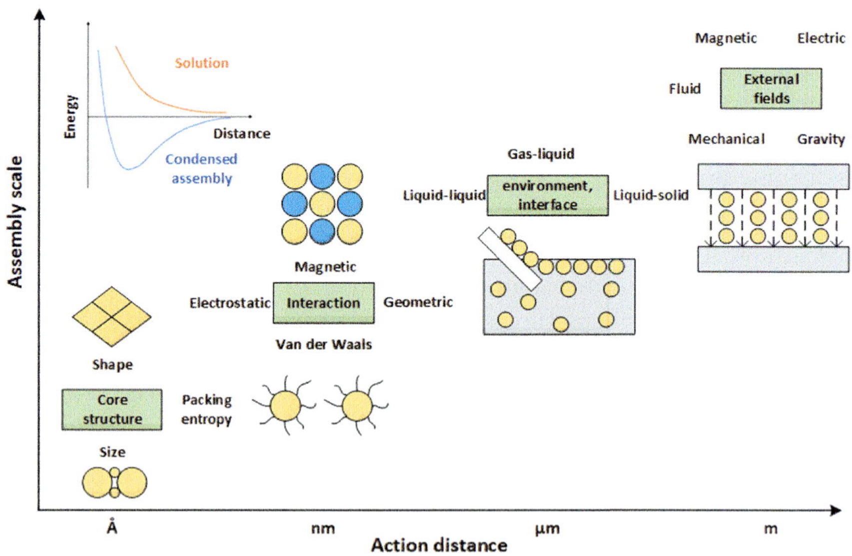

Figure 3.8 The principle of self-assembly. From hard-sphere model upon contact to nanoscale interaction of building blocks, microscale interaction between environments and interface, as well as the macroscale external field effect.

aggregations of atoms, so that the control over the low-dimensional structures, such as dimensionality, scale, shape, crystal structure and surface chemistry, is the prerequisite for the assembly. The interaction refers to the forces between the building blocks, including hydrogen bonding, vdW forces, electrostatic interactions, and surface tensions. Entropy is also the important part of free energy of a system, which contains configuration entropy and free volume entropy. Self-assembly usually occurs in solution environments, and the building blocks of traditional supermolecules are colloidal particles. Inorganic materials such as nanocrystals require surface ligands to form a stably dispersed solution, in which the interactions of core−ligand structure in a solvent need to be considered.

3.3.2 The assembly in liquid phase

The assembly in liquid phase is the simplest method for the assembly of low-dimensional materials. Generally, low-dimensional materials are stably dispersed in solution, forming appropriate repulsive force between them against the attraction of vdW forces. The repulsion forces are mainly steric hindrance of surface ligands and electrostatic forces. Once the equilibrium of repulsive forces and attractive forces is disrupted, the destabilization of solutions occurs through processes such as precipitation, aggregation and assembly. For instance, solvent evaporation results in an increase in the concentration of particle volume fraction, which shortens the distance between each other and increases attractive forces. On the other hand, the introduction of new charged ions or other low-dimensional structures with dissimilar charges will also disrupt the equilibrium of repulsive forces.

During the evaporation-based self-assembly, the nanocrystalline three-dimensional-ordered assemblies can be precipitated by controlling the temperature of water or oil bath and evaporation rate of solvents. Under the effect of isotropic vdW forces, the equilibrium structures of single-component superlattices are densely packed nanocrystals, such as FCC and hexagonal close-packed structures (Boles et al., 2016).

For the preparation of heterogeneous nanocrystalline superlattices, on the other hand, it is necessary to take into account the influence of structural parameters such as the shape and aspect ratio of building blocks on the filling ratio, configuration entropy, and free volume entropy (Chen & O'Brien, 2008; Murray et al., 2000). For example, for binary assemblies of hard ferromagnets FePt and soft ferromagnetic Fe_3Pt nanocrystals, "exchange spring magnets" can be formed by adjusting the ratio and size of the building blocks and optimizing the exchange interaction (Kneller & Hawig, 1991). They combined the high coercivity (H_c) of the hard magnetic phase with the high saturation magnetization strength (M_s) of the soft magnetic phase. The magnetic energy density (magnetic energy product) of the nanocomposite from an $Fe_3O_4-Fe_{58}Pt_{42}$ assembly is 20.1 MGOe, which exceeds the value for single-phase $Fe_{58}Pt_{42}$, reflecting the emergent phenomenon of the assembly (Zeng et al., 2002). Another example, Pd and Pt nanocrystals self-assembled by air–liquid interfacial evaporation to form a binary AB_{13}-type superlattice. The assembly shows a high activity toward electrocatalysis of the oxygen reduction reaction, with the current densities two and four times as high as those on Pt octahedra and random mixtures, respectively. This is attributed to the shape effects of Pt nanocrystals and synergistic effects of Pt–Pd (Kang et al., 2013).

Mixing heterogeneously charged low-dimensional structures to produce flocculation is a soft-chemical method for rapidly obtaining macroscopic superlattice structures. Sasaki's group at the National Institute for Materials Science in Japan used this approach to obtain a variety of two-dimensional material superlattices, including graphene, oxide, and hydroxide nanosheets. The formation of composite structures at the molecular level enables the optimization of their catalytic activity, electrical conductivity, and ion-diffusion channels for superior performance in energy storage and conversion (Ma et al., 2015; Wang, Wang et al., 2011).

3.3.3 The assembly of liquid crystal materials

Liquid crystal ordering is an important assembly process for the preparation of novel materials. A liquid crystal transition from an isotropic solution to a high oriented nematic structure can take place spontaneously when anisotropic one- and two-dimensional structures are suspended in a solution at a sufficiently high concentration. This phenomenon has been observed in solutions of low-dimensional materials such as CNTs and reduced graphene oxide (RGO) (Zakri et al., 2013).

The aligned nanocarbon materials on a large scale have been used to make high-strength and high-conductivity fiber materials. Pasquali's group at Rice University found that CNTs in strong acids were concentrated and transformed to liquid crystals by centrifugation, with the critical transition concentration increasing as the acidity increases (Davis et al., 2009; Rai et al., 2006). In ultra-strong chlorosulfonic acid solution, CNTs were dispersed by protonation and the dispersioin concentration can reach 2%–6% by weight. After the as-formed liquid crystal was extruded through a spinneret, continuous CNT fibers were obtained by extracting the acid solvents with a coagulant such as acetone and water. The best tensile strength of the as-made individual filament is 1.3 GPa, the optimum Young modulus is 200 GPa, the average electrical conductivity is 2.9 ± 0.3 MS/m and the average thermal conductivity is 380 ± 15 W/m K (Behabtu et al., 2013).

Gao's group at Zhejiang University, China, has proposed that single-layered graphene oxides (GO) can be well dispersed in aqueous solution and exhibit liquid crystal transition after treatments by high concentrations of sulfuric acid and potassium permanganate. The critical concentration of GO sheets in aqueous solution to form liquid crystal is 0.025%, and the layers are oriented and have long-range chiral characteristics (Xu & Gao, 2011a, 2011b). Xu et al. (2016) prepared RGO fibers by wet spinning method (Liu et al., 2017; Xu & Gao, 2011b, 2014), which reached a tensile strength of 2.2 GPa and a modulus of 440 GPa. After doping, the electrical conductivity reached 2.24×10^7 S/m, close to the metal conductivity (Liu et al., 2016).

3.3.4 Liquid-phase interface assembly

Driven by surface energy and interface energy, low-dimensional structures tend to assembly at surface or interface (Kutuzov et al., 2007). Park's group at the University of Chicago integrated a layer-by-layer (LBL) assembly of organic–inorganic two-dimensional superlattices (Zhong et al., 2019). They injected molecular precursors into the pentane/water interface to form a laminar flow, and the polymerization reaction self-assembles to form a large-area two-dimensional polymer. Two-dimensional polymer hybrid heterostructures can be prepared by controlling the injection order of monomer molecules. The monomer molecules are periodically linked into a two-dimensional structure through covalent or coordination bonds to form a two-dimensional covalent organic framework structure or MOF structure, which can be transferred onto a quartz substrate or fabricated into a self-supported thin film.

3.3.5 Gas–liquid interface assembly

In 1935, Langmuir and Blodgett, at General Electric Company, invented a method for assembling monomolecular films at gas–liquid interface, that is, Langmuir—Blodgett (L—B) technique (Langmuir & Blodgett, 1935; Blodgett, 1935; Zhong et al., 2019). L—B assembly is the key technology to prepare large-scale homogeneous assembled structures, including the molecular assembly process at air—water interface (Langmuir-film) and the deposition of the film on solid substrate (L—B film).

The gas—liquid interface assembly of low-dimensional structures such as nanocrystals can be seen as a two-dimensional phase transition process. Nanocrystals usually carry hydrophobic ligands on their surfaces, which form stable solutions or colloids in nonpolar solvents, and when added to polar liquids, monolayer molecules can form on the surface of the polar liquid. With the compression of a controlled barrier, the distance and interactions between the molecules can be adjusted, leading to a two-dimensional phase transition in the assembly structure, similar to the three-dimensional gaseous, liquid, and solid state (Hussain et al., 2018). When the inter molecular distance is large and the interaction is negligible, the resulting floating monolayer can be considered as a two-dimensional gas. As the distance decreases, the surface area decreases and the surface energy increases, the molecules are in a liquid state. When the distance decreases and the surface energy rises steeply, a compact and well-ordered two-dimensional film is formed at gas-liquid interface. Then, the gas—liquid interface assembly of nanocrystal structures can be transferred by lifting the substrate to form an ordered monolayer structure on the substrate.

The L—B method is widely used to assemble low-dimensional materials such as zero-, one-, and two-dimensional materials (Chen et al., 2007; Cote et al., 2009; Li et al., 2007; Tao et al., 2003). Weller's group at the University of Hamburg, Germany, used the L—B method to assemble Co—Pt nanocrystals into large-area, continuous and tightly arranged monolayers. By fabricating gold electrodes on the monolayer film with an interelectrode distance and a width of 1 μm, its room-temperature conductance is 6.5 nS, as measured by direct current electrical measurements from 80 to 300K. It is found that the film follows a thermally activated charge transport model with a thermal activation energy of 18 meV (Aleksandrovic et al., 2008). Dai's group at Stanford University reported the L—B method of assembling single-walled CNT films, where CNTs were functionalized with PmPV polymers noncovalently. The conjugated backbones and single-walled CNTs formed $\pi-\pi$ interactions, stably dispersed in 1,2-dichloroethane (DCE) solution and assembled on water surface by L—B process. Field effect transistors (FETs) fabricated in this way achieve a current of 3 mA at a source-drain voltage of 3 V (Li et al., 2007).

In 2013 Cao et al. at International Business Machines (IBM) achieved high-density CNT thin-film assembly using the developed Langmuir–Schaefer approach. CNTs were dispersed in DCE solution, which spreads and volatilizes on the water surface, and guides CNTs to be aligned by mechanical extrusion. The aligned CNT film was then transferred onto the substrate to fabricate the device, achieving a current density of 120 μA/μm (Cao et al., 2013). In 2020 Peng's research group at Peking University used multiple-dispersion sorting process to increase the purity of semiconducting single-walled CNTs to over 99.9999%. They also developed a dimension-limited self-alignment procedure. That is, a 2-butene-1,4-diol upper liquid layer was dropped on top of the 1,1,2-trichloroethane solution of polymer-coated CNTs to form a liquid-layered interface on a vertical substrate, where CNTs were confined between the upper and lower solutions. As the substrate was slowly pulled out, the aligned CNT arrays assembled onto a 4-in. wafer. Finally, the orientation of the CNTs within 9 degrees of alignment and the tunable density ranging from 100 to 20 CNTs/μm can be prepared on the substrate with a full wafer coverage. The fabricated FET achieves a current density exceeding 1.3 mA/μm under a bias voltage of −1 V and a low bias on/off ratio greater than 10^5. The CNT array-based ring oscillators show a maximum oscillating frequency of 8 GHz. The integrated circuits exhibit the performance metrics exceeded those of conventional Si transistors with similar dimensions (Liu et al., 2020).

3.3.6 Electrostatic layer-by-layer assembly

Low-dimensional nanostructures stabilized in solutions are usually charged, and Decher and Hong (1991) invented the LBL method utilizing the molecular electrostatic properties. By either dipping a substrate in solutions or spaying solution onto a substrate, the substrate can be charged. Due to the electrostatic effect, molecules with the same charge are assembled into a monolayer structure on the substrate. LBL assembly can be achieved sequentially by spraying molecular solutions with opposite charges (Richardson et al., 2015, 2016). This method requires only electrostatic interaction and enables the assembly of low-dimensional materials with very different structures and functions. Kotov et al. synthesized a composite multilayer of polymers with inorganic nanoparticles and nanowires by LBL method (Srivastava & Kotov, 2008). Sasaki's group assembled ultrathin films of oxides or hydroxide nanosheets by this method (Osada et al., 2007; Renzhi & Takayoshi, 2012; Sasaki et al., 2001), and it was found that titanium oxide nanoplates had a very high dielectric constant (125) and low drain current (10^{-7} A/cm^2) (Osada et al., 2007). The assembled homologous perovskite nanosheets have a high dielectric constant of 470, which is the highest value for dielectric materials with a thickness of less than 10 nm, and have the potential to be used in high-density capacitors (Li et al., 2017).

3.3.7 Guided assembly by external field

The self-assembly of low-dimensional materials is mainly about the interactions between materials, but the roles of environments and external fields cannot be ignored. Except for the solution environment and solvent effects, temperature and gravity fields, which are already included in the scope of self-assembly, externally applied electric, magnetic, and stress fields can also effectively regulate the assembly process and final structures of low-dimensional materials.

Dielectrophoresis is a method of assembling low-dimensional materials by arranging them using the electric field effects. Under the action of a nonuniform electric field, polarized particles will be subjected to a directional dielectric force, moving in the direction of the largest electric field gradient and assembling on the electrode surface. The polarization of a particle related to its shape, etc. has a characteristic frequency. Maximum dielectrophoresis is produced when the applied alternating current frequency is comparable to the particle polarization frequency, thus enabling screening of the particle structure. Freer et al. (2010) assembled up to 16,000 groups of single nanowires by dielectrophoresis method. Sarker et al. (2011) assembled horizontally aligned enriched semiconducting single-walled CNT films using dielectrophoresis, and the density of CNTs could be tuned from 1 to 25 CNTs/μm by frequency and concentration.

Ren's group has developed a continuous centrifugal casting method to rapidly produce highly aligned and compact two-dimensional nanosheet films and vertical heterostructures. Taking GO as an example, its dispersion fluid in the dispersion is subjected to both centrifugal force and shear force simultaneously in high-speed rotating. Shear forces orient two-dimensional materials and centrifugal forces lead to densification. By this method, 10-μm-thick GO films can be synthesized within 1 minute, and scalable synthesis of meter-scale films is demonstrated. The RGO films show superhigh strength (~660 MPa) and conductivity (~650 S/cm). The RGO/CNT hybrid-film-based all-solid-state flexible supercapacitors exhibit ultrahigh volumetric capacitance (407 F/cm^3) and energy density (~10 mW h/cm^3) (Fig. 3.9) (Zhong et al., 2018).

3.3.8 Dynamic self-assembly

The previous sections focus on the static self-assembly of low-dimensional materials, which is an assembly process approaching to an equilibrium state under the influence of interactions and external fields. Maintaining its stable configuration after assembly is important for the reliable use of the materials. In addition, dynamic self-assembly is very common in nature, and the living organisms can be viewed as a self-assembly out of equilibrium. Unlike static assembly, dynamic self-assembly, accompanied by energy transfer and material transport, changes over time and environment (Self-assembling life, 2016).

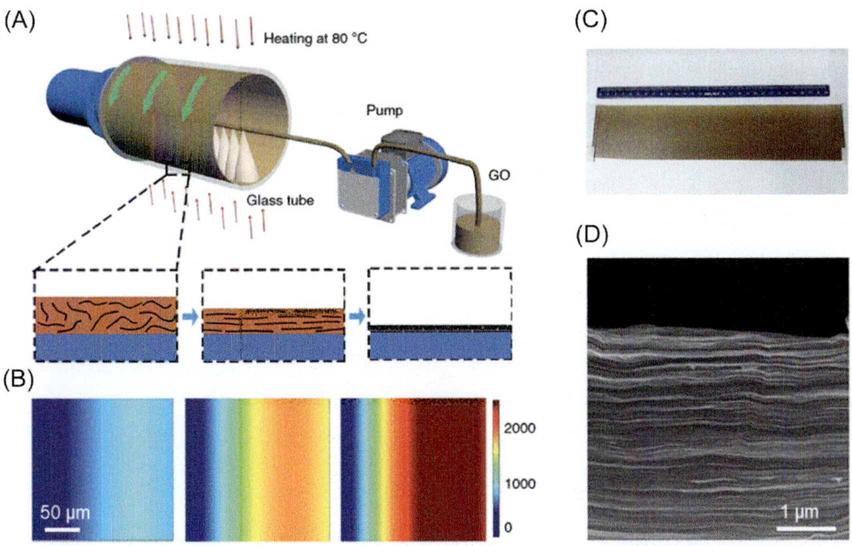

Figure 3.9 Production and characterization of highly aligned and compact GO films by CCC. (A) Schematic of the CCC production process. (B) Simulated transient shear rate field at different rotating rate. (C) A GO film with a thickness of ~100 μm and size of ~30 × 10 cm^2. (D) Cross-sectional SEM image of a GO film (Zhong et al., 2018). *CCC*, Continuous centrifugal casting; *GO*, graphene oxides; *SEM*, scanning electron microscopy.

In 1952, the British scientist Turing explored mathematically how reaction−diffusion evolves into patterns in nature (Turing, 1952). In 2007 the Nobel Prize-winner in Chemistry, Gerhard Ertl, summarized the pervasive phenomenon of dynamic self-organization in terms of surface chemical reactions, where the coupling between different parts of the surface by the reaction−diffusion systems leads to the formation of complex dynamic patterns and temporal oscillations. Similar self-organization can be found not only in atomic-scale surface chemical reaction, but also in the patterns of animal fur in nature and even in Van Gogh's painting *The Starry Night* (Ertl, 1991, 2008). Overall, dynamic self-assembly has been relatively poorly studied, and it will be a very important subject in the emerging fields of micro/nano robotics and smart materials.

3.4 Processing of low-dimensional materials

Processing is a top-down approach to prepare low-dimensional materials through the interaction of external forces and the material. These forces include mechanical and fluidic shear forces, temperature, gravity, light, electrons, and ion beams (Fig. 3.10). From a dimensional perspective, processing is dimensionality reduction. This is achieved by physically, chemically, and mechanically interacting with three-dimensional bulk materials to obtain low-dimensional materials and devices. Material processing is fundamentally a process of controlled destructive process. The crucial issue in it is how to effectively and accurately remove the excess parts while preventing any structural damage to the useful parts. This section will briefly introduce the basic principles, control accuracy, and research progress of representative processing methods.

3.4.1 Lithography processing

Lithography serving as the foundation of the microelectronics industry, has now reached the nanoscale level. The fundamental aspect of photolithography is to create photoresist pattern through optical exposure. The electronic devices are then produced through a number of reciprocal processes such as exposure, etching, coating, and ion implantation. The resolution limit of optical exposure is limited by the diffraction effect. To

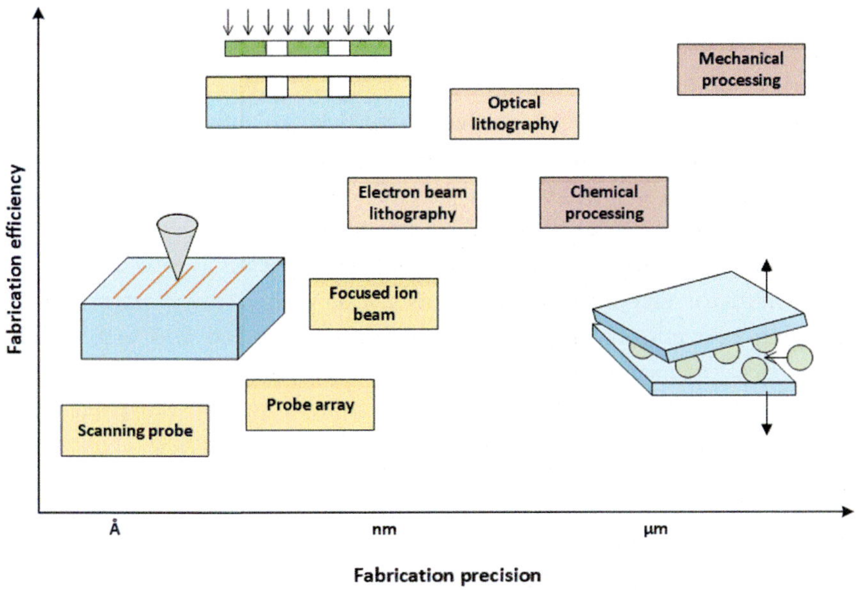

Figure 3.10 Processing methods for low-dimensional materials present a trade-off between fabrication precision and efficiency. They can be classified into micro-probe processing, FIB processing, EBL, optical lithography, chemical and mechanical processing. *FIB, Focused ion beam; EBL, electron beam lithography.*

break through this limitation, the industry is constantly improving the design of optical paths and the use of light sources with shorter wavelength, including deep ultraviolet light, X-rays, electron beams, and so on.

The Rayleigh limit of photolithography: $R = k\frac{\lambda}{NA}$, where k is a constant related to the optical path system, λ is the optical wavelength, and NA is the numerical aperture. The diffraction limit is similar to the imaging limit of a lens. Imaging refers to seeing clearly, whereas processing precision is defined by the characteristic scale of changes in the material created by light. In improving resolution and pushing the limitations, processing and imaging share similarities but also differences in some aspects.

The resolution of lithography processing is primarily dependent on the wavelength of the light. The diffraction limit for a fixed wavelength is 0.61λ. As the characteristic size of semiconductor electronics decreases from microns to below 10 nm, lithography has shifted from the 365 nm "i-line" to deep ultraviolet light at 248–193 nm, and even to extreme ultraviolet at 13.5 nm (Wagner & Harned, 2010).

For photolithography, the value of k in the above equation is related to the aberrations of the optical lens, and also to the light–resist interaction. In semiconductor industry, it is crucial to adjust the nonlinear response of the photoresist to beam exposure to enhance the exposure pattern accuracy. Another critical parameter is the refractive index of the light medium, which is contained in NA, and immersive exposures with a high refractive coefficient is based on this principle.

Lithography plays a significant role in fabricating low-dimensional semiconductor electronic devices and assessing their transport properties. Utilizing conventional lithography methods, oxidation and wet etching, Chen et al. managed to produce silicon nanowire arrays with a diameter ranging from 200 to 10–20 nm and a length up to 100 μm, whose wafer-scale yields exceed 90% (Fig. 3.11) (Chen et al., 2009).

3.4.2 Beam processing

All kinds of moving particles have associated wavelength, known as de Broglie wavelength. The smaller the wavelength, the higher the resolution. Since the wavelength of the electron and ion beam is much shorter

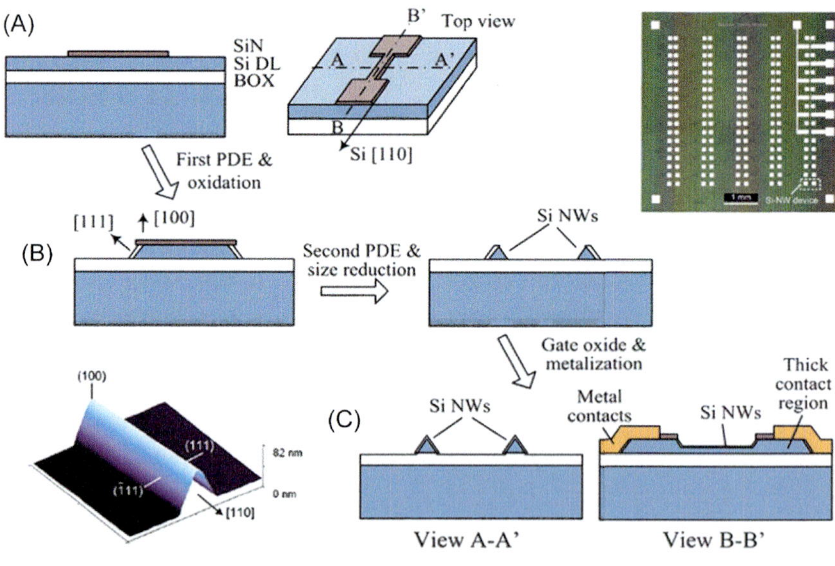

Figure 3.11 Photolithographic processing of device arrays comprised of silicon nanowires. (A, B) Multistep selective etch-oxidation, (C) metal deposition processes. The illustration is an optical photography of device arrays and atomic force microscopy image of individual silicon nanowire devices (Chen et al., 2009).

than the wavelength of light, far less than 1 nm, it is a logical preference to replace light with electron and ion beams for the exposure and processing of delicate structures. The direct interaction of electrons and ions as particles with the material is notably stronger than that of light, making it possible to simplify the procedure and manufacture the material directly. During the processing with electron and ion beam, the excited secondary electrons can also be collected for imaging, so that three-dimensional complex structures can be processed. Besides, as incident electron energy increases, there is a proportional increase in the volume of interaction between the particle beam and material. In 2000 Vieu et al. (2000) reported that the resolution limits of the electron beam lithography (EBL) technique can be pushed below 10 nm in the conventional polymethylmethacrylate organic resist at high incident electron energy. With the advancements in spherical aberration—correction technology, the electron beam generated by low acceleration voltage can also reach the atomic level, which is expected to reach the single-atom level with low irradiation damage (Manfrinato et al., 2013).

EBL has become a common method for the fabrication of microsized devices. Mirza et al. used EBL to fabricate silicon nanowires with mean widths reducing from 12 to 4 nm. They observed that the electron transport properties exhibited a gradual transition from three-dimensional to two-dimensional and one-dimensional characteristics. Furthermore, the rise in channel resistivity indicated the detrimental effect of surface roughness on device performance (Mirza et al., 2014). Peng et al. from Peking University utilized EBL to fabricate graphene-contacted CNT FETs with a channel length of 5 nm. The subthreshold slope was found to be only 73 mV/dec at an operating voltage of 0.4 V, almost reaching theoretical limits. Moreover, the gate delay is just 43 fs, surpassing that of comparable silicon complementary metal oxide semiconductor FETs (Qiu et al., 2017).

Compared to EBL, the focused ion beam (FIB) has higher particle kinetic energy and can cut materials directly. Its low lateral exposure results in more precise processing at smaller scale, regarded as an "ion knife." The gas precursor can be induced to form three-dimensional complex shapes by FIB, similar to nano horizontal engraving (Tseng, 2005). Li et al. (2001) fabricated holes in Si_3N_4 thin films using FIBs, with pore size reducing from 60 to 1.8 nm after the ion beam exposure, which could be used as a single DNA molecule detector.

3.4.3 Mechanical processing

Traditional materials processing, such as forging and cutting, utilizes mechanical forces for machining. The precision is related to the size of the tool used. It is obvious that ordinary tools are too large to controllably machine individual low-dimensional structures. By using micromechanical exfoliation, a wide range of two-dimensional materials could be prepared, representing an unexpected breakthrough in mechanical processing (Novoselov et al., 2004, 2005). Moreover, single-atom-thickness materials could also be achieved by tape approach, due to the weak interlayer chemical bonding of two-dimensional crystals. It also indicates that the choice of processing methods should take the full account of the structural characteristics of materials. The followings are four kinds of fast developing mechanical processing methods, all of which are based on scanning probes, except for mechanical exfoliation.

3.4.3.1 Scanning probe manipulation

It is now available to manipulate surface or subsurface of materials based on a scanning probe microscope (SPM) at the nanometer or even single-atom scale. Direct contact approach based on SPM can avoid the diffraction limit of light-wave-based processing, and more can be found in Chapter 4. So that as long as the tip of a needle is sharp enough, it can be manipulated at the individual atom or molecule level (Custance et al., 2009). Different from the lithography methods introduced in Section 3.4.1, SPM-based scanning probe lithography (SPL), using a sharp scanning probe to pattern materials by produce local modification on a surface, is a relatively simple process with sube-10-nm resolution. SPL usually does not need subsequent development steps and a vacuum condition. The SPL is a nanofabrication tool to mainly destruct a surface through oxidation, etching, shaving, or grafting and subsequent absorption of the patterning materials. In addition to direct contact modification between a probe tip and a sample surface, material patterning and device fabrication can also be achieved by local changes of properties of a sample surface by means of voltage, thermal effects, chemical reactions, etc. (Garcia et al., 2014).

Simmons' group at the University of New South Wales, Australia, has developed scanning tunneling microscopy-based lithography under a gaseous dopant source, for preparing an atomic-scale dopant Si nanowire on an H-passivated single-crystalline Si substrate (Schofield et al., 2003). Then they in situ fabricated a device by creating a 1.5-nm-wide nanowire with a length of 106 nm and doping two electrodes, which shows an exceptionally low resistivity and the current-carrying capabilities of copper. It is also found that the phosphorous-doped wires achieved a diameter-independent resistivity, demonstrating ohmic scaling to the atomic limit (Weber et al., 2012). They further fabricated a single-atom transistor by deterministically placing an individual phosphorus atom between highly phosphorus-doped source and drain electrodes in a silicon transport device. By measuring the transport measurement at liquid helium temperatures, a quantum confinement of the solitary phosphorus dopant was found (Fuechsle et al., 2012).

Pires et al. at IBM patterned a thin film of organic resist into tens of nanometers with a heated scanning probe tip. It is important to choose a resist material for the processing, in which organic molecules are weakly interacted by hydrogen-bonding into a glassy state. Then a heated tip was used to locally desorb materials from the thin film to resist above glass transition temperature. This method can be used for the processing of three-dimensional micro—nanostructures, such as the replica of the Swiss mountain Matterhorn in a 100-nm-thick molecular glass film with a height of 25 nm (Pires et al., 2010). Wei et al. at the U.S. Naval Research Laboratory used atomic force microscopy tip to locally heat GO and then fabricate a conductive graphene nanoribbon embedded in GO insulators with a width of only 12 nm (Wei et al., 2010).

The atomic-scale manipulation capabilities are not practical for large-scale patterning or device fabrication, and not all applications require precise control over the position of single atoms. Imprint lithography methods, however, allow for a rapid patterning at 25-nm resolution (Chou et al., 1996). Guo's group at the University of Michigan reported a large-area roll-to-roll and roll-to-plate nanoimprint lithography to continuously produce patterns with a line width of 300 nm on different substrates (Ahn & Guo, 2009).

3.4.3.2 Dip-pen microengraving

A scanning probe can be used as a shaving knife or a pen for the processing of low-dimensional materials at the individual atom or molecule level. Mirkin's group at Northwestern University invented the dip-pen nanolithography method (DPN), an SPM-based nanofabrication technique, which uses the tip of an atomic force microscope as a "pen" to deliver the coated chemical compound such as molecules as an "ink" onto a solid substrate as a "paper", then patterns with submicrometer resolution can be created (Piner et al., 1999; Salaita et al., 2007). Different from the indirect SPL method, DPN is a direct-write lithography process without destructing surfaces. By regulating the ambient humidity to control the size of water meniscus between the tip of the needle and the substrate, the DPN resolution and the transport rate of molecules from the tip surface onto the surface of the substrate by capillary force can be adjusted. Thereby stable nanostructures with pristine properties, such as patterns of small organic molecules, polymers, DNA, and colloidal nanoparticles, can be directly written. For example, they used this method to generate patterns of DNA on both metallic and insulating substrates (Demers et al., 2002). It is also possible to rapidly generate chemical templates with tailored composition and structure by DPN, which are then used for the assembly of low-dimensional structures.

It is inefficient to process materials by direct-write DPN, if only an individual scanning probe is used. Mirkin's group has improved the DPN technology by designing an array of cantilevers or hard silicon tips. By using tens of thousands of pens, it is possible to simultaneously generate patterned structures of different chemical compositions with 50-nm lateral resolution over a centimeter-scale area (Salaita et al., 2006; Shim et al., 2011). Especially, to keep each hard silicon tip in close contact with a substrate over large areas, they introduced a spring-like elastomer layer underlying ultra-sharp Si tips, which provides mechanical flexibility to each tip and reduce the cost of fabrication of cantilevers. Based on the method, a large array of Au dots with the minimum diameter of 34 ± 5 nm was produced (Salaita et al., 2006).

For the high-throughput lithography method, instead of using hard tips mounted on individual cantilevers, a soft elastomeric tip array can also be utilized to deliver materials in a direct-write way. By this polymer pen lithography method, Mirkin's group prepared arrays with 11 million pens to contact-print uniform patterns. For instance, a representative region of $\sim$15,000 miniaturized duplicates of the 2008 Beijing Olympic logo with a feature size of $\sim$90 nm was prepared within 40 minutes, demonstrating the low-cost and scalability of this method (Huo et al., 2008). By using this DPN method, it is also able to realize the high-throughput and rapid screening of compositional combinations of catalysts and high-entropy alloys. For example, when block polymers with different metal ion precursors were used as an ink and selectively deposited on a substrate, they can be decomposed into individual site-isolated nanoparticles upon thermal annealing, and then nanoparticle libraries with every combination of five metallic elements (Au, Ag, Co, Cu, and Ni) can be prepared (Chen et al., 2016).

3.4.3.3 Mechanical exfoliation/cleavage

Geim, Novoselov, et al. from Manchester University used the mechanical exfoliation method to obtain a single-layered graphene by repeatedly peeling of the mesa of highly oriented monocrystalline graphite films with transparent adhesive tape (Novoselov et al., 2004). Mechanical exfoliation/cleavage of layered crystals into two-dimensional crystals of atomic-level thickness is now a well-established top-down method, leading to the discovery of a family of two-dimensional materials. Since the obtained two-dimensional structures are usually free of impurities, they have exhibited excellent mechanical, electrical, and thermal properties, opening up a new field of research on two-dimensional materials and physical properties.

Tang et al., the authors of this book, investigated the cleavage processes and associated mechanical behaviors of atomically thin MoS_2 layers as a model material, by a direct in situ TEM probing technique. A tungsten tip (~5 nm in diameter) was moved to contact the edge of the MoS_2 crystal, then tip movements were controlled by a piezoelectric motor to approach and selectively contact the atomic edge steps under high-resolution TEM imaging. Then atomically thin flakes with a clean surface are peeled off. The mechanical behaviors of the two-dimensional crystals during bending, stretching, and sliding are found strongly dependent on the numbers of layers (Fig. 3.12) (Tang et al., 2014).

Mechanical exfoliation for two-dimensional materials is essentially a controlled rupture process under external force. It utilized the

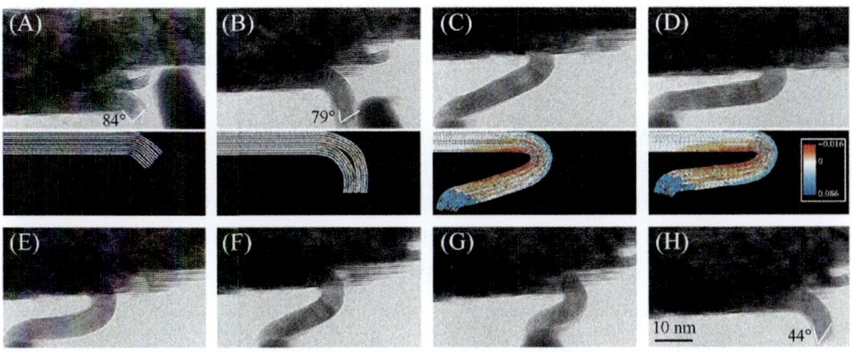

Figure 3.12 Nanomechanical cleavage of a layered MoS_2 by in situ TEM probing method. The strain distribution was simulated by molecular dynamics (Tang et al., 2014). (A–D) The peeling process of an 11-layer MoS_2 along the surface of the MoS_2 single crystal with a tungsten tip. The simulated structures show the strain distribution. (E–H) Reverse process of the peeling process. *TEM*, transmission electron microscopy.

characteristics of weak interlayer bonding and strong intralayered covalent bonding of layered materials, allowing the extension of cracks along the weakly bonded layers to form a two-dimensional structure. Shim et al. from MIT designed a layer-resolved splitting technique as a wafer-scale 2D material exfoliation and transfer method. They demonstrated that various monolayers of 2D materials, including hexagonal boron nitride, tungsten disulfide, tungsten diselenide, molybdenum disulfide, and molybdenum diselenide, can be produced by splitting single stacks of multilayers of two-dimensional material crystals grown on a sapphire substrate. This is based on the differences in the interfacial toughness of these materials, nickel film that takes as an atomic-scale adhesive, and sapphire wafer. The wafer-scale multilayer vdW heterostructures were then constructed to fabricate FETs, showing uniform device performances across the wafer (Shim et al., 2018).

"Lego" is a Latin word, originally from Denmark, meaning fun to put together. Recently, a variety of vdW heterostructures are synthesized by stacking atomically thin layers together via weak vdW forces, similar to how Lego bricks attach to each other, which result in the emergence of unusual properties that are not inherent in the individual two-dimensional structures. Numerous possible combinations of Lego-vdW heterojunctions open a larger degree of freedom for new designer-materials and novel devices in the field of materials science and physics (Geim & Grigorieva, 2013). The electronic spectrum of the graphene placed on h-BN in a high magnetic field shows a fractal pattern of Landau levels, referred to as the *Hofstadter butterfly* pattern (Yu et al., 2014). Except for heterostructures of different two-dimensional materials, two-dimensional materials of the same component but layers overlapped and twisted at a certain angle, also form new periodic conditions and band structures due to the coupling between the layers. Jarillo-Herrero's group at MIT has obtained bilayer graphene with different twist angles through vdW engineering. They found an unusual quantum transport phenomenon in magic-angle graphene superlattices, that is, Mott-like insulator state appears when the twist angle of bilayer graphene is close to the "magic" angle (Cao et al., 2018).

3.4.3.4 Thermomechanical processing

Mechanical processing of nanomaterials during Joule heating or electromigration is also an effective means to engineer low-dimensional materials. CNTs can be viewed as one-dimensional macromolecules with special chiral structures, which determine their electrical properties as metallic or semiconducting (Saito et al., 1992). Theoretically, the change of the CNT chirality causes the corresponding variations of electrical properties (Yakobson, 1998). However, the covalently bonded CNTs are brittle at room temperature and difficult to be processed in a controlled manner (Nardelli et al., 1998), Huang et al. (2006) found that stretching CNTs at high temperatures can activate the kink (a kind of defects) motion and exhibits a superplasticity.

Tang et al., the authors of this book, in situ observed the plastic deformation of Joule-heated CNTs by thermomechanical processing in a controlled manner. Metal-to-semiconductor transitions along with the chirality changes are observed, as confirmed by both the electron diffraction and electrical measurements. They found a trend for the chiral angles to increase to the larger angle region near the armchair. A CNT intramolecular transistor with a channel length as short as 2.8 nm was then created by this method. The in situ measurement results demonstrated a quantum transport at room temperature (Tang et al., 2018, 2021).

3.4.4 Peeling and thinning

Two-dimensional materials with near-perfect crystal structures can be obtained by mechanical exfoliation, which are suitable for exploring extreme physical properties. However, they also suffer from low yield and difficulties in controlling the number and size of layers. Liquid exfoliation is a "soft chemistry" method for obtaining two-dimensional materials in liquids, including ion intercalation, ion exchange, and sonication-assisted exfoliation methods. These methods usually first weaken interlayer attractions of layered materials and then separate the layers by agitation, such as ultrasonication and thermal treatments. Liquid exfoliation method allows the formation of thin films, hybrids and composites for a range of application (Nicolosi et al., 2013). While weakening the interlayer attractions, these methods also suffer the difficulty in the inevitable damage to the structure. Therefore different exfoliation strategies are required for different layered compounds, and relatively mild conditions are also required to maintain the integrity of the in-plane chemical bonding and crystal structure of weakly interlayer-attracted materials.

In the conventional Hummers method (Hummers & Offeman, 1958), graphite is oxidized in a strongly acidic medium, so that oxygen-containing functional groups are introduced to weaken the interlayer attraction, GO sheets are obtained upon ultrasonication separation. RGO sheets can then be obtained via reduction of GO. This method is inexpensive and can be scaled up, but the strong oxidation process destroys the graphite-layered structures and inevitably introduces a large number of defects on GO sheets meanwhile produces serious environmental pollution, such as waste water containing mixed acids and heavy metal ions. Pei et al. (2018) from Institute of Metal Research, Chinese Academy of Sciences reported a scalable and green electrochemical method to synthesize clean GO sheets. The water electrolytic oxidation process of graphite can be completed within a few seconds, and dilute acid can be recycled mainly acting as a control agent to tune the anodic electrocatalytic oxygen evolution reaction of water (Fig. 3.13).

For sonication-assisted exfoliation, the choice of solvent is important, because only by utilizing the appropriate surface tension of the solvent, the interlayer attractions can be weakened and retained. Coleman's group at Trinity College Dublin obtained the mass production of high-quality two-dimensional materials such as few-layer graphene, h-BN, and MoS_2. Taking graphene as an example, graphene dispersions were produced by dispersion and exfoliation of graphite in many organic solvents such as *N*-methyl-pyrrolidone. They proposed that only the exfoliation energetic cost is very small, the 2D materials can be prepared and stabilized against

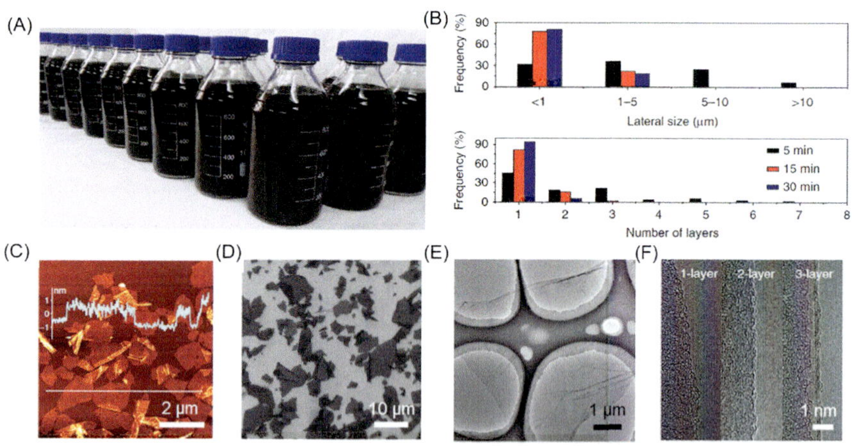

Figure 3.13 Synthesis and characterization of electrochemically synthesized GO. (A) Mass-produced GO aqueous solution (5 mg/mL). (B) Lateral size and number of layers distribution of GO sheets. (C) Typical AFM, (D) SEM, and (E,F) TEM images of GO sheets (Pei et al., 2018). *SEM*, Scanning electron microscopy. *GO*, graphene oxide; *AFM*, atomic force microscopy; *TEM*, transmission electron microscopy.

reaggregation by the liquid exfoliation method. Because in good solvents, whose surface energies match that of graphene, the exfoliation energy is balanced by the solvent–graphene interaction (Coleman, 2013; Hernandez et al., 2008; Nicolosi et al., 2013).

For ion-exchange methods, the ions in-between the layers of layered compounds, such as TMO and layered double hydroxide, can be exchanged in a liquid environment. Thereby the interlayer interaction can be adjusted. In 1996, Sasaki et al. at the National Institute for Materials Science, Japan, reported that the layered protonic titanate with weak electrostatic interlayer interaction, $H_xTi_{2-x/4}\square_{x/4}O_4 \cdot H_2O$ ($x \sim 0.7$, $\square$, vacancy), was substantially swelled upon inserting tetrabutylammonium ions and then exfoliated into single layers by shaking (Sasaki et al., 1996). Based on this method, a series of oxide and hydroxide nanosheets were obtained (Ma & Sasaki, 2010; Minoru & Takayoshi, 2012).

3.5 Summary

This chapter introduces the basic principles, methods, and structural control strategies for the fabrication, assembly and processing of low-dimensional materials from both bottom-up and top-down approaches. Bottom-up approach utilizes interactions between structural units to form an ordered structure, which is a local process, while top-down approach utilizes external actions for processing, which can control over the entire process. In the real fabrication process, a combination of the two approaches is usually required to realize the multiscale control, including the well-ordered arrangements at macroscopic scale, the control over the interfacial bondings and atomic crystal structures at microscopic scale.

Bottom-up molecular synthesis and top-down scanning probe processing are performed with atomic-level precision. Fine structural features that determine the electrical properties of low-dimensional materials, such as the size and shape of nanocrystals, the chirality of CNTs, and the edges of two-dimensional materials, are all controlled to some extent during synthesis process. Using low-dimensional structures as building blocks, self-assembly methods under weak chemical bonding interactions can be used to obtain ordered structures at a higher scale. In addition, structural control of low-dimensional materials faces many challenges, such as control accuracy, mechanism understanding, defect control, production scale, and cost.

It is worth noting that a well-arranged structure does not necessarily have a better property. Functions often arise from defect and interface modulation, such as defect luminescence centers and interfacial electron transport. For practical applications, it is not only necessary to control the structure of individual low-dimensional materials, but also to tailor specific functions at multiple levels by regulating composition, crystal structure, defects, size, shape, surface, interface and other aspects. This requires the development of an integrated fabrication system that considers the entire chain of production and application, as well as green manufacturing and material recycling throughout the entire production-consumption-recycling- environment process.

Furthermore, the fine structural control of low-dimensional materials depends on the in-depth understanding of their preparation processes and mechanisms. With the development and application of various in situ research techniques and theoretical calculation methods, the understanding of the growth mechanism of low-dimensional materials has been deepened in recent years. In the future, it is necessary to pay attention to the study of growth mechanisms at different levels and to establish relatioinships between growth conditions and structures at multiple scales.

References

Ahn, S. H., & Guo, L. J. (2009). Large-Area Roll-to-Roll and Roll-to-Plate Nanoimprint Lithography: A Step toward High-Throughput Application of Continuous Nanoimprinting. *ACS Nano*, *3*(8), 2304–2310.

Aleksandrovic, V., Greshnykh, D., Randjelovic, I., Frömsdorf, A., Kornowski, A., Roth, S. V., Klinke, C., & Weller, H. (2008). Preparation and Electrical Properties of Cobalt – Platinum Nanoparticle Monolayers Deposited by the Langmuir – Blodgett Technique. *ACS Nano*, *2*(6), 1123–1130.

Behabtu, N., Young, C. C., Tsentalovich, D. E., Kleinerman, O., Wang, X., Ma, A. W. K., Bengio, E. A., ter Waarbeek, R. F., de Jong, J. J., Hoogerwerf, R. E., Fairchild, S. B., Ferguson, J. B., Maruyama, B., Kono, J., Talmon, Y., Cohen, Y., Otto, M. J., & Pasquali, M. (2013). Strong, Light, Multifunctional Fibers of Carbon Nanotubes with Ultrahigh Conductivity. *Science (New York, N.Y.)*, *339*(6116), 182.

Bethune, D. S., Kiang, C. H., de Vries, M. S., Gorman, G., Savoy, R., Vazquez, J., & Beyers, R. (1993). Cobalt-catalysed growth of carbon nanotubes with single-atomic-layer walls. *Nature*, *363*(6430), 605–607.

Bieri, M., Treier, M., Cai, J., Aït-Mansour, K., Ruffieux, P., Gröning, O., Gröning, P., Kastler, M., Rieger, R., Feng, X., Müllen, K., & Fasel, R. (2009). Porous graphenes: two-dimensional polymer synthesis with atomic precision. *Chemical Communications* (45), 6919–6921.

Blodgett, K. B. (1935). Films Built by Depositing Successive Monomolecular Layers on a Solid Surface. *Journal of the American Chemical Society*, *57*(6), 1007–1022.

Boles, M. A., Engel, M., & Talapin, D. V. (2016). Self-Assembly of Colloidal Nanocrystals: From Intricate Structures to Functional Materials. *Chemical Reviews*, *116*(18), 11220–11289.

Bolhuis, P. G., Frenkel, D., Mau, S.-C., & Huse, D. A. (1997). Entropy difference between crystal phases. *Nature*, *388*(6639), 235–236.

Burger, B., Maffettone, P. M., Gusev, V. V., Aitchison, C. M., Bai, Y., Wang, X., Li, X., Alston, B. M., Li, B., Clowes, R., Rankin, N., Harris, B., Sprick, R. S., & Cooper, A. I. (2020). A mobile robotic chemist. *Nature*, *583*(7815), 237–241.

Cai, J., Ruffieux, P., Jaafar, R., Bieri, M., Braun, T., Blankenburg, S., Muoth, M., Seitsonen, A. P., Saleh, M., Feng, X., Müllen, K., & Fasel, R. (2010). Atomically precise bottom-up fabrication of graphene nanoribbons. *Nature*, *466*(7305), 470–473.

Cao, Q., Han, S.-j, Tulevski, G. S., Zhu, Y., Lu, D. D., & Haensch, W. (2013). Arrays of single-walled carbon nanotubes with full surface coverage for high-performance electronics. *Nature Nanotechnology*, *8*, 180.

Cao, Y., Fatemi, V., Demir, A., Fang, S., Tomarken, S. L., Luo, J. Y., Sanchez-Yamagishi, J. D., Watanabe, K., Taniguchi, T., Kaxiras, E., Ashoori, R. C., & Jarillo-Herrero, P. (2018). Correlated insulator behaviour at half-filling in magic-angle graphene superlattices. *Nature*, *556*(7699), 80–84.

Chen, L., Hernandez, Y., Feng, X., & Müllen, K. (2012). From Nanographene and Graphene Nanoribbons to Graphene Sheets: Chemical Synthesis. *Angewandte Chemie International Edition*, *51*(31), 7640–7654.

Chen, P.-C., Liu, X., Hedrick, J. L., Xie, Z., Wang, S., Lin, Q.-Y., Hersam, M. C., Dravid, V. P., & Mirkin, C. A. (2016). Polyelemental nanoparticle libraries. *Science (New York, N.Y.)*, *352*(6293), 1565.

Chen, S., Bomer, J. G., van der Wiel, W. G., Carlen, E. T., & van den Berg, A. (2009). Top-Down Fabrication of Sub-30 nm Monocrystalline Silicon Nanowires Using Conventional Microfabrication. *ACS Nano*, *3*(11), 3485–3492.

Chen, X., Lenhert, S., Hirtz, M., Lu, N., Fuchs, H., & Chi, L. (2007). Langmuir–Blodgett Patterning: A Bottom–Up Way To Build Mesostructures over Large Areas. *Accounts of Chemical Research*, *40*(6), 393–401.

Chen, Z., & O'Brien, S. (2008). Structure Direction of II – VI Semiconductor Quantum Dot Binary Nanoparticle Superlattices by Tuning Radius Ratio. *ACS Nano*, *2*(6), 1219–1229.

Cheng, H. M., Li, F., Su, G., Pan, H. Y., He, L. L., Sun, X., & Dresselhaus, M. S. (1998). Large-scale and low-cost synthesis of single-walled carbon nanotubes by the catalytic pyrolysis of hydrocarbons. *Applied Physics Letters*, *72*(25), 3282–3284.

Chou, S. Y., Krauss, P. R., & Renstrom, P. J. (1996). Imprint Lithography with 25-Nanometer Resolution. *Science (New York, N.Y.)*, *272*(5258), 85.

Coleman, J. N. (2013). Liquid Exfoliation of Defect-Free Graphene. *Accounts of Chemical Research*, *46*(1), 14–22.

Cote, L. J., Kim, F., & Huang, J. (2009). Langmuir – Blodgett Assembly of Graphite Oxide Single Layers. *Journal of the American Chemical Society*, *131*(3), 1043–1049.

Crommie, M. F., Lutz, C. P., & Eigler, D. M. (1993). Confinement of Electrons to Quantum Corrals on a Metal Surface. *Science (New York, N.Y.)*, *262*(5131), 218.

Custance, O., Perez, R., & Morita, S. (2009). Atomic force microscopy as a tool for atom manipulation. *Nature Nanotechnology*, *4*(12), 803–810.

Dasgupta, N. P., Sun, J., Liu, C., Brittman, S., Andrews, S. C., Lim, J., Gao, H., Yan, R., & Yang, P. (2014). 25th Anniversary Article: Semiconductor Nanowires – Synthesis, Characterization, and Applications. *Advanced Materials*, *26*(14), 2137–2184.

Davis, V. A., Parra-Vasquez, A. N. G., Green, M. J., Rai, P. K., Behabtu, N., Prieto, V., Booker, R. D., Schmidt, J., Kesselman, E., Zhou, W., Fan, H., Adams, W. W., Hauge, R. H., Fischer, J. E., Cohen, Y., Talmon, Y., Smalley, R. E., & Pasquali, M. (2009). True solutions of single-walled carbon nanotubes for assembly into macroscopic materials. *Nature Nanotechnology*, *4*(12), 830–834.

Decher, G., & Hong, J.-D. (1991). Buildup of ultrathin multilayer films by a self-assembly process, 1 consecutive adsorption of anionic and cationic bipolar amphiphiles on charged surfaces. *Makromolekulare Chemie. Macromolecular Symposia*, *46*(1), 321–327.

Demers, L. M., Ginger, D. S., Park, S. J., Li, Z., Chung, S. W., & Mirkin, C. A. (2002). Direct Patterning of Modified Oligonucleotides on Metals and Insulators by Dip-Pen Nanolithography. *Science (New York, N.Y.)*, *296*(5574), 1836.

Duan, X., Wang, C., Shaw, J. C., Cheng, R., Chen, Y., Li, H., Wu, X., Tang, Y., Zhang, Q., Pan, A., Jiang, J., Yu, R., Huang, Y., & Duan, X. (2014). Lateral epitaxial growth of two-dimensional layered semiconductor heterojunctions. *Nature Nanotechnology*, *9*(12), 1024–1030.

Eigler, D. M., & Schweizer, E. K. (1990). Positioning single atoms with a scanning tunnelling microscope. *Nature*, *344*(6266), 524–526.

Ek, M., & Filler, M. A. (2018). Atomic-Scale Choreography of Vapor–Liquid–Solid Nanowire Growth. *Accounts of Chemical Research*, *51*(1), 118–126.

Ertl, G. (1991). Oscillatory Kinetics and Spatio-Temporal Self-Organization in Reactions at Solid Surfaces. *Science (New York, N.Y.)*, *254*(5039), 1750.

Ertl, G. (2008). Reactions at Surfaces: From Atoms to Complexity (Nobel Lecture). *Angewandte Chemie International Edition*, *47*(19), 3524–3535.

Fanourgakis, G. S., Gkagkas, K., Tylianakis, E., & Froudakis, G. E. (2020). A Universal Machine Learning Algorithm for Large-Scale Screening of Materials. *Journal of the American Chemical Society*, *142*(8), 3814–3822.

Feynman, R. P. (1960). Plenty of Room at the Bottom. *California Institute of Technology Journal of Engineering and Science*, *4*(2), 23.

Freer, E. M., Grachev, O., Duan, X., Martin, S., & Stumbo, D. P. (2010). High-yield self-limiting single-nanowire assembly with dielectrophoresis. *Nature Nanotechnology*, *5* (7), 525–530.

Fuechsle, M., Miwa, J. A., Mahapatra, S., Ryu, H., Lee, S., Warschkow, O., Hollenberg, L. C. L., Klimeck, G., & Simmons, M. Y. (2012). A single-atom transistor. *Nature Nanotechnology*, 7, 242.

Garcia, R., Knoll, A. W., & Riedo, E. (2014). Advanced scanning probe lithography. *Nature Nanotechnology*, *9*, 577.

Geim, A. K., & Grigorieva, I. V. (2013). Van der Waals heterostructures. *Nature*, *499*, 419.

Gleiter, H. (1989). Nanocrystalline materials. *Progress in Materials Science*, *33*(4), 223–315.

Gong, Y., Lei, S., Ye, G., Li, B., He, Y., Keyshar, K., Zhang, X., Wang, Q., Lou, J., Liu, Z., Vajtai, R., Zhou, W., & Ajayan, P. M. (2015). Two-Step Growth of Two-Dimensional WSe2/MoSe2 Heterostructures. *Nano Letters*, *15*(9), 6135–6141.

Gong, Y., Lin, J., Wang, X., Shi, G., Lei, S., Lin, Z., Zou, X., Ye, G., Vajtai, R., Yakobson, B. I., Terrones, H., Terrones, M., Tay., Beng, K., Lou, J., Pantelides, S. T., Liu, Z., Zhou, W., & Ajayan, P. M. (2014). Vertical and in-plane heterostructures from WS2/MoS2 monolayers. *Nature Materials*, *13*(12), 1135–1142.

Grill, L., Dyer, M., Lafferentz, L., Persson, M., Peters, M. V., & Hecht, S. (2007). Nano-architectures by covalent assembly of molecular building blocks. *Nature Nanotechnology*, *2*(11), 687–691.

Hata, K., Futaba, D. N., Mizuno, K., Namai, T., Yumura, M., & Iijima, S. (2004). Water-Assisted Highly Efficient Synthesis of Impurity-Free Single-Walled Carbon Nanotubes. *Science (New York, N.Y.)*, *306*(5700), 1362–1364.

Hernandez, Y., Nicolosi, V., Lotya, M., Blighe, F. M., Sun, Z., De, S., McGovern, I. T., Holland, B., Byrne, M., Gun'Ko, Y. K., Boland, J. J., Niraj, P., Duesberg, G., Krishnamurthy, S., Goodhue, R., Hutchison, J., Scardaci, V., Ferrari, A. C., & Coleman, J. N. (2008). High-yield production of graphene by liquid-phase exfoliation of graphite. *Nature Nanotechnology, 3*, 563.

Hines, M. A., & Guyot-Sionnest, P. (1996). Synthesis and Characterization of Strongly Luminescing ZnS-Capped CdSe Nanocrystals. *The Journal of Physical Chemistry, 100*(2), 468–471.

Hong, Y.-L., Liu, Z., Wang, L., Zhou, T., Ma, W., Xu, C., Feng, S., Chen, L., Chen, M.-L., Sun, D.-M., Chen, X.-Q., Cheng, H.-M., & Ren, W. (2020). Chemical vapor deposition of layered two-dimensional $MoSi_2N_4$ materials. *Science (New York, N.Y.), 369*(6504), 670.

Huang, J. Y., Chen, S., Wang, Z. Q., Kempa, K., Wang, Y. M., Jo, S. H., Chen, G., Dresselhaus, M. S., & Ren, Z. F. (2006). Superplastic carbon nanotubes. *Nature, 439*(7074), 281.

Hummers, W. S., & Offeman, R. E. (1958). Preparation of Graphitic Oxide. *Journal of the American Chemical Society, 80*(6), 1339.

Huo, F., Zheng, Z., Zheng, G., Giam, L. R., Zhang, H., & Mirkin, C. A. (2008). Polymer Pen Lithography. *Science (New York, N.Y.), 321*(5896), 1658.

Hussain, S. A., Dey, B., Bhattacharjee, D., & Mehta, N. (2018). Unique supramolecular assembly through Langmuir – Blodgett (LB) technique. *Heliyon, 4*(12)e01038.

Iijima, S. (1991). Helical microtubules of graphitic carbon. *Nature, 354*(6348), 56–58.

Iijima, S., & Ichihashi, T. (1993). SINGLE-SHELL CARBON NANOTUBES OF 1-NM DIAMETER. *Nature, 363*(6430), 603–605.

Ji, Z.-H., Zhang, L., Tang, D.-M., Chen, C.-M., Nordling, T. E. M., Zhang, Z.-D., Ren, C.-L., Da, B., Li, X., Guo, S.-Y., Liu, C., & Cheng, H.-M. (2021). High-throughput screening and machine learning for the efficient growth of high-quality single-wall carbon nanotubes. *Nano Research, 14*, 4610–4615.

Jiang, K., Li, Q., & Fan, S. (2002). Spinning continuous carbon nanotube yarns. *Nature, 419*(6909), 801.

Jourdain, V., & Bichara, C. (2013). Current understanding of the growth of carbon nanotubes in catalytic chemical vapour deposition. *Carbon, 58*(0)), 2–39.

Kaiser, K., Scriven, L. M., Schulz, F., Gawel, P., Gross, L., & Anderson, H. L. (2019). An sp-hybridized molecular carbon allotrope, cyclo[18]carbon. *Science (New York, N.Y.)*, eaay1914.

Kang, Y., Ye, X., Chen, J., Cai, Y., Diaz, R. E., Adzic, R. R., Stach, E. A., & Murray, C. B. (2013). Design of Pt–Pd Binary Superlattices Exploiting Shape Effects and Synergistic Effects for Oxygen Reduction Reactions. *Journal of the American Chemical Society, 135*(1), 42–45.

Kitagawa, S. (2017). Future Porous Materials. *Accounts of Chemical Research, 50*(3), 514–516.

Kluender, E. J., Hedrick, J. L., Brown, K. A., Rao, R., Meckes, B., Du, J. S., Moreau, L. M., Maruyama, B., & Mirkin, C. A. (2019). Catalyst discovery through megalibraries of nanomaterials. *Proceedings of the National Academy of Sciences, 116*(1), 40.

Kneller, E. F., & Hawig, R. (1991). The exchange-spring magnet: a new material principle for permanent magnets. *IEEE Transactions on Magnetics, 27*(4), 3588.

Kong, J., Soh, H. T., Cassell, A. M., Quate, C. F., & Dai, H. (1998). Synthesis of Individual Single-Walled Carbon Nanotubes on Patterned Silicon Wafers. *Nature, 395*(6705), 878–881.

Krätschmer, W., Lamb, L. D., Fostiropoulos, K., & Huffman, D. R. (1990). Solid C60: a new form of carbon. *Nature, 347*, 354.

Krogstrup, P., Jørgensen, H. I., Johnson, E., Madsen, M. H., Sørensen, C. B., Morral, A. F. I., Aagesen, M., Nygård, J., & Glas, F. (2013). Advances in the theory of III–V nanowire growth dynamics. *Journal of Physics D: Applied Physics*, *46*(31), 313001.

Kroto, H. W., Heath, J. R., O'Brien, S. C., Curl, R. F., & Smalley, R. E. (1985). C60: Buckminsterfullerene. *Nature*, *318*, 162–163.

Kutuzov, S., He, J., Tangirala, R., Emrick, T., Russell, T. P., & Böker, A. (2007). On the kinetics of nanoparticle self-assembly at liquid/liquid interfaces. *Physical Chemistry Chemical Physics*, *9*(48), 6351–6358.

LaMer, V. K., & Dinegar, R. H. (1950). Theory, Production and Mechanism of Formation of Monodispersed Hydrosols. *Journal of the American Chemical Society*, *72* (11), 4847–4854.

Langmuir, I., & Blodgett, K. B. (1935). Über einige neue Methoden zur Untersuchung von monomolekularen Filmen. *Kolloid-Zeitschrift*, *73*(3), 257–263.

Li, B.-W., Osada, M., Kim, Y.-H., Ebina, Y., Akatsuka, K., & Sasaki, T. (2017). Atomic Layer Engineering of High-κ Ferroelectricity in 2D Perovskites. *Journal of the American Chemical Society*, *139*(31), 10868–10874.

Li, J., Stein, D., McMullan, C., Branton, D., Aziz, M. J., & Golovchenko, J. A. (2001). Ion-beam sculpting at nanometre length scales. *Nature*, *412*(6843), 166–169.

Li, S., Hong, J., Gao, B., Lin, Y.-C., Lim, H. E., Lu, X., Wu, J., Liu, S., Tateyama, Y., Sakuma, Y., Tsukagoshi, K., Suenaga, K., & Taniguchi, T. (2021). Tunable Doping of Rhenium and Vanadium into Transition Metal Dichalcogenides for Two-Dimensional Electronics. *Advanced Science*, *8*(11), 2004438.

Li, S., Lin, Y.-C., Zhao, W., Wu, J., Wang, Z., Hu, Z., Shen, Y., Tang, D.-M., Wang, J., Zhang, Q., Zhu, H., Chu, L., Zhao, W., Liu, C., Sun, Z., Taniguchi, T., Osada, M., Chen, W., Xu, Q.-H., et al. (2018). Vapour–liquid–solid growth of monolayer MoS2 nanoribbons. *Nature Materials*, *17*(6), 535–542.

Li, S., Wang, S., Tang, D.-M., Zhao, W., Xu, H., Chu, L., Bando, Y., Golberg, D., & Eda, G. (2015). Halide-assisted atmospheric pressure growth of large WSe2 and WS2 monolayer crystals. *Applied Materials Today*, *1*(1), 60–66.

Li, W., Liu, J., & Zhao, D. (2016). Mesoporous materials for energy conversion and storage devices. *Nature Reviews Materials*, *1*, 16023.

Li, X., Cai, W., An, J., Kim, S., Nah, J., Yang, D., Piner, R., Velamakanni, A., Jung, I., Tutuc, E., Banerjee, S. K., Colombo, L., & Ruoff, R. S. (2009). Large-Area Synthesis of High-Quality and Uniform Graphene Films on Copper Foils. *Science (New York, N. Y.)*, *324*(5932), 1312.

Li, X., Zhang, L., Wang, X., Shimoyama, I., Sun, X., Seo, W.-S., & Dai, H. (2007). Langmuir – Blodgett Assembly of Densely Aligned Single-Walled Carbon Nanotubes from Bulk Materials. *Journal of the American Chemical Society*, *129*(16), 4890–4891.

Li, Y., Peng, S., Mann, D., Cao, J., Tu, R., Cho, K. J., & Dai, H. (2005). On the Origin of Preferential Growth of Semiconducting Single-Walled Carbon Nanotubes. *The Journal of Physical Chemistry. B*, *109*(15), 6968–6971.

Lin, L., Zhang, J., Su, H., Li, J., Sun, L., Wang, Z., Xu, F., Liu, C., Lopatin, S., Zhu, Y., Jia, K., Chen, S., Rui, D., Sun, J., Xue, R., Gao, P., Kang, N., Han, Y., Xu, H. Q., et al. (2019). Towards super-clean graphene. *Nature Communications*, *10*(1), 1912.

Liu, B., Tang, D.-M., Sun, C., Liu, C., Ren, W., Li, F., Yu, W.-J., Yin, L.-C., Zhang, L., Jiang, C., & Cheng, H.-M. (2011). Importance of Oxygen in the Metal-Free Catalytic Growth of Single-Walled Carbon Nanotubes from SiOx by a Vapor – Solid – Solid Mechanism. *Journal of the American Chemical Society*, *133*(2), 197–199.

Liu, G., Yang, H. G., Pan, J., Yang, Y. Q., Lu, G. Q., & Cheng, H.-M. (2014). Titanium Dioxide Crystals with Tailored Facets. *Chemical Reviews*, *114*(19), 9559–9612.

Liu, L., Han, J., Xu, L., Zhou, J., Zhao, C., Ding, S., Shi, H., Xiao, M., Ding, L., Ma, Z., Jin, C., Zhang, Z., & Peng, L.-M. (2020). Aligned, high-density semiconducting carbon nanotube arrays for high-performance electronics. *Science (New York, N.Y.)*, *368*(6493), 850.

Liu, Y., Xu, Z., Gao, W., Cheng, Z., & Gao, C. (2017). Graphene and Other 2D Colloids: Liquid Crystals and Macroscopic Fibers. *Advanced Materials*, *29*(14), 1606794.

Liu, Y., Xu, Z., Zhan, J., Li, P., & Gao, C. (2016). Superb Electrically Conductive Graphene Fibers via Doping Strategy. *Advanced Materials*, *28*(36), 7941–7947.

Ma, R., & Sasaki, T. (2010). Nanosheets of Oxides and Hydroxides: Ultimate 2D Charge-Bearing Functional Crystallites. *Advanced Materials*, *22*(45), 5082–5104.

Ma, T., Liu, Z., Wen, J., Gao, Y., Ren, X., Chen, H., Jin, C., Ma, X.-L., Xu, N., Cheng, H.-M., & Ren, W. (2017). Tailoring the thermal and electrical transport properties of graphene films by grain size engineering. *Nature Communications*, *8*14486.

Ma, T., Ren, W., Liu, Z., Huang, L., Ma, L.-P., Ma, X., Zhang, Z., Peng, L.-M., & Cheng, H.-M. (2014). Repeated Growth–Etching–Regrowth for Large-Area Defect-Free Single-Crystal Graphene by Chemical Vapor Deposition. *ACS Nano*, *8* (12), 12806–12813.

Ma, T., Ren, W., Zhang, X., Liu, Z., Gao, Y., Yin, L.-C., Ma, X.-L., Ding, F., & Cheng, H.-M. (2013). Edge-controlled growth and kinetics of single-crystal graphene domains by chemical vapor deposition. *Proceedings of the National Academy of Sciences*, *110*(51), 20386–20391.

Ma, W., Ma, R., Wang, C., Liang, J., Liu, X., Zhou, K., & Sasaki, T. (2015). A Superlattice of Alternately Stacked Ni–Fe Hydroxide Nanosheets and Graphene for Efficient Splitting of Water. *ACS Nano*, *9*(2), 1977–1984.

Magnin, Y., Amara, H., Ducastelle, F., Loiseau, A., & Bichara, C. (2018). Entropy-driven stability of chiral single-walled carbon nanotubes. *Science (New York, N.Y.)*, *362*(6411), 212.

Magnin, Y., Zappelli, A., Amara, H., Ducastelle, F., & Bichara, C. (2015). Size Dependent Phase Diagrams of Nickel-Carbon Nanoparticles. *Physical Review Letters*, *115*(20), 205502.

Manfrinato, V. R., Zhang, L., Su, D., Duan, H., Hobbs, R. G., Stach, E. A., & Berggren, K. K. (2013). Resolution Limits of Electron-Beam Lithography toward the Atomic Scale. *Nano Letters*, *13*(4), 1555–1558.

Manoharan, V. N. (2015). Colloidal matter: Packing, geometry, and entropy. *Science (New York, N.Y.)*, *349*(6251), 1253751.

Minoru, O., & Takayoshi, S. (2012). Two-Dimensional Dielectric Nanosheets: Novel Nanoelectronics From Nanocrystal Building Blocks. *Advanced Materials*, *24*(2), 210–228.

Mirza, M. M., MacLaren, D. A., Samarelli, A., Holmes, B. M., Zhou, H., Thoms, S., MacIntyre, D., & Paul, D. J. (2014). Determining the Electronic Performance Limitations in Top-Down-Fabricated Si Nanowires with Mean Widths Down to 4 nm. *Nano Letters*, *14*(11), 6056–6060.

Moliner, M., Román-Leshkov, Y., & Corma, A. (2019). Machine Learning Applied to Zeolite Synthesis: The Missing Link for Realizing High-Throughput Discovery. *Accounts of Chemical Research*, *52*(10), 2971–2980.

Murray, C. B., Kagan, C. R., & Bawendi, M. G. (2000). Synthesis and Characterization of Monodisperse Nanocrystals and Close-Packed Nanocrystal Assemblies. *Annual Review Of Materials Science*, *30*(1), 545–610.

Murray, C. B., Norris, D. J., & Bawendi, M. G. (1993). Synthesis and characterization of nearly monodisperse CdE (E = sulfur, selenium, tellurium) semiconductor nanocrystallites. *Journal of the American Chemical Society*, *115*(19), 8706–8715.

Naguib, M., Kurtoglu, M., Presser, V., Lu, J., Niu, J., Heon, M., Hultman, L., Gogotsi, Y., & Barsoum, M. W. (2011). Two-Dimensional Nanocrystals Produced by Exfoliation of Ti3AlC2. *Advanced Materials*, *23*(37), 4248–4253.

Najmaei, S., Liu, Z., Zhou, W., Zou, X., Shi, G., Lei, S., Yakobson, B. I., Idrobo, J.-C., Ajayan, P. M., & Lou, J. (2013). Vapour phase growth and grain boundary structure of molybdenum disulphide atomic layers. *Nature Materials*, *12*, 754.

Nardelli, M. B., Yakobson, B. I., & Bernholc, J. (1998). Brittle and Ductile Behavior in Carbon Nanotubes. *Physical Review Letters*, *81*(21), 4656–4659.

Nicolosi, V., Chhowalla, M., Kanatzidis, M. G., Strano, M. S., & Coleman, J. N. (2013). Liquid Exfoliation of Layered Materials. *Science (New York, N.Y.)*, *6139*, 340.

Nikolaev, P., Hooper, D., Perea-López, N., Terrones, M., & Maruyama, B. (2014). Discovery of Wall-Selective Carbon Nanotube Growth Conditions via Automated Experimentation. *ACS Nano*, *8*(10), 10214–10222.

Novoselov, K. S., Geim, A. K., Morozov, S. V., Jiang, D., Zhang, Y., Dubonos, S. V., Grigorieva, I. V., & Firsov, A. A. (2004). Electric Field Effect in Atomically Thin Carbon Films. *Science (New York, N.Y.)*, *306*(5696), 666–669.

Novoselov, K. S., Jiang, D., Schedin, F., Booth, T. J., Khotkevich, V. V., Morozov, S. V., & Geim, A. K. (2005). Two-dimensional atomic crystals. *Proceedings of the National Academy of Sciences of the United States of America*, *102*(30), 10451–10453.

Osada, M., Akatsuka, K., Ebina, Y., Funakubo, H., Kiguchi, T., Takada, K., & Sasaki, T. (2007). Solution-Based Fabrication of High-κ Dielectric Nanofilms Using Titania Nanosheets as a Building Block. *Japanese Journal of Applied Physics*, *46*(10B), 6979–6983.

Pei, S., Wei, Q., Huang, K., Cheng, H.-M., & Ren, W. (2018). Green synthesis of graphene oxide by seconds timescale water electrolytic oxidation. *Nature Communications*, *9*(1), 145.

Peng, X., Manna, L., Yang, W., Wickham, J., Scher, E., Kadavanich, A., & Alivisatos, A. P. (2000). Shape control of CdSe nanocrystals. *Nature*, *404*, 59.

Peng, X., Wickham, J., & Alivisatos, A. P. (1998). Kinetics of II-VI and III-V Colloidal Semiconductor Nanocrystal Growth: "Focusing" of Size Distributions. *Journal of the American Chemical Society*, *120*(21), 5343–5344.

Peng, Z. A., & Peng, X. (2001). Formation of High-Quality CdTe, CdSe, and CdS Nanocrystals Using CdO as Precursor. *Journal of the American Chemical Society*, *123*(1), 183–184.

Peterson, J. J., & Krauss, T. D. (2006). Fluorescence Spectroscopy of Single Lead Sulfide Quantum Dots. *Nano Letters*, *6*(3), 510–514.

Piner, R. D., Zhu, J., Xu, F., Hong, S., & Mirkin, C. A. (1999). Dip-Pen" Nanolithography. *Science (New York, N.Y.)*, *283*(5402), 661.

Pires, D., Hedrick, J. L., De Silva, A., Frommer, J., Gotsmann, B., Wolf, H., Despont, M., Duerig, U., & Knoll, A. W. (2010). Nanoscale Three-Dimensional Patterning of Molecular Resists by Scanning Probes. *Science (New York, N.Y.)*, *328*(5979), 732.

Qin, H., Niu, Y., Meng, R., Lin, X., Lai, R., Fang, W., & Peng, X. (2014). Single-Dot Spectroscopy of Zinc-Blende CdSe/CdS Core/Shell Nanocrystals: Nonblinking and Correlation with Ensemble Measurements. *Journal of the American Chemical Society*, *136* (1), 179–187.

Qiu, C., Zhang, Z., Xiao, M., Yang, Y., Zhong, D., & Peng, L.-M. (2017). Scaling Carbon Nanotube Complementary Transistors to 5-nm Gate Lengths. *Science (New York, N.Y.)*, *355*(6322), 271.

Rai, P. K., Pinnick, R. A., Parra-Vasquez, A. N. G., Davis, V. A., Schmidt, H. K., Hauge, R. H., Smalley, R. E., & Pasquali, M. (2006). Isotropic – Nematic Phase Transition of Single-Walled Carbon Nanotubes in Strong Acids. *Journal of the American Chemical Society*, *128*(2), 591–595.

Renzhi, M., & Takayoshi, S. (2012). Synthesis of LDH Nanosheets and their Layer-by-Layer Assembly. *Recent Patents on Nanotechnology*, *6*(3), 159–168.

Richardson, J. J., Björnmalm, M., & Caruso, F. (2015). Technology-driven layer-by-layer assembly of nanofilms. *Science (New York, N.Y.)*, *348*(6233), aaa2491.

Richardson, J. J., Cui, J., Björnmalm, M., Braunger, J. A., Ejima, H., & Caruso, F. (2016). Innovation in Layer-by-Layer Assembly. *Chemical Reviews*, *116*(23), 14828–14867.

Saito, R., Fujita, M., Dresselhaus, G., & Dresselhaus, M. S. (1992). Electronic structure of chiral graphene tubules. *Applied Physics Letters*, *60*(18), 2204–2206.

Salaita, K., Wang, Y., Fragala, J., Vega, R. A., Liu, C., & Mirkin, C. A. (2006). Massively Parallel Dip–Pen Nanolithography with 55 000-Pen Two-Dimensional Arrays. *Angewandte Chemie International Edition*, *45*(43), 7220–7223.

Salaita, K., Wang, Y., & Mirkin, C. A. (2007). Applications of dip-pen nanolithography. *Nature Nanotechnology*, *2*, 145.

Sanchez-Valencia, J. R., Dienel, T., Groning, O., Shorubalko, I., Mueller, A., Jansen, M., Amsharov, K., Ruffieux, P., & Fasel, R. (2014). Controlled synthesis of single-chirality carbon nanotubes. *Nature*, *512*(7512), 61–64.

Sarker, B. K., Shekhar, S., & Khondaker, S. I. (2011). Semiconducting Enriched Carbon Nanotube Aligned Arrays of Tunable Density and Their Electrical Transport Properties. *ACS Nano*, *5*(8), 6297–6305.

Sasaki, T., Ebina, Y., Tanaka, T., Harada, M., Watanabe, M., & Decher, G. (2001). Layer-by-Layer Assembly of Titania Nanosheet/Polycation Composite Films. *Chemistry Of Materials*, *13*(12), 4661–4667.

Sasaki, T., Watanabe, M., Hashizume, H., Yamada, H., & Nakazawa, H. (1996). Macromolecule-like Aspects for a Colloidal Suspension of an Exfoliated Titanate. Pairwise Association of Nanosheets and Dynamic Reassembling Process Initiated from It. *Journal of the American Chemical Society*, *118*(35), 8329–8335.

Schofield, S. R., Curson, N. J., Simmons, M. Y., Rueß, F. J., Hallam, T., Oberbeck, L., & Clark, R. G. (2003). Atomically Precise Placement of Single Dopants in Si. *Physical Review Letters*, *91*(13), 136104.

Self-assembling life. (2016). *Nature Nanotechnology*, *11*(11), 909.

Shim, J., Bae, S.-H., Kong, W., Lee, D., Qiao, K., Nezich, D., Park, Y. J., Zhao, R., Sundaram, S., Li, X., Yeon, H., Choi, C., Kum, H., Yue, R., Zhou, G., Ou, Y., Lee, K., Moodera, J., Zhao, X., et al. (2018). Controlled crack propagation for atomic precision handling of wafer-scale two-dimensional materials. *Science (New York, N.Y.)*, *362*(6415), 665.

Shim, W., Braunschweig, A. B., Liao, X., Chai, J., Lim, J. K., Zheng, G., & Mirkin, C. A. (2011). Hard-tip, soft-spring lithography. *Nature*, *469*, 516.

Slater, A. G., & Cooper, A. I. (2015). Function-led design of new porous materials. *Science (New York, N.Y.)*, *348*(6238).

Srivastava, S., & Kotov, N. A. (2008). Composite Layer-by-Layer (LBL) Assembly with Inorganic Nanoparticles and Nanowires. *Accounts of Chemical Research*, *41*(12), 1831–1841.

Steigerwald, M. L., & Brus, L. E. (1990). Semiconductor crystallites: a class of large molecules. *Accounts of Chemical Research*, *23*(6), 183–188.

Takata, T., Jiang, J., Sakata, Y., Nakabayashi, M., Shibata, N., Nandal, V., Seki, K., Hisatomi, T., & Domen, K. (2020). Photocatalytic water splitting with a quantum efficiency of almost unity. *Nature*, *581*(7809), 411–414.

Tang, D.-M., Erohin Sergey, V., Kvashnin Dmitry, G., Demin Victor, A., Cretu, O., Jiang, S., Zhang, L., Hou, P.-X., Chen, G., Futaba Don, N., Zheng, Y., Xiang, R., Zhou, X., Hsia, F.-C., Kawamoto, N., Mitome, M., Nemoto, Y., Uesugi, F., Takeguchi, M., et al. (2021). Semiconductor nanochannels in metallic carbon nanotubes by thermomechanical chirality alteration. *Science (New York, N.Y.)*, *374* (6575), 1616–1620.

Tang, D.-M., Kvashnin, D. G., Cretu, O., Nemoto, Y., Uesugi, F., Takeguchi, M., Zhou, X., Hsia, F.-C., Liu, C., Sorokin, P. B., Kawamoto, N., Mitome, M., Cheng, H.-M., Golberg, D., & Bando, Y. (2018). Chirality transitions and transport properties of individual few-walled carbon nanotubes as revealed by in situ TEM probing. *Ultramicroscopy*, *194*, 108–116.

Tang, D.-M., Kvashnin, D. G., Najmaei, S., Bando, Y., Kimoto, K., Koskinen, P., Ajayan, P. M., Yakobson, B. I., Sorokin, P. B., Lou, J., & Golberg, D. (2014). Nanomechanical cleavage of molybdenum disulphide atomic layers. *Nature Communications*, *5*, 3631.

Tao, A., Kim, F., Hess, C., Goldberger, J., He, R., Sun, Y., Xia, Y., & Yang, P. (2003). Langmuir – Blodgett Silver Nanowire Monolayers for Molecular Sensing Using Surface-Enhanced Raman Spectroscopy. *Nano Letters*, *3*(9), 1229–1233.

Thompson, D. A. W. (1992). *On Growth and Form*. Cambridge: Cambridge University Press.

Treier, M., Pignedoli, C. A., Laino, T., Rieger, R., Müllen, K., Passerone, D., & Fasel, R. (2011). Surface-assisted cyclodehydrogenation provides a synthetic route towards easily processable and chemically tailored nanographenes. *Nature Chemistry*, *3*(1), 61–67.

Tseng, A. A. (2005). Recent Developments in Nanofabrication Using Focused Ion Beams. *Small (Weinheim an der Bergstrasse, Germany)*, *1*(10), 924–939.

Tshitoyan, V., Dagdelen, J., Weston, L., Dunn, A., Rong, Z., Kononova, O., Persson, K. A., Ceder, G., & Jain, A. (2019). Unsupervised word embeddings capture latent knowledge from materials science literature. *Nature*, *571*(7763), 95–98.

Turing, A. M. (1952). The chemical basis of morphogenesis. *Philosophical Transactions of the Royal Society of London. Series B, Biological Sciences*, *237*(641), 37–72.

VahidMohammadi, A., Rosen, J., & Gogotsi, Y. (2021). The world of two-dimensional carbides and nitrides (MXenes). *Science (New York, N.Y.)*, *372*(6547), eabf1581.

Vieu, C., Carcenac, F., Pépin, A., Chen, Y., Mejias, M., Lebib, A., Manin-Ferlazzo, L., Couraud, L., & Launois, H. (2000). Electron beam lithography: resolution limits and applications. *Applied Surface Science*, *164*(1), 111–117.

Voznyy, O., Levina, L., Fan, J. Z., Askerka, M., Jain, A., Choi, M.-J., Ouellette, O., Todorović, P., Sagar, L. K., & Sargent, E. H. (2019). Machine Learning Accelerates Discovery of Optimal Colloidal Quantum Dot Synthesis. *ACS Nano*, *13*(10), 11122–11128.

Wagner, C., & Harned, N. (2010). Lithography gets extreme. *Nature Photonics*, *4*, 24.

Wagner, R. S., & Ellis, W. C. (1964). Vapor-liquid-solid mechanism of single crystal growth. *Applied Physics Letters*, *4*(5), 89–90.

Wang, C., Chi, M., Li, D., Strmcnik, D., van der Vliet, D., Wang, G., Komanicky, V., Chang, K.-C., Paulikas, A. P., Tripkovic, D., Pearson, J., More, K. L., Markovic, N. M., & Stamenkovic, V. R. (2011). Design and Synthesis of Bimetallic Electrocatalyst with Multilayered Pt-Skin Surfaces. *Journal of the American Chemical Society*, *133*(36), 14396–14403.

Wang, J., Jin, X., Liu, Z., Yu, G., Ji, Q., Wei, H., Zhang, J., Zhang, K., Li, D., Yuan, Z., Li, J., Liu, P., Wu, Y., Wei, Y., Wang, J., Li, Q., Zhang, L., Kong, J., Fan, S., et al. (2018). Growing highly pure semiconducting carbon nanotubes by electrotwisting the helicity. *Nature Catalysis*, *1*, 326.

Wang, L., Wang, D., Dong, X. Y., Zhang, Z. J., Pei, X. F., Chen, X. J., Chen, B., & Jin, J. (2011). Layered assembly of graphene oxide and Co–Al layered double hydroxide nanosheets as electrode materials for supercapacitors. *Chemical Communications*, *47*(12), 3556–3558.

Wang, X., Gong, Y., Shi, G., Chow, W. L., Keyshar, K., Ye, G., Vajtai, R., Lou, J., Liu, Z., Ringe, E., Tay, B. K., & Ajayan, P. M. (2014). Chemical Vapor Deposition Growth of Crystalline Monolayer MoSe2. *ACS Nano*, *8*(5), 5125–5131.

Wang, Y., Qiu, L., Zhang, L., Tang, D.-M., Ma, R., Wang, Y., Zhang, B., Ding, F., Liu, C., & Cheng, H.-M. (2020). Precise Identification of the Active Phase of Cobalt Catalyst for Carbon Nanotube Growth by In Situ Transmission Electron Microscopy. *ACS Nano*, *14*(12), 16823–16831.

Weber, B., Mahapatra, S., Ryu, H., Lee, S., Fuhrer, A., Reusch, T. C. G., Thompson, D. L., Lee, W. C. T., Klimeck, G., Hollenberg, L. C. L., & Simmons, M. Y. (2012). Ohm's Law Survives to the Atomic Scale. *Science (New York, N.Y.)*, *335*(6064), 64–67.

Wei, Z., Wang, D., Kim, S., Kim, S.-Y., Hu, Y., Yakes, M. K., Laracuente, A. R., Dai, Z., Marder, S. R., Berger, C., King, W. P., de Heer, W. A., Sheehan, P. E., & Riedo, E. (2010). Nanoscale Tunable Reduction of Graphene Oxide for Graphene Electronics. *Science (New York, N.Y.)*, *328*(5984), 1373.

Weidman, M. C., Smilgies, D.-M., & Tisdale, W. A. (2016). Kinetics of the self-assembly of nanocrystal superlattices measured by real-time in situ X-ray scattering. *Nature Materials*, *15*(7), 775–781.

Xia, Y., Xiong, Y., Lim, B., & Skrabalak, S. E. (2009). Shape-Controlled Synthesis of Metal Nanocrystals: Simple Chemistry Meets Complex Physics? *Angewandte Chemie International Edition*, *48*(1), 60–103.

Xiang, R., Inoue, T., Zheng, Y., Kumamoto, A., Qian, Y., Sato, Y., Liu, M., Tang, D., Gokhale, D., Guo, J., Hisama, K., Yotsumoto, S., Ogamoto, T., Arai, H., Kobayashi, Y., Zhang, H., Hou, B., Anisimov, A., Maruyama, M., et al. (2020). One-dimensional van der Waals heterostructures. *Science (New York, N.Y.)*, *367*(6477), 537.

Xiong, Y., & Xia, Y. (2007). Shape-Controlled Synthesis of Metal Nanostructures: The Case of Palladium. *Advanced Materials*, *19*(20), 3385–3391.

Xu, Z., & Gao, C. (2011). Graphene chiral liquid crystals and macroscopic assembled fibres. *Nature Communications*, *2*, 571.

Xu, Z., & Gao, C. (2011). Aqueous Liquid Crystals of Graphene Oxide. *ACS Nano*, *5*(4), 2908–2915.

Xu, Z., & Gao, C. (2014). Graphene in macroscopic order: liquid crystals and wet-spun fibers. *Accounts of Chemical Research*, *47*(4), 1267–1276.

Xu, Z., Liu, Y., Zhao, X., Peng, L., Sun, H., Xu, Y., Ren, X., Jin, C., Xu, P., Wang, M., & Gao, C. (2016). Ultrastiff and strong graphene fibers via full-scale synergetic defect engineering. *Advanced Materials*, *28*(30), 6449–6456.

Yakobson, B. I. (1998). Mechanical relaxation and "intramolecular plasticity" in carbon nanotubes. *Applied Physics Letters*, *72*(8), 918–920.

Yan, Z., Peng, Z., & Tour, J. M. (2014). Chemical Vapor Deposition of Graphene Single Crystals. *Accounts of Chemical Research*, *47*(4), 1327–1337.

Yang, F., Wang, X., Zhang, D., Yang, J., LuoDa., Xu, Z., Wei, J., Wang, J.-Q., Xu, Z., Peng, F., Li, X., Li, R., Li, Y., Li, M., Bai, X., Ding, F., & Li, Y. (2014). Chirality-specific growth of single-walled carbon nanotubes on solid alloy catalysts. *Nature*, *510* (7506), 522–524.

Yang, H. G., Sun, C. H., Qiao, S. Z., Zou, J., Liu, G., Smith, S. C., Cheng, H. M., & Lu, G. Q. (2008). Anatase TiO2 single crystals with a large percentage of reactive facets. *Nature*, *453*, 638.

Yu, G. L., Gorbachev, R. V., Tu, J. S., Kretinin, A. V., Cao, Y., Jalil, R., Withers, F., Ponomarenko, L. A., Piot, B. A., Potemski, M., Elias, D. C., Chen, X., Watanabe, K., Taniguchi, T., Grigorieva, I. V., Novoselov, K. S., Fal'ko, V. I., Geim, A. K., & Mishchenko, A. (2014). Hierarchy of Hofstadter states and replica quantum Hall ferromagnetism in graphene superlattices. *Nature Physics*, *10*, 525.

Zakri, C., Blanc, C., Grelet, E., Zamora-Ledezma, C., Puech, N., Anglaret, E., & Poulin, P. (2013). Liquid crystals of carbon nanotubes and graphene. *Philosophical Transactions of the Royal Society A: Mathematical, Physical and Engineering Sciences, 371*(1988), 20120499.

Zeng, H., Li, J., Liu, J. P., Wang, Z. L., & Sun, S. (2002). Exchange-coupled nanocomposite magnets by nanoparticle self-assembly. *Nature, 420*(6914), 395–398.

Zhang, J., Sun, L., Jia, K., Liu, X., Cheng, T., Peng, H., Lin, L., & Liu, Z. (2020). New Growth Frontier: Superclean Graphene. *ACS Nano, 14*(9), 10796–10803.

Zhao, T., Xu, C., Ma, W., Liu, Z., Zhou, T., Liu, Z., Feng, S., Zhu, M., Kang, N., Sun, D.-M., Cheng, H.-M., & Ren, W. (2019). Ultrafast growth of nanocrystalline graphene films by quenching and grain-size-dependent strength and bandgap opening. *Nature Communications, 10*(1), 4854.

Zherebetskyy, D., Scheele, M., Zhang, Y., Bronstein, N., Thompson, C., Britt, D., Salmeron, M., Alivisatos, P., & Wang, L.-W. (2014). Hydroxylation of the surface of PbS nanocrystals passivated with oleic acid. *Science (New York, N.Y.), 344*(6190), 1380.

Zhong, J., Sun, W., Wei, Q., Qian, X., Cheng, H.-M., & Ren, W. (2018). Efficient and scalable synthesis of highly aligned and compact two-dimensional nanosheet films with record performances. *Nature Communications, 9*(1), 3484.

Zhong, Y., Cheng, B., Park, C., Ray, A., Brown, S., Mujid, F., Lee, J.-U., Zhou, H., Suh, J., Lee, K.-H., Mannix, A. J., Kang, K., Sibener, S. J., Muller, D. A., & Park, J. (2019). Wafer-scale synthesis of monolayer two-dimensional porphyrin polymers for hybrid superlattices. *Science (New York, N.Y.)*, eaax9385.

Zhou, J., Lin, J., Huang, X., Zhou, Y., Chen, Y., Xia, J., Wang, H., Xie, Y., Yu, H., Lei, J., Wu, D., Liu, F., Fu, Q., Zeng, Q., Hsu, C.-H., Yang, C., Lu, L., Yu, T., Shen, Z., et al. (2018). A library of atomically thin metal chalcogenides. *Nature, 556* (7701), 355–359.

Zhu, C., Du, D., Eychmüller, A., & Lin, Y. (2015). Engineering Ordered and Nonordered Porous Noble Metal Nanostructures: Synthesis, Assembly, and Their Applications in Electrochemistry. *Chemical Reviews, 115*(16), 8896–8943.

CHAPTER FOUR

Structural characterization of low-dimensional materials

4.1 Principles of structural characterization

Structural characterization is to collect, process, and analyze signals generated by probe—matter interactions, which is the basis for understanding the structure and properties of materials. The probe can be physical probes used in scanning probing microscopes, electromagnetic waves such as visible light and X-rays, and particles such as electrons and neutrons. The probe—matter interaction can be mainly classified into elastic and inelastic scattering of the probing particles or fields. The probing energy determines the type of interaction and the depth of characterization, the probe size and the interaction volume determine the spatial resolution, and the characteristic time of interaction determines the temporal resolution. In addition, the sensitivity and resolution of the detector, the speed of data acquisition and real-time analysis, are also critical to enhance the efficiency and to understand fast transition processes. Structural characterization is essentially an "image" of the object through the "lens" of the interaction. Rigorous analysis is required so that the image shows the true structural characteristics of the material. Paradoxically, the higher the resolution, the higher the excitation/probing energy required, which may drastically modify the structure and thus limit the characterization of fine structures. In addition, in quantum mechanics, the measurement of a quantum state will cause a "collapse" of the wave function from a superposition state to an eigenstate and to obtain the physical quantities corresponding to this eigenstate. How to characterize quantum states and obtain quantum information in quantum materials and devices is becoming an active research field.

4.1.1 Classification of characterization techniques

Characterization techniques for low-dimensional material can be classified into four categories based on the types of information obtained: imaging, diffraction, spectroscopy, and measurement, which correspond to the

Low-Dimensional Materials.
DOI: https://doi.org/10.1016/B978-0-443-29046-6.00010-9

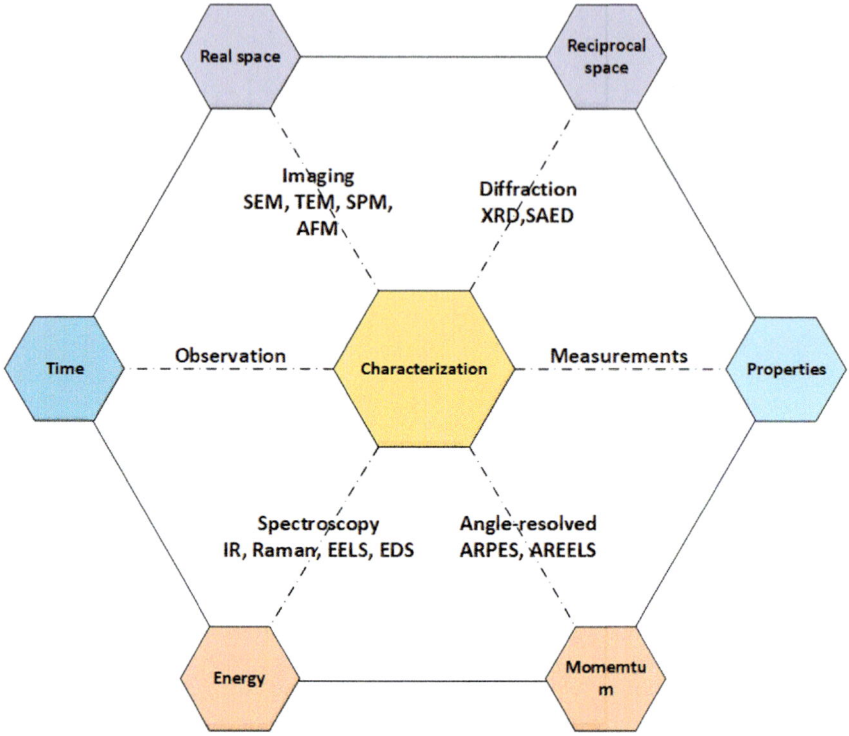

Figure 4.1 ***Characterization techniques for low-dimensional materials.*** Probing the interaction between signals and low-dimensional materials to obtain structural information such as real-space images, reciprocal-space diffraction patterns, and energy–momentum relationships. The relationships between structure, processing, and properties of a material can also be established by real-time observation and measurement under operating conditions.

structures in real, reciprocal, and energy spaces, and are associated with their physicochemical properties (Fig. 4.1).

By imaging, it is possible to obtain a visual image to get information on the shape, surface, crystalline structure, atomic structure, pore structure, defects, and interfaces. Imaging techniques include optical microscopy, X-ray imaging, scanning electron microscopy (SEM), transmission electron microscopy (TEM), scanning tunneling microscopy (STM), atomic force microscopy (AFM).

Diffraction techniques, on the other hand, reflect the overall periodicity of the material and are complementary to imaging techniques which focus more on local structures. Mathematically, diffraction reciprocally related to the corresponding imaging, can be expressed in terms of the Fourier transform, including X-ray, electron, and neutron diffraction.

Spectroscopy is to analyze the changes of energy and momentum during inelastic scattering processes, to obtain the energy-related information such as the energy states and composition. Spectroscopy techniques include infrared (IR) absorption spectroscopy, Raman scattering spectroscopy, electron energy loss spectroscopy (EELS), X-ray photoelectron spectroscopy (XPS), energy-dispersive X-ray spectroscopy (EDS), and angle-resolved photoemission spectroscopy (ARPES).

By using different characterization techniques, we can obtain structure information of the materials from different aspects to get a comprehensive understanding. For example, scanning probe microscopy (SPM) gives microstructured features on a flat surface, SEM shows the surface morphology, while TEM can image the internal structure of a thin specimen.

By analyzing probe–matter interaction processes from different perspectives, it is possible for researchers to obtain different structure information of a sample. Taking electron beam probe as an example, after interaction with matter, the produced secondary or backscattered electrons are used for SEM imaging, the transmitted electrons contribute to TEM imaging, the elastically scattered electrons interfere with each other to form selective area electron diffraction (SAED), the inelastically scattered electrons form EELS, and the characteristic X-rays are detected for EDS. Thus, to have a comprehensive understanding of the structure, it is necessary to use a combination of characterization methods.

The goal of advanced structural characterization is to establish the structure–property relationship of a material. In particular, the structural evolution of a material during the growth process is the key to understanding the growth mechanism and controlling the structures. And the structural degradation under the working environment is important to understand the failure mechanism and to design robust structures. In this chapter, we will also discuss techniques for in situ observations of structural evolutions and property measurements.

4.1.2 Diffraction limits and breakthroughs in resolution

Around 1015 Arab scientist Ibn al-Haytham published the Book of Optics, which described the principles of reflection and refraction of light for magnified imaging. According to the Abbey principle, there is a resolution limit determined by diffraction: $d = 0.61\frac{\lambda}{\alpha}$, where λ is the wavelength of the light wave, α is the incident half angle corresponding to the limiting aperture. The shortest wavelength of visible light is about 400 nm. Even if the lenses of optical microscopes are nearly perfect, the corresponding diffraction resolution limit is more than 200 nm. For low-dimensional materials with feature sizes of less than 100 nm, or even down to the atomic level, it is necessary to break the limit.

There are two paths to break the resolution limit, one is to shorten the wavelength and the other is to avoid diffraction (Fig. 4.2). Following the first path, electron microscopes were invented, because the wavelength of electrons is much shorter than that of the light. At an accelerating voltage of 100 kV, the wavelength of electrons is about 3.7 pm, which is much smaller than the typical size of atoms (~100 pm). Therefore the theoretical resolution limit of either SEM or TEM is much better than that of optical microscopes. The resolution of SEM is about 1 nm, which is two orders of magnitude smaller than that of optical microscopes. With the advancement of electromagnetic lens technology, especially the invention of spherical aberration correctors, the resolution of TEM has reached deep atomic level. It has become one of the most powerful tools for the structural characterization of low-dimensional materials.

Another way to break the diffraction limit is to avoid diffraction. Scientists have developed SPM by using tiny probes to scan across the

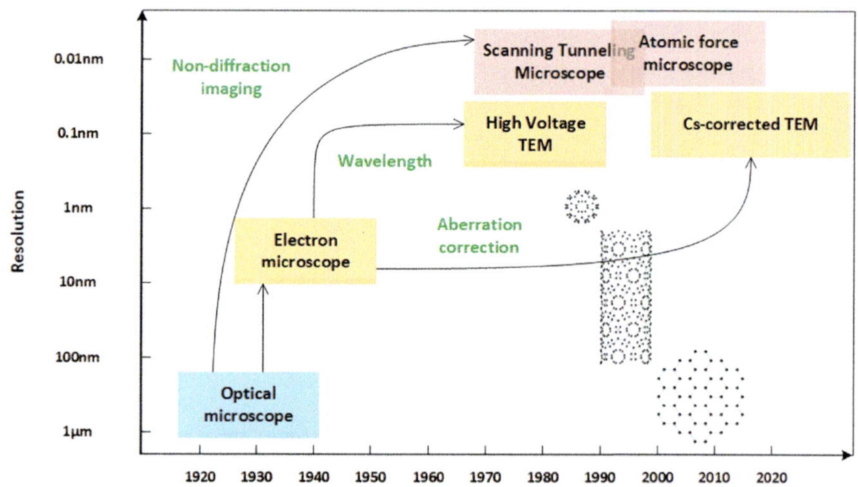

Figure 4.2 ***Diffraction limits and breakthroughs of imaging resolution.*** Optical microscopes are limited by diffraction and have a resolution limit of about 200 nm. It has been broken by the invention of electron microscopy, which employs electrons of shorter wavelengths as "light." On the other hand, scanning probe microscope such as scanning tunneling microscope and atomic force microscope is based on tip-surface interaction, thus avoiding the far-field diffraction. The improvement of microscopy is indispensable for the discovery of low-dimensional structures, such as fullerene, carbon nanotubes, and graphene.

surface of a sample and measure the interaction to form images as a function of the probe positions. It is essentially a near-field process that avoids the far-field diffraction. The resolution depends on the size of the probe tip and interaction of the probe tip with the sample surface. Based on the exponential relationship between the tunneling current and the distance between the tip and the sample, STM has been invented to obtain the surface morphology, atomic arrangement, and electronic structure of the material. AFM has been invented using the electrostatic, magnetic, and van der Waals forces between the tip—surface interaction to allow atomic-scale observation of the surface, including insulating surfaces.

4.2 Transmission electron microscopy

TEM is an electron-based characterization technique to "see" the structure of a material. It utilizes the interaction of incident electron beam with matter and generates signals for characterization of a thin specimen. In 1931 Ernst Ruska developed electron microscope and demonstrated electron images for the first time, which brought him the 1986 Nobel Prize in Physics. TEM is an effective tool for characterizing the structural and chemical properties of low-dimension materials, resolving the crystal structure, defect structure, and chemical composition of a specimen at atomic scale. With the recent development of TEM-based techniques, such as tomography and in situ techniques, three-dimensional (3D) structural information, and physical properties of materials, can also be characterized.

A TEM usually consists of the following components: electron gun, illumination, interaction, magnification, and observation. It can be operated in two different modes, either TEM or scanning TEM (STEM). In TEM mode, electrons emitted from electron source are accelerated by a voltage ranging from tens to hundreds of kV and are turned into a parallel beam by the condenser lens system. The parallel electron beam interacts with specimens, passes through the objective, intermediate, and projector lenses, and then the transmitted beam is recorded by the charge-coupled device (CCD) camera. In STEM mode, electron beam is focused to a probe by condenser system and scanned over the specimen. During scanning, the electron probe is scattered by the specimen and recorded by the electron detectors at different scatter angles.

For the electron gun, the beam of electrons is generated mainly by thermionic emission or field emission sources. The electron source is either heated to a high temperature or applied at a high voltage to overcome barriers for the electron emission. The emitted electrons are then accelerated to a speed approaching that of light. With the correction of relativity, the electron wavelength is determined by the accelerating voltage (V) and can be calculated as follows: $\lambda = \frac{h}{\left[2m_0eV\left(1+\frac{eV}{2m_0c^2}\right)\right]^{\frac{1}{2}}}$. When an accelerating voltage of 100 kV is applied, the electron wavelength is 0.0037 nm, which is far less than the diameter of atoms (0.1 nm).

When an electron beam interacts with specimens, the phase and amplitude of the electron wave function change, carrying structural information of materials. The electrons that pass through the specimen are named transmitted beam that contains elastic scattering and inelastic scattering electrons. Elastic scattering (especially Bragg diffraction) can be used for electron diffraction, diffraction contrast imaging, and high-resolution TEM (HRTEM) imaging. Inelastic scattering process causes electron adsorption, secondary electrons, characteristic X-rays emission, electron energy loss, cathodoluminescence (CL), etc., which can be used for chemical analysis. Among these, the incoherent scattering over a large angle shows a roughly proportional scattering intensity to the square of the atomic number, the contrast is known as Z-contrast, which is the basic principle of high-angle annular dark-field (HAADF) imaging.

The key component of a TEM is electromagnetic lens, where the electron beam is deflected and focused by the Lorentz force in a round and axially gradient-distributed magnetic field. As electromagnetic lens can only realize the function of a convex lens, Scherzer pointed out that it is not possible to eliminate spherical aberrations by round magnetic lenses. Thus the resolution of TEM has been hindered for many decades (Scherzer, 1947). In the 1990s Rose et al. invented spherical aberration correctors, which make use of nonround quadrupoles to compensate the spherical aberration of objective lens and improve TEM and STEM resolution to subangstrom levels (Haider et al., 1998).

4.2.1 High-resolution imaging

4.2.1.1 Discovery of new materials

HRTEM imaging plays an important role in discovering new materials. HRTEM methods based on field-emission were developed by John Cowley and Sumio Iijima in the 1970s (Cowley & Iijima, 1976; Iijima,

1971). In 1991 Iijima published a paper entitled *Helical microtubules of graphitic carbon*. In this chapter, HRTEM images clearly showed double- and multi-shell hollow tubular structures, i.e., carbon nanotubes (CNTs). Based on the symmetry of the electron diffraction pattern, he proposed a structural model of rolled-up tube made by graphite sheets (Iijima, 1991). In 1993 he discovered single-walled carbon nanotubes (SWCNTs) rolled from a single layer of graphite and used electron diffraction method to analyze the helical angle and chiral index (Iijima & Ichihashi, 1993).

According to thermodynamic theory (Mermin & Wagner, 1966), it had long been believed that thin film of two-dimensional (2D) materials cannot exist stably. However, in 2004, Novoselov and Geim at the University of Manchester discovered that single-layer graphene could be separated by mechanical exfoliation method, that is, repeated peeling of graphite by using scotch tapes. Interestingly, the single-layer graphene was proved by an optical microscope (Novoselov et al., 2004). Meyer et al. (2007) at the Max Planck Institute used TEM to directly observe the structure of suspended graphene and identified its stable existence and long-range ordered crystal structure by electron diffraction. In 2020 Ren et al. at the Institute of Metal Research, Chinese Academy of Sciences synthesized a new type of layered 2D crystal, $MoSi_2N_4$, by chemical vapor deposition method (Hong et al., 2020). As shown in Fig. 4.3, the atomically resolved HAADF image clearly shows the positions and types of atoms in the hexagonal arrangement, which was verified by the calculated atomic model.

4.2.1.2 High resolution TEM at low accelerating voltage

Low-dimension materials are more vulnerable to the electron radiation damage because of these relatively small sizes compared to the bulk

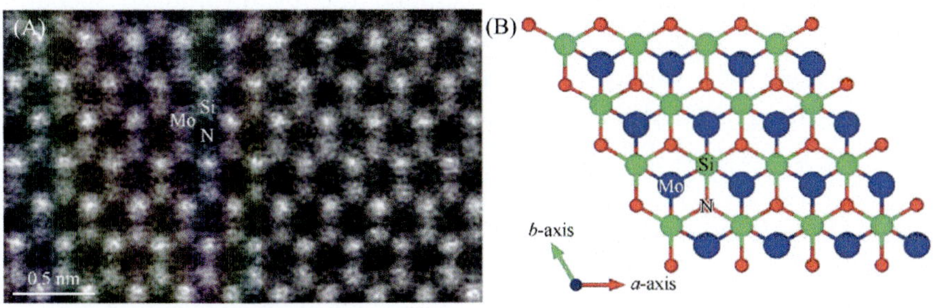

Figure 4.3 ***STEM characterization of 2D structure of $MoSi_2N_4$.*** (A) HAADF-STEM image of $MoSi_2N_4$, (B) the atomic model showing the in-plane crystal structure of the monolayer $MoSi_2N_4$ (Hong et al., 2020). *2D*, Two-dimensional; *HAADF*, high-angle annular dark-field; *STEM*, scanning TEM.

materials. As a result, the characterization of the structure of low-dimensional materials is a big challenge. Electron radiation damage includes knock-on damage, specimen heating, and ionization damage (radiolysis). It is mainly associated with the types of elements, chemical bonding, thermal and electrical properties of materials.

For nanocarbon materials, which are composed of lightweight carbon atoms, the damage mechanism is mainly about the kinetic energy transfer (related to the acceleration voltage) of knock-on damage. With the availability of spherical- and chromatic-aberration correction techniques, atomic resolution TEM images can be realized at low accelerating voltage. The Sub-Angstrom Low Voltage Electron Microscopy project led by Thermo Fisher Scientific Inc. (United States), Corrected Electron Optical Systems GmbH (Germany), and Ulm University (Germany) has developed a spherical- /chromatic-aberration corrector, which achieves a TEM resolution of 0.139 nm at an acceleration voltage of 20 kV (Linck et al., 2016). With the "Delta" correctors jointly developed by JEOL Ltd. and National Institute of Advanced Industrial Science and Technology (AIST), Japan, high-order spherical aberration can be corrected. Using an acceleration voltage of 30 kV and atomic-resolution STEM imaging with a large electron beam convergence angle up to 48 mrad, the dumbbell structures and single-carbon atoms of graphene could be resolved (Sawada et al., 2015).

For materials such as zeolites that are composed of ionic bonds, the main electron beam damage is ionization. To solve this problem, Han et al. at King Abdullah University of Science and Technology developed a low electron dose imaging method. They developed a program for one-step alignment of the zone axis, which reduced the total electron dose during sample tilting to $1e/\text{Å}^2$. In addition, by using the state-of-the-art direct-detection electron-counting camera, the beam-induced specimen vibration and image blur can be avoided. These methods enable atomic-level imaging of a variety of electron beam-sensitive materials, including organic–inorganic hybrid perovskite $CH_3NH_3PbBr_3$ and metal-organic framework materials (Zhang et al., 2018).

4.2.1.3 Single-atom catalysts

In the STEM mode, the electron probe is controlled to scan a sample point by point, and scattered electrons are collected at different scattering angles for bright field (BF), annular BF (ABF), and HAADF imaging. HAADF imaging is suitable for characterizing heavy atoms on light atomic number substrates, as the contrast of the images is approximately proportional to the square of the atomic number (*Z*-contrast). Together with EELS under STEM mode, chemical analysis is realized at single-atomic scale. Suenaga et al. at AIST characterized the six-membered ring arrangement and chemical bonding of carbon atoms in single-layer graphene with aberration-corrected STEM–EELS. Thus the fine differences on the electronic structure between edge and internal carbon atoms were distinguished (Suenaga & Koshino, 2010).

HAADF imaging is an effective method for single-atom characterization, because heavy metal atoms can be clearly distinguished from oxides supporting materials by *Z*-contrast (Qin et al., 2020). In 2011 Zhang et al. at Dalian Institute of Chemical Physics, Chinese Academy of Sciences reported the synthesis of Pt_1/FeO_x single-atom catalysts with excellent conversion efficiency and stability for the oxidation reaction of CO. The HAADF image clearly showed the single-atom distribution of Pt atoms on the FeO_x substrate (Qiao et al., 2011). Zelenay et al. at Los Alamos National Laboratory demonstrated the distribution of Fe atoms and FeN_4 coordination structure on nitrogen-doped graphene by HAADF–EELS (Chung et al., 2017).

4.2.1.4 Imaging light atoms

The energy storage materials are usually composed of light elements, such as hydrogen and lithium. The small scattering cross sections of light elements lead to the difficulties for characterizing the compositions and structures of energy storage materials by traditional TEM methods. In 2009 Okunishi et al. (2009) from JEOL Ltd. proposed the ABF-STEM technique. The electrons with scattering angle between 11 and 22 mrad are collected, and the phase contrast in the ABF-STEM images is approximately proportional to $Z^{1/3}$ (where Z is atomic number). As a result, light and heavy atom columns are visible simultaneously with relatively high contrasts (Findlay et al., 2010). Abe et al. at the University of Tokyo resolved H atom columns in the YH_2 lattice with ABF method (Ishikawa et al., 2011). Ikuhara et al. revealed Li ions in $LiFePO_4$ and found that in partially delithiated $LiFePO_4$, Li ions preferably occupied every second layer. This result disclosed the evolution picture between $LiFePO_4$ and $FePO_4$ (Gu et al., 2011).

4.2.2 Phase reconstruction

After electron beam–matter interaction, the structural information carried by the electron wave is contained in the amplitude and phase of the electron exit wave. However, only the amplitude/intensity of electron wave can be recorded for imaging, whatever the detectors are photographic film, CCD camera, or recently developed direct electron-detection camera. For retrieving phase information of electron waves, techniques such as Lorentz microscopy, electron holography, and differential phase contrast (DPC), have been developed.

Lorentz microscopy was developed for imaging internal magnetic field of specimens with special objective lens. Tonomura et al. at Hitachi used Lorentz microscopy to characterize vortex lattices in superconductor (Harada et al., 1992). Yu et al. (2010) at National Institute of Materials Science (NIMS), Japan observed a 2D skyrmion crystal in an $Fe_{0.5}Co_{0.5}Si$ thin film by Lorentz imaging.

Electron holography was proposed for characterizing nanoscale electrostatic and magnetic fields of materials. In electron holography images, interference fringes are recorded from two coherent electron beams in and outside of the specimen (Tanji et al., 1996). In the 1980s Tonomura et al. (1980) used electron holography to investigate the structure of domain walls in plate-shaped cobalt particles. Dunin-Borkowski et al. (2004) at the University of Cambridge observed the magnetic field distribution in magnetic nanowires and nanoparticle aggregates and identified the annular-shaped magnetic field in self-assembled cobalt nanoparticles. Cumings et al. at the University of California, Berkeley used this technique to study the magnitude and spatial distribution of the electric field surrounding individual field-emitting CNTs. It was determined that the electric field is concentrated at the tips of the CNTs (Cumings et al., 2002).

Besides the Lorentz and holography imaging in TEM modes, the recently developed DPC method under STEM mode reveals phase information by measuring the deflection of electron beams by the electromagnetic fields in specimens. Ikuhara et al. at the University of Tokyo and JEOL Ltd. demonstrated DPC STEM imaging of crystals to visualize the potential and electric field at atomic resolution by using aberration-corrected TEM and a segmented-type annular detector (Shibata et al., 2012). Since the resolution of DPC is similar to that of STEM at atomic scale, it can be used to characterize single atoms, atomic-scale defects, and the distribution of magnetic fields and electric fields in crystals (Fig. 4.4) (Ishikawa et al., 2018).

Four-dimensional (4D) STEM is a newly developed method on the basis of the segmented detectors, where each pixel of the detector record information independently. Each point scanned by the electron beam on

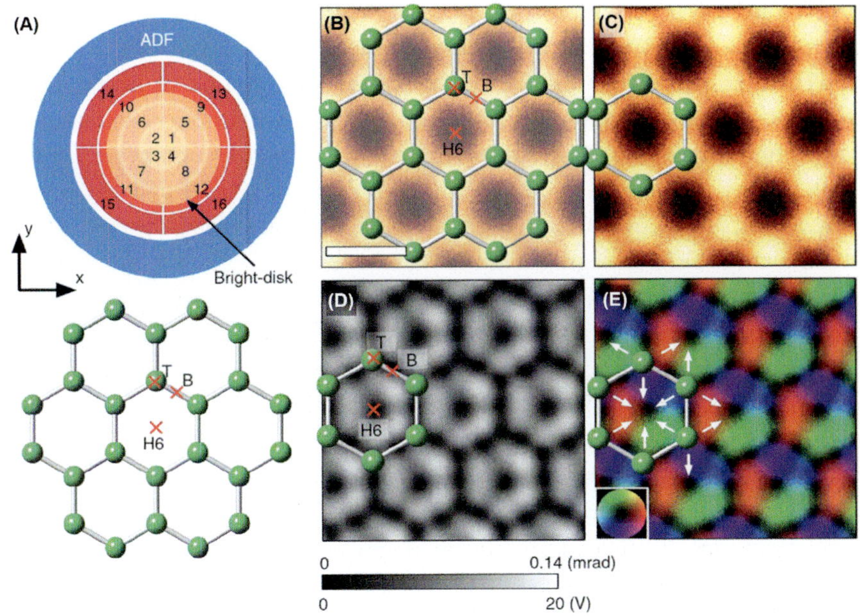

Figure 4.4 ***Atomic-resolution STEM-DPC imaging of monolayer graphene.*** (A) Segmented/ADF detector and the orientation relation with graphene. (B and C) STEM-ADF images of graphene and the corresponding atomic positions. (D and E) The projected field strength map and vector map of electric field in graphene (Ishikawa et al., 2018). *ADF*, Annular dark field; *DPC*, differential phase contrast; *STEM*, scanning TEM.

the specimen corresponds not to the intensity of an image, but to the electron diffraction patterns. By careful analysis with this method, more structural information of specimens can be obtained. Pan et al. at the University of California, Irvine, resolved the charge distributions by analyzing convergent beam electron diffraction patterns obtained by 4D STEM. They have successfully imaged the local electric field between anions and cations in $BiFeO_3$ crystal, etc. (Gao et al., 2019). Fang et al. at Harvard University used 4D STEM to analyze charge distributions in 2D materials such as MoS_2 and revealed the presence of metallic one-dimensional line defects in the 2D semiconductor lattice (Fang et al., 2019).

4.2.3 Three-dimensional tomography

TEM imaging is usually a 2D projection of a 3D specimen, with a loss of information in depth. The 3D reconstruction technique is based on an iterative algorithm invented by Randon in 1917, which uses multiple projections to inversely reconstruct the high dimensional space. X-ray computed tomography allows millimeter-level 3D imaging of the human body and has become an indispensable diagnostic method in hospitals. By using X-ray ptychography technique, nondestructive 3D imaging at a ~10 nm spatial resolution of integrated circuits was achieved (Holler et al., 2017). Focused ion beam (FIB) together with SEM dual-beam system allows 3D reconstruction with 5 nm resolution by FIB layer-by-layer etching and SEM imaging (Motta et al., 2019). The 3D atom probe tomography, developed from field ion microscopy, can be used to obtain 3D structures with near-atomic resolution by tracking the trajectories of field-evaporated ions on a specimen with a needle tip (Gault et al., 2012).

The application of 3D reconstruction techniques of TEM in resolving low-dimensional materials structures is briefly described (Friedrich et al., 2009; Weyland & Midgley, 2004). Due to the development of techniques such as spherical aberration correctors, high-sensitivity detectors, and high-precision automatically controlled goniophotometers, as well as the improvement of computing power, the 3D reconstruction technique has approached atomic resolution. The electron tomography 3D reconstruction technique is capable of obtaining 3D structural information of materials, including the morphology of nanoparticles and the distribution of catalysts in mesoporous materials.

In heterogeneous catalysis, interfacial bindings between metal catalyst nanoparticles and supports are crucial for the activity and stability of catalytic reactions. Midgley's group at the University of Cambridge, UK used the HAADF-STEM tomography to provide the 3D information about the location of PtRu bimetallic particles on and within mesoporous silica. By analyzing size, volume, shape, and surface area of pores, it was found that the mesoporous silica is a fractal structure with a dimension of 2.4, and the catalyst particles are preferentially adsorbed on saddle-like regions within the interior of supports (Ward et al., 2007). The information on the surface crystallography of a support and location of the catalyst is important to understand the metal/support system. González et al. quantitatively analyzed support surface and the distribution of gold catalyst particles on the oxide support by 3D reconstruction. They found that the preferential adsorption site is at the facet vertices and nanocrystal boundaries of the support. High-resolution characterization revealed an epitaxial relationship between the (1 1 1) planes of metal particles and support, which helps to understand the stabilization of the catalyst nanoparticles (González et al., 2009).

Combining the spherical aberration corrector, fast and sensitive detector, high-throughput data acquisition method, and superior computing power, Miao's group at the University of California, Los Angeles developed an atomically resolved 3D reconstruction technique (Chen et al., 2013; Miao et al., 2016). Using spherical aberration-corrected TEM to obtain subangstrom-resolved images, precise alignment around the center-of-mass during rotation, and iterative reconstruction in Fourier space to establish an atomic model and the 3D coordinates of each atom in a nanoparticle can be obtained. Fine structures, such as multiple twin crystals inside gold nanoparticles, atomic steps on grain boundaries, and dislocations, were resolved using this technique (Chen et al., 2013; Scott et al., 2012). By using an in situ liquid cell with a graphene window, Alivisatos' group at the University of California, Berkeley achieved 3D reconstruction of individual platinum nanoparticles in solution at near-atomic resolution (Park et al., 2015).

4.2.4 In situ transmission electron microscopy

The microstructural characterization of materials aims to establish relationships among structure, processing and property in the field of materials science. Structural control of low-dimensional materials relies on an understanding of their growth mechanisms, which require direct observations of their formation and evolution processes during growth. The measurements of the intrinsic properties of low-dimensional materials also need to be performed in a microscope due to the tiny size of nanomaterials. In situ TEM is a powerful technology that can provide real-time observation during the growth and measurement process of materials.

4.2.4.1 Time-resolved transmission electron microscopy

The spatial resolution of TEM has now reached the atomic scale, while the time resolution remains limited by several factors such as the brightness of the electron gun, the shutter speed of camera, the sensitivity of camera sensors, and the data transfer rate. In recent years, the invention of direct electron-detection camera has improved the time resolution to milliseconds and can acquire 1000 frames per second owing to their high sensitivity. Ahmed Zewail, the Nobel laureate in Chemistry at the California Institute of Technology, invented a four-dimensional electron microscope, which enables a resolution of picoseconds (10^{-12} seconds) (Zewail, 2010). The main enhancement is the integration of a two-laser system on a TEM, which is utilized to stimulate the electron gun and samples, respectively. Photoelectric effect was used to generate electron packets, while the laser was used to induce structural changes in the samples. The time intervals of observations can be well defined by variable delay between these two laser beams. This technology has been used to acquire ultrafast dynamic information on thermal vibrations, liquid fluidity, phase changes, and mechanical vibrations (Barwick et al., 2008; Lorenz & Zewail, 2014).

Ropers' group at the University of Göttingen studied the phase change of charge density in 1T TaS_2 nanosheets by using time-resolved electron microscopy (Danz et al., 2021). They designed a unique aperture array based on the expected phase structure and initiated the structural phase transition with ultrashort laser pulses. By using pump–probe ultrafast dark-field electron microscopy, they observed the dynamic movement of domain configurations during phase transitions with a femtosecond temporal resolution of 5 nm.

Time-resolved TEM represents a promising development in electron microscopy. Li et al. at the Institute of Physics, Chinese Academy of Sciences have developed a new generation of field emission cathode capable of achieving a spatial resolution of 0.27 nm and a temporal resolution of femtoseconds. The study investigated the anisotropic lattice dynamics of hexagonal-boron nitride (h-BN) upon laser excitation. The fast and slow lattice dynamics processes were distinguished, with the corresponding time constants of approximately 8 and 100–300 ps, respectively (Li, Sun, et al., 2017).

4.2.4.2 Observations of in situ growth

In general, electron microscopy is often performed in a high vacuum to preserve the performance of electron gun and prevent sample contamination, etc. However, many applications, such as catalytic reactions and material growth, are often carried out in gaseous or liquid conditions. Environmental TEM (ETEM) was then conceived to bridge the gap. With the adaption of small-bore apertures and differential pumping system to a TEM, a gaseous atmosphere of the order of 1000 Pa near the polepieces or samples can be provided, while the rest of the microscope can be kept at low pressure or remained in the high vacuum (Hansen et al., 2001). Another adaption to the holder is the development of the environmental cell (E-cell), which separates the reaction from the vacuum of the microscope column with electron-transparent windows. The gas pressure of E-cell is typically 1–2 atm. Similarly, introducing liquid into the reactor allows for in situ observations of the liquid reactions and the growth process in real time (De Jonge & Ross, 2011).

By means of ETEM, Frances Ross et al. from Purdue University in the United States investigated the growth processes of various semiconductor nanowires such as Si, Ge, and GaP. They revealed that a periodically changing morphology of the gas–liquid–solid interface, which corresponded to the changing of catalyst supersaturation (Wen et al., 2011). The formation of the Pd_xSi alloy phase was found during the palladium-catalyzed growth of silicon nanowires, and the Si atoms precipitated layer-by-layer at the catalyst–nanowire interface was observed (Hofmann et al., 2008).

Helveg et al. (2004) studied the state of the catalyst nanoparticles and the location of the precipitated carbon layers during the nickel-catalyzed growth of CNTs in ETEM, and found that the fluid catalyst particles were periodically elongated, deformed, and extruded during the precipitation of carbon layers. Yoshida et al. (2008) from Osaka University found that the Fe_3C catalysts maintained a solid crystalline structure during the nucleation and growth process of a SWCNT. Sharma et al. from the National Institute of Standards and Technology observed the precipitation and anchoring process of graphene sheets on cobalt catalyst nanoparticles, in which the graphene sheets ultimately lifted off and converted into a cap (Picher et al., 2014).

The phase and structure of catalyst nanoparticles are key issues in the growth mechanism of CNTs. It is generally believed that the catalyst nanoparticles have many possible compositions and phases in the complex growth environments. Cobalt and carbon, for example, can form two carbide phases (i.e., Co_2C and Co_3C). The mixed phases of cobalt carbide and metallic cobalt are reported to catalyze the growth of CNTs (Lin et al., 2017). The precise identification of the active phase of catalyst nanoparticles is crucial for understanding the growth mechanism of CNTs but faces significant challenges due to the tiny size effect of nanoscale particles, etc. The authors of this book investigated the growth of CNTs from Co catalyst at near-atmospheric pressure using E-cell TEM and proposed an in situ method to determine the phase of catalyst nanoparticles accurately. The aberration-corrected TEM was used for the in situ observation of the dynamic growth process. By performing the fast Fourier transformation of the HRTEM images of the catalyst nanoparticles, together with simulated diffraction patterns, limited error criteria of the lattice spacing ($<5\%$) and the angle (<2 degrees), the phases of Co_2C

and Co_3C can be distinguished clearly. It was found that the active phase of Co catalyst for CNT growth was unambiguously maintained as a Co_3C single-phase (Fig. 4.5) (Wang et al., 2020).

The atomic-scale studies by ETEM can provide insights into material reaction mechanisms. DeRita et al. (2019) at the University of California, Santa Barbara found that isolated Pt species can adopt a range of local coordination environments and oxidation states, which evolved in accordance with different environmental conditions by atomic-resolution ETEM and spectroscopy-based characterization. The spatial distribution of CO molecules on triangular gold nanocatalysts was also characterized by STEM–EELS. Sharma's group found that the CO molecules preferred to absorb at the junctions of the triangle rather than the flat surfaces. Meanwhile, the intensity of the localized surface plasmon modes was also the highest at the junction of the triangle, which confirmed that the surface-plasmon assisted the activation of catalytic reactions (Yang et al., 2019). Zhang and Wang's group at Zhejiang University revealed that water molecules preferred to absorb on the highly ordered TiO_2 (1×4) surface, which led to the formation of twinned active sites (Yuan et al., 2020). The interfacial structures of Au–TiO_2 and the rotation of gold nanoparticles toward TiO_2 support were observed under different reaction gases (Yuan et al., 2021).

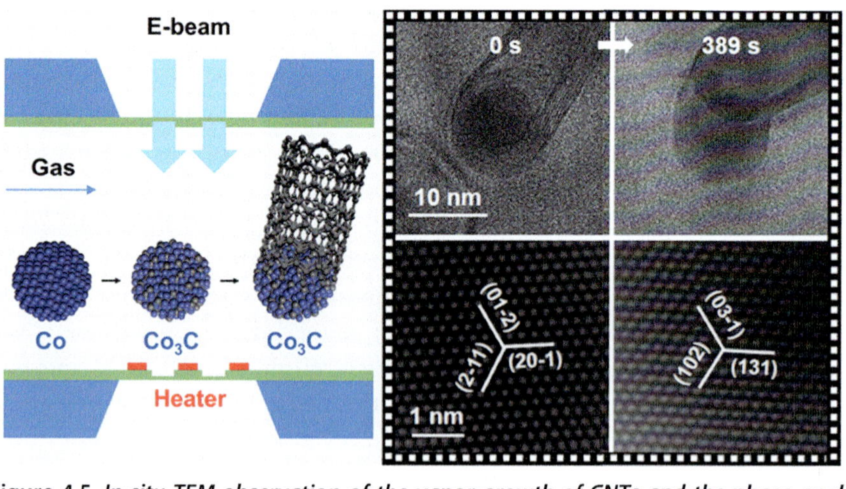

Figure 4.5 ***In situ TEM observation of the vapor-growth of CNTs and the phase analysis of catalyst nanoparticles.*** The active phase of Co nanoparticles was determined as cobalt carbide (Co_3C) based on the precise identification and statistical analysis (Wang et al., 2020). *CNTs*, Carbon nanotubes.

Compared to gas environments, liquid environments are more challenging for in situ TEM. The density of liquids is 1000 times higher than gases, which leads to the 1000 times stronger scattering of electron beams. Bubbles are easily formed in the liquid layer under the e-beam irradiation, which may cause serious damage to sample and cell window. Therefore it is necessary to develop ultrathin window with good chemical inertness, and electron beam transparency (De Jonge & Ross, 2011; Liao et al., 2012). Graphene is a 2D carbon material with atomic-scale thickness, together with its high electron beam transparency, high mechanical strength, and chemical inertness, making it a promising window material for an in situ liquid cell (Textor & de Jonge, 2018). By using this kind of E-cell, Wang et al. at the Institute of Physics, Chinese Academy of Sciences found that the cubic NaCl can grow into a peculiar hexagonal morphology, thus revealing the impact of confinement on nonclassical kinetics and thermodynamics of crystallization (Wang et al., 2021).

4.2.4.3 Property measurement by in situ transmission electron microscopy

In situ TEM measurements of the properties of low-dimensional materials are important methods to establish a dynamic structure–property relationship. Over the last decades, numerous studies of in situ measurements involving electrical, mechanical, thermal, and optical properties have been reported, among them, some of the author's work will be introduced in this section.

By in situ measurement, we studied the electrical properties of atomic-level heterostructures and devices. Metal atomic chain, which is an ultimate one-dimensional structure, shows great potential as possible components in new electronic devices. Electron irradiation and electrical heating were used to fabricate the metal-filled CNTs. An electron beam was focused to peel off the carbon shells and to thin the exposed metal nanorod until several atomic layers; thus the CNT-clamped metal atomic chain was fabricated. By simultaneous measurement of electrical conductance during processing of the structure, we observed the quantized conductance (Tang et al., 2010). An SWCNT can be either metallic or semiconducting, depending on its chirality. Joule heating and mechanical strain were applied to initial the plastic deformation of an SWCNT and alter the local chirality. A controlled metal-to-semiconductor transition was realized by monitoring the electrical properties as a feedback signal during the repeated thermomechanical processing. Additionally, quantum transport at room temperature was

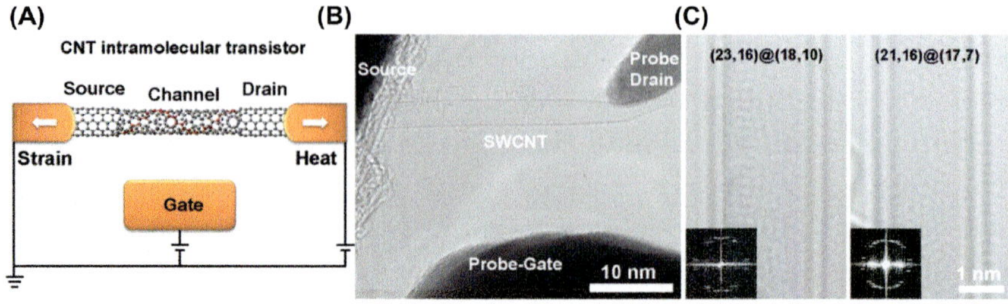

Figure 4.6 ***Fabrication and characterization of CNT intramolecular transistors.*** (A) Schematic of a CNT intramolecular transistor with local chirality altered by mechanical strain and Joule heating. (B) TEM image of an SWCNT transistor with a fixed source electrode, a probe as drain electrode, and another probe as gate electrode. (C) Lattice-resolved TEM images and corresponding fast Fourier transformation patterns of a double-walled CNT before and after the chirality transition (Tang et al., 2021). *CNT*, Carbon nanotube; *SWCNT*, single-walled CNTs.

demonstrated for the fabricated nanotube transistors with a ultrashort channel length (Fig. 4.6) (Tang et al., 2021).

In situ TEM technique can also be used to measure the elastic, plastic, bending, and fracture properties of individual low-dimensional nanostructures. The resonant frequency of a single silicon nanowire was measured with periodic electrostatic forces generated by an alternating current electric field. For the same nanowire, up to 10 resonance modes were observed from 1.1 to 137.5 MHz, corresponding to the first- to fifth-order resonance frequencies for normal and parametric modes. According to the in situ characterization, an $Si-SiO_2$ core–shell nanowire model was constructed to predict its resonance frequency. The predicted values are very close to the experimental measurements, and the deviation decreases from 3.14% at the first order to 0.25% at the fifth order. The elastic modulus of silicon was calculated, and it was found that at the 20 nm scale, Young's modulus along its <1 1 0> direction is 169 GPa, which is consistent with that of bulk Si crystals (Hsia et al., 2019).

The AFM–TEM holder is also used for bending and stretching mechanical measurements of single silicon nanowires. By in situ TEM measurement, it was found that nanowires exhibit brittle fracture under tensile conditions, and their fracture strength is size-dependent. The strength of small-diameter nanowires reaches 11.3 GPa, which is close to its theoretical limit. Under bending conditions, considerable plastic deformation of nanowires can be observed, by dislocation activities and local amorphous transition. This behavior can be understood through a stress state parameter, the ratio of maximum tensile stress to maximum shear stress, which corresponds to the driving forces for brittle fracture and plastic deformation, respectively (Tang et al., 2012).

Two-dimensional materials are characterized by an atom-thick structure with strong intralayer covalent bonds and weak interlayer van der Waals bonds, making them suitable for flexible electronic devices. We used STM probes to mechanically exfoliate, stretch, and bend MoS_2 crystals with different layer numbers in TEM. It is found that the relatively thick MoS_2 crystal, for example, 20-layer MoS_2, tends to form kinks and bend with interlayer sliding, corresponding to an abrupt force drop. When the number of layers is less than 10, the interlayer structure remains stable even if it was bent to the ultimate curvature radius of 1.3 nm. The equilibrium shape of a single-layer MoS_2 under bending condition was analyzed, and the interlayer force was calculated to be 0.11 N/m. This layer-dependent bending behavior can be understood in terms of the competition between the van der Waals forces of the layered lattice and the layer-dependent bending energy (Tang et al., 2014).

4.3 Scanning electron microscopy

TEM described in the previous section uses "transmitted" electrons to *see* and analyze internal structure of thin specimens, while SEM uses "reflected" electrons to *see* and analyze the surface, or near surface, structure of bulk specimens. The imaging process of SEM is similar to that of STEM, in which a focused electron beam is used as a probe to scan across a selected area, and the signal at each point is measured and synchronized to an SEM image. All SEM instruments normally have facilities for detecting secondary electrons, backscattered electrons, or X-rays. The secondary electron energy can give a better imaging of surface topography information, whereas compositional contrast is better imaged by backscattering electrons. Besides, with electron backscatter diffraction, the crystal structures, such as grain size and shape, orientations, and phase distributions, can be obtained. The electron-excited X-rays can be used for chemical analysis by obtaining the composition and its distribution of materials, this technique is known as EDS.

4.3.1 Approaching atomic resolution

The resolution of SEM is not only dependent on the size of the scanning electron beam but also limited by the sampling volume of the signal used. Secondary electrons have the smallest sampling volume, compared to other signals, so that secondary electron imaging is the most popular mode for high resolution characterization in SEM. Due to the nonlocalization nature of secondary electron excitation, the resolution of SEM is usually at the nanometer level. In 1980 Pennycook and Howie showed that secondary electrons produced by electron excitation in the inner layers of heavy element atoms contain localization and crystallographic information. Thus it is possible for heavy element atoms to acquire atomic-level secondary electron images (Pennycook & Howie, 2006).

In 2009 Zhu from Brookhaven National Laboratory, USA and Inada et al. from Hitachi High Technologies Corp, Japan reported that secondary electron imaging with atomic-level resolution was obtained by using a subnanometer probe in an aberration-corrected TEM, which captures the surface information of the sample using secondary electrons (Fig. 4.7) (Inada et al., 2011). The full width at half maximum of the intensity profile of individual U atoms supported on the carbon film was around 0.08 nm. The lattice spacing of 0.14 nm for the $YBa_2Cu_3O_{7-x}$ crystal was also clearly resolved (Zhu et al., 2009). Compared with TEM, secondary-electron imaging is highly surface sensitive with only a few-nanometer depth. This technology is suitable for compositional and structural characterization of superficial layers and helpful for mechanistic studies such as catalytic active sites (Krumeich et al., 2011).

Atomic resolution secondary electron imaging has promoted the development of material characterization for a more comprehensive structure and composition information, such as 2D heterostructure. However, the imaging mechanism is complicated, since it involves complex secondary electron scattering processes as the change of incident energy and material, etc. Ding from the University of Science and Technology of China (USTC) and Da et al. from the NIMS, Japan proposed a new understanding of underlying mechanism of secondary electron imaging. They developed a simulation approach by combining the quantum trajectory and Monte Carlo models, so that lattice information and the complex inelastic scattering processes, such as valence excitation and inner layer ionization, have been considered. Therefore this

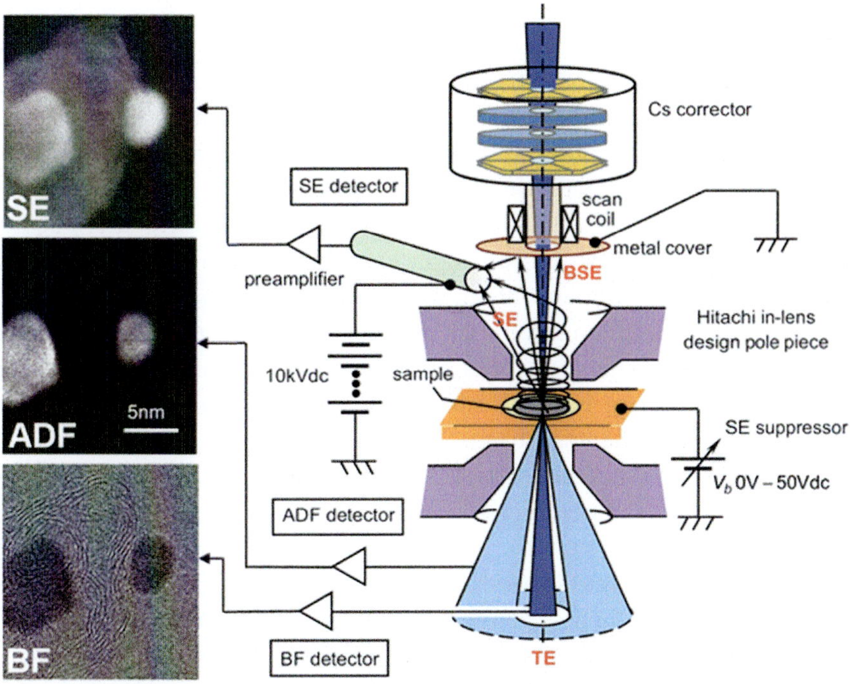

Figure 4.7 ***High-resolution STEM for simultaneous acquisition and imaging of SE, ADF, and BF electron signals*** (Inada et al., 2011). *ADF*, Annular dark-field; *BF*, bright field; *SE*, secondary electrons; STEM, scanning TEM.

technology with an ultrahigh surface resolution can even distinguish the atoms of same elements among the few atomic layers near the surface (Cheng et al., 2021).

4.3.2 Contrast mechanisms

SEM is capable of yielding information not only on the surface topography but also on the composition of a sample, that is, topographic and atomic number contrasts. In addition, material properties or their variations can also be imaged by using electron beam-induced current/voltage signals, which is known as voltage contrast. As incident electrons can generate a huge amount of "electron–hole pairs" from the outer shells of the atoms of a sample, the majority recombine extremely fast within 10^{-12} seconds. When a voltage is applied across a semiconductor sample, the recombination of electron–hole pairs can be suppressed, and a current can flow in the external circuit and take as the signal for imaging.

Jiang's group at Tsinghua University has used this contrast of SEM to effectively distinguish and separate metallic or semiconducting SWCNTs. The contribution of secondary electrons is mainly considered for imaging under a low accelerating voltage (1 kV). They show that the contrast of SEM image has a relationship with the electrical properties of CNTs due to surface charging effect, which was also verified using transfer characteristics measurements of a single-CNT field effect transistor (Li et al., 2012). The metallic nanotubes show bright contrast, and the semiconducting ones show darker contrast. When an external high voltage is applied on CNT samples, metallic ones on an insulator substrate retain their potential equal to that of metal electrodes and even form potential gradient with the positive charging insulator surface. In contrast, semiconducting ones show almost no potential change as external voltage. This is because of the conductivity difference between metallic and semiconducting nanotubes and the surface charging effect of insulator substrates under electron irradiation or external voltage. Later on, they have developed the method to evaluate effectively the bandgap distribution of semiconducting nanotubes and realize a high-throughput analysis on the ratio of metallic to semiconducting nanotubes in CNT arrays through a single-SEM image (Zhao et al., 2020). Using this method, they also investigated the changes in structural and electrical properties during the growth of CNTs under the action of an electric field. It was found that metallic CNTs can transform to semiconducting ones with the reversal of an applied electric field, finally obtaining horizontal arrays of CNTs with a semiconducting ratio of more than 99.9% (Wang et al., 2018).

4.3.3 In situ measurement

By SEM imaging combined with in situ measurements, it is possible to correlate the properties of materials with their morphology, crystal structure, etc. at the nanoscale. SEM owns a larger specimen chamber and can be extended with more functions, compared with TEM. CL can be another mode of SEM imaging, which detects the light emitted from materials under electron bombardment. It can appear as a band of wavelengths and correlate to the energy gap between the filled valence and empty conduction bands of semiconducting and insulting materials. Therefore the factors that influence energy gap, such as temperature, crystal structure, or impurity, can alter CL results. The lower the specimen temperature, the intense and sharper the CL emission bands.

Recently, it has proved to be a powerful technique for measuring the optical properties of materials. Takashi's group from NIMS built a nanoprobe-CL system based on Schottky SEM. The system equips an SEM specimen He or N_2 cooling stage, piezoelectric nanomanipulator, and optical signal acquisition system (Watanabe et al., 2016). They used this system to visualize the carrier concentration map of free-standing ZnO nanorods at high sensitivity and tell the history of local transient growth events within each growth sector. They also used the SEM-CL spectroscopy to study the relationship between the bandgap and applied uniaxial strain of ZnO nanorods. Point-bending stress was loaded to a single ZnO nanorod using a nanoprobe until its elastic limit ($\approx 4\%$), the local bandgap was evaluated by the peak energy in a CL spectrum, and the nanorod-bending deformation is observed by SEM. It was found that the regulation effect of strain on the bandgap decreases with the amount of strain (Watanabe et al., 2015).

Wei et al. from Peking University used in situ SEM technique to study the mechanical and electrical properties of 2D materials. When measuring interlayer electrical transport of graphite island in different interlayer rotation angles, it was found that the resistance in the *c*-axis direction increased linearly as the rotation angle increased (Li et al., 2018). They also measured the interlayer binding energy of hexagonal MoS_2 by in situ peeling-to-fracture method in an SEM, which was measured as $0.55 \pm 0.13\,J/m^2$ (Fang et al., 2020). The frictional properties of MoS_2 layers were measured by manipulating an Si nanowire probe to make an MoS_2 nanoflake slide on the MoS_2 substrate, and the Si nanowire was used as a cantilever force sensor to detect the friction force. The friction coefficient was in the order of 10^{-4}, which was in the regime of superlubricity (Li, Wang, et al., 2017).

4.3.4 In situ reaction

Like ETEM, environmental SEM (ESEM) that operates at a high pressure of 1−10 Torr has also been developed. Several electron beam apertures and pumping systems are used to keep a high vacuum in the microscope gun and column, and a higher pressure in the region of the specimen chamber. ESEM enables the imaging of wet samples and in situ observations of reactions involving gas or liquid phases. Willinger et al. at the Max Planck Institute, Germany achieved in situ growth of graphene on a copper substrate under ESEM. They realized the observation of the stacking processes of few-layer graphene during in situ growth (Wang et al., 2016).

Wei's group took the advantage of the large space of the SEM chamber to set up a multifunctional in situ platform, integrating optical, electrical, mechanical, and other functions. They realized the whole-journey nanomaterial research from material synthesis, composition and structure characterizations, property measurement to device construction, and test in one equipment (Tang et al., 2016). Taking tungsten oxide nanowires as an example, they established the relationships between crystal structures, physical properties and device performances. The $W_{18}O_{49}$ nanowire arrays were in situ grown by irradiating a tungsten filament in oxygen atmosphere, and the chemical composition and distribution of single nanowires were analyzed by EDS. By mechanical resonance and tensile force on as single nanowire, the Young modulus of 128 GPa and the breaking strength of 2.2 GPa were obtained. With the in situ construction of the electronic nanodevices by using the as-synthesized nanowire, the electric conductivity was measured to be 241.1 S/cm, and the $W_{18}O_{49}$ nanowire photodetectors, and gas sensors were also constructed and evaluated in ESEM as shown in Fig. 4.8.

4.4 Scanning probe microscopy

In 1959 Feynman named *arrange atoms one by one the way we want* the ultimate challenge in his "Plenty of Room at the Bottom" lecture (Feynman, 1960). In 1982 Gerd Binnig and Heinrich Rohrer invented scanning tunneling microscope (STM) and realized the observation of monoatomic steps and resolution orders of magnitude better than SEM (Binnig et al., 1982), so that they shared the 1986 Nobel Prize in Physics. In 1990 Eigler and Schweizer (1990) demonstrated the ability of STM to position individual atoms on a single-crystal nickel surface, realizing the dream of manipulating individual atoms. The history has shown that the breakthrough from "zero to one" can only be realized through bold imagination and relentless exploration within the scope of physical principles.

TEM, described in the previous section, can resolve individual atoms and normally requires hundreds of kilovolts to accelerate the electrons for a high-energy electron wave. In contrast, the bias voltage is usually about 1 V for STM, which is considerably lower than those of the TEM. As aforementioned, one way to break the resolution limit is to shorten the

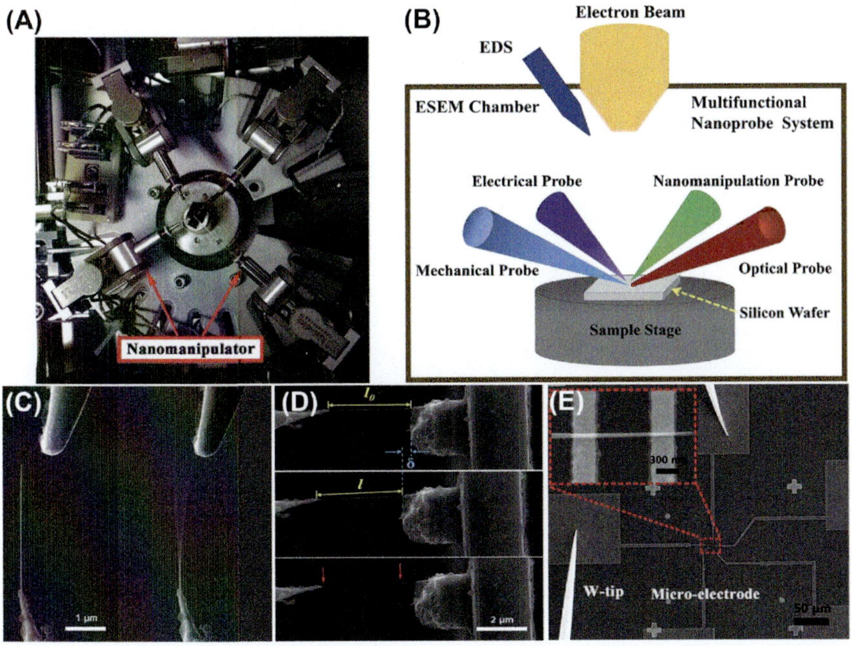

Figure 4.8 ***The whole-journey research of nanomaterials in an SEM from material synthesis, composition characterization, and physical property measurement to device construction and tests.*** (A and B) A picture and schematic drawing of the multifunctional ESEM platform; (C) mechanical resonance measurements; (D) mechanical tensile measurements; and (E) electrical properties and sensor performance measurements (Tang et al., 2016). ESEM, Environmental SEM.

wavelength, and the other way is to avoid diffraction. The TEM overcomes the first limitation, while the STM uses the second way to obtain local structural information on a sample surface by using tunneling currents.

In addition to manipulating individual atoms, STM combined with spectroscopy enables near-field optical and spectroscopic analyses of single-atom, single-molecule energetic processes. Many variations of probe-based microscopies have been developed. SPM, a generic name for this family, includes STM, AFM, magnetic force microscopy (MFM), scanning near-field optical microscopy, and scanning thermal microscopy (SThM) (Fig. 4.9). It enables the study of probe-sample thermal and magnetic interactions and quantitative measurement of local physicochemical properties on the surface of samples. Advanced SPMs are also capable of operating at high temperatures and in atmospheric environments, making

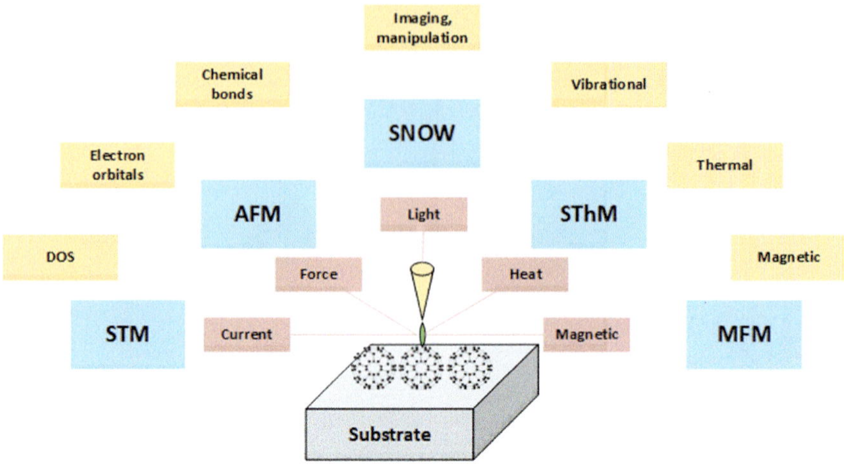

Figure 4.9 Basic principles of SPM. *SPM*, Scanning probe microscopy.

them an important tool for characterizing the surface atomic structure, electronic structure, reaction processes, and property measurement of low-dimensional materials (Bonnell et al., 2012; Park et al., 2010).

4.4.1 Scanning tunneling microscopy

According to quantum mechanics, the sufficient proximity of two objects leads to an overlap of the wave functions, which generates a tunneling current at a certain bias voltage. This is described in the one-dimensional model by the following equation: $I_t \propto V\rho_S(E_F)e^{-1.025\sqrt{\Phi}z}$, where V is the bias voltage, $\rho_S(E_F)$ is the density of electronic states in the Fermi surface, Φ is the electronic work function, and z is the distance from the tip of the probe to the surface. It can be seen that the tunneling current decays exponentially with increasing spacing, and it is the extreme sensitivity to distance that gives the STM its extremely high vertical resolution. The vertical resolution of the STM is about 0.01 nm, and the horizontal resolution is about 0.1 nm.

When the atomically sharp tip of an STM is brought progressively closer to the sample surface, tunneling current with electrons transferring from one metal to the other without going over potential barrier can be measured. It forms a six-dimensional information space, that is, time, 3D coordinates, tunneling current, and bias voltage, which can be combined to produce a variety of imaging modes. In the constant current mode, a tip is moved slowly on the sample surface to maintain the tunneling current constant between the tip and the sample. It takes as feedback to adjust the height of the tip z toward sample surface, so as to yield surface topography images of sample surface. Alternatively, in the constant height mode, a tip can be scanned across a sample surface at a constant voltage and height of the tip relative to the sample surface, and then the variation of the current can be measured for imaging during the scanning process. This allows the study of the differences in the electrical properties of the sample surface. In scanning tunneling spectroscopy (STS) mode, the sample coordinates are fixed to measure the tunneling currents at different bias voltages, the energy level distribution and electronic density of states near the Fermi surface of the sample, that is, its electronic structure, can be obtained.

Generally, a lot of structure information of low-dimensional materials can be obtained by STM characterization, which includes (1) surface and defect structures (Binnig & Rohrer, 1987; Binnig et al., 1982), (2) single-atom and single-molecule manipulation, as well as assembly processes (Bartels et al., 1997; Eigler & Schweizer, 1990), (3) chemical information, such as chemical bonds and their variations (Lee & Ho, 1999), reaction processes (Hla et al., 2000), molecular orbitals (Repp et al., 2005), and intermolecular noncovalent bonding (Weiss et al., 2010), and (4) localized physical properties, such as quantized conductivity (Pascual et al., 1993), single-atom magnetization curves (Meier et al., 2008), and thermoelectric power factors (Lyeo et al., 2004). The following are typical examples of STM characterization (Fig. 4.10).

4.4.1.1 Surface topography and electronic structure

Over the last few decades, the surface topography and electronic nature of low-dimensional materials have been widely investigated by using STM. Taking carbon nanomaterials as examples. Lu et al. at the University of California, Berkeley observed constant current topograph of individual C_{60} molecules with different orientations. The energy-resolved local density of states on the Ag (1 0 0) surface, for example, the electron density distribution of the highest occupied molecular orbital (HOMO) and lowest unoccupied molecular orbital (LUMO), can be obtained by STS (Lu et al., 2003). Néel from University of Kiel, Germany observed the transition from quantum tunneling to quantized conductance (G_0) when the tip is close to and in contact with a single C_{60} molecule (Néel et al., 2007). Charles Lieber's group at Harvard University used STM to obtain atomically resolved images of individual SWCNTs. Their individual six-membered

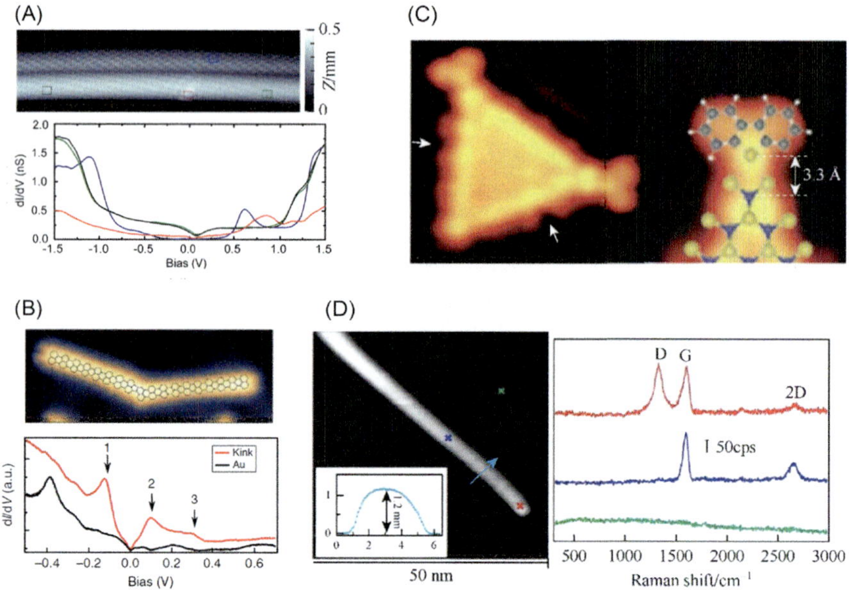

Figure 4.10 ***STM characterization of low-dimensional materials.*** (A) Atomic structure of nitrogen-substituted carbon defects in CNTs with current-voltage spectra embodying the localized electronic density of states (Tison et al., 2013). (B) STM image and current-voltage spectra of graphene nanoribbon torsion structures (Kimouche et al., 2015). (C) STM image and atomic model of catalytically active sites in molybdenum disulfide nanosheets (Tuxen et al., 2010). (D) STM controlled tip-enhanced Raman scattering spectra of CNTs with spatial distributions of their G and D vibrational modes (Liao et al., 2016). *STM*, scanning tunneling microscopy; *CNT*, carbon nanotube.

ring structures can be resolved and the chirality can be directly determined. The energy band structures obtained by STS reflect van Hove singularities and the measured bandgaps are inversely proportional to nanotube diameters, which is in line with theoretical predictions (Odom et al., 1998).

CNTs may change their chirality during growth and form intramolecular junctions, which can be used to assemble all-CNT nanoelectronics. Lieber's group observed CNT intramolecular junctions formed by different combinations of chirality using STM, and STM/STS results showed that localized states do not exist on metal–semiconductor junctions, whereas metal–metal junctions form localized states near the Fermi surface (Ouyang et al., 2001). Tison et al. at the Université Paris Diderot-Paris 7 observed multiple atomic configurations of doping sites in nitrogen-doped SWCNTs. These point defects can induce extended shallow levels or localized deep levels in a semiconducting SWCNT, and donor levels close to the first van Hove singularity can be observed in metallic SWCNTs (Fig. 4.10A) (Tison et al., 2013).

Kimouche et al. from the University of Aalto, Finland characterized the topograph and local density of states of the ultranarrow armchair graphene nanoribbons by STM/STS (Fig. 4.10B). It was found that the atomically well-defined graphene nanoribbons exhibit narrow bandgap and its electronic states depend on the ribbon length. The near-metallic behavior can be reached for a ribbon longer than 5 nm with the bandgap of about 100 meV (Kimouche et al., 2015). LeRoy's group at the University of Arizona characterized graphene on h-BN with STM/STS and found the appearance of Moore's patterns. The surface roughness and the electron−hole charge fluctuations of graphene on h-BN are reduced by two orders of magnitude as compared with those on silicon oxide substrates (Xue et al., 2011), which helps to understand the excellent electrical transport properties of graphene on h-BN substrates (Dean et al., 2010).

The laminated transition metal disulfides are well known as typical 2D semiconducting materials, which can take as building block for van der Waals heterojunctions and fabricate a variety of functional devices. Heinz's group at Columbia University measured the electronic structure of MoS_2/WS_2 heterostructures and found that the bandgaps of MoS_2 and WS_2 are around 2.16−2.38 eV, respectively. This heterostructure constitutes a type II band alignment that has an interfacial bandgap of 1.45 eV (Hill et al., 2016). Renner's group at the University of Geneva characterized vertically aligned heterostructures composed of MoS_2 and $NbSe_2$ with STM. The exfoliated $NbSe_2$ crystals tend to degrade when exposed to ambient conditions, which can be protected by the design of chemically stable MoS_2 as capping layer. When performing STM imaging and spectroscopy of encapsulated heterostructures, it was found that the electronic coupling between them is misalignment angle dependent, and they have strong coupling at smaller angles. While at larger angles, they are weakly coupled and are able to probe the intrinsic properties both of MoS_2 and $NbSe_2$, demonstrating the encapsulation is compatible with the STM/STS investigation of artificial 2D heterostructure materials (Martinez-Castro et al., 2018).

4.4.1.2 Catalytic reaction mechanism

STM is capable of providing electronic structures with respect to the local atomic structure of low-dimensional materials, which is valuable for understanding of chemical reaction processes, because it can provide important information such as preferential adsorption on active sites, charge transfer during reactions, and intermediate configurations.

The characterization of low-dimensional materials used as catalysts is their small size, high ratio of surface and edge atoms. These catalysts exhibit high reactivity in many reactions, and a variety of catalyst systems have been developed, including single-atom catalysts (Qiao et al., 2011), MoS_2 (Voiry et al., 2013), nanogold (Haruta et al., 1989), and nanocarbon (Zhang et al., 2008). In catalytic reactions, the identification of the structure and the nature of active sites are essential for understanding the reaction mechanism. It is difficult for general research tools to characterize the structure of catalysts with atomic-level resolution at high temperatures and atmospheric pressures. Wintterlin's group at the University of Munich, Germany investigated Fischer–Tropsch synthesis on Co (0 0 0 1) surfaces using an operando STM. STM imaging under H_2/CO gas mixtures at pressures of up to 950 mbar (1 bar $= 10^5$ Pa) and a temperature of 500 K, STM imaging results revealed the fluctuations of atomic steps on the catalyst surface as the reaction proceeds. They used surface sputtering method to control the density of monoatomic steps and measured the turnover frequency by a customized gas chromatography, the catalytic activity was then found to be proportional to the density of monoatomic steps, confirming that the atomic steps are the active sites of the catalytic reaction (Böller et al., 2019).

Besenbacher's group at Aarhus University, Denmark selectively probed and atomically resolved the active sites of single-layer MoS_2 nanoclusters by means of STM. They revealed that the adsorption properties of dibenzothiophene (DBT) molecule in the hydrodesulfurization reactions depend on the size of MoS_2 nanoclusters. The single-layer nanoclusters were well-defined triangular, and they prefer to expose the Mo-edge termination when the number of Mo atoms along the edges is greater than six, while S-edge termination when the number of Mo atoms is less than 6. When the size of nanoclusters exceeds a threshold value of 1.5 nm, MoS_2 reacts with hydrogen to generate sulfur vacancies on the edges and binds weakly to the reactant DBT. In contrast, when the size of the nanoclusters is below the threshold, the sulfur vacancies are formed mainly at the top corners of the triangles with strong adsorption to the reactant molecules, which explains the unique catalytic properties of very small MoS_2 nanoclusters (Fig. 4.10C) (Tuxen et al., 2010).

The Sykes's group at Tufts University, USA has investigated the oxidation of CO by single-Pt atoms at low temperatures by STM.

The atomic-scale geometry and electronic structure of the active sites were probed and correlated to CO oxidation activity. CO molecules were found to preferentially adsorb on a Cu_2O film supporting single Pt atoms, in which lattice oxygen is involved in the oxidation process (Therrien et al., 2018). Parkinson et al. from the Technical University of Vienna, Austria investigated how local environment of single-atom catalysts toward supports affect their stability and reactivity. It was found that the transition group metals of the Cu, Ag, Au, Ni–Pd, Pd, Pt, Ru, and Ir series all occupy the same twofold-coordinated adsorption site of Fe_3O_4 (Scherzer, 1947) characterized by STM. In addition, charge transfer occurs between the support and the d-states of metal atoms may either strengthen or weaken the catalyst–CO bonds. They demonstrated that binding metal atoms to an oxide support can alter their electronic structure and activities, and for most metals, the metal–CO bond strength can be increased. According to Sabatier principle, it is possible to have higher reactivity for the non-Pt-group but stabilized metal adatoms (Hulva et al., 2021).

4.4.1.3 Tip-enhanced spectroscopy

Photovoltaic devices, such as solar cells, contain optoelectronic processes such as light absorption, emission, transport, and attenuation. Understanding the relationship between these processes and the electronic structure of semiconductor materials is the key to improving the energy conversion efficiency. Spectral analysis can obtain the macroscopic energy distribution and conversion process, but the spatial resolution is limited by the optical diffraction limit. STM can overcome the optical diffraction limit by combining with spectroscopies and using the tip of the needle to enhance the local electric field, turning far-field optics into near-field optics. It can also image the light–matter interaction process, which is directly related to local features such as grain boundaries, vacancies, doping, and interfaces (Wieghold & Nienhaus, 2020).

The team of Hou and Dong at the USTC has done a series of pioneering work in single-molecule optics. They obtained single-molecule-resolved Raman scattering spectra and spatial distributions of intramolecular vibrational modes (Zhang, Zhang, et al., 2013), observed dipole electroluminescence corresponding to different numbers of molecules (Zhang et al., 2016), and probed the spatial distributions of luminescent sites within single molecules (Yang et al., 2020). They combined three techniques, STM, AFM, and tip-enhanced Raman scattering spectroscopy (TERS), to comprehensively characterize individual *p*-pentacene molecules adsorbed on Ag (1 1 0) surfaces, and to investigate the geometrical configurations, electronic structures, and vibrational properties of single molecules at the level of individual chemical bonds. Among them, STM measures the electrical properties and induces the conformational transitions, TERS determines the positional changes in the C−H bond vibrational peak and identifies the breakage of the C−H bond in dehydrogenation reactions, and noncontact AFM resolves the process of changes in the molecular backbone structure. The combination of the three techniques is able to give comprehensive structural and chemical information to distinguish between different molecular isomers (Xu et al., 2021). Fig. 4.10D shows the STM-controlled TERS spectra of CNTs, and it can be seen that the spatial distributions of G-band associated with C = C stretching vibrations and D band associated with a dispersive photo mode are enhanced. With a spatial resolution of about 0.7 nm, the spatial extent of the defect-induced D-band scattering can be visualized in real space, and tiny impurities attached to the CNT can be chemically identified (Liao et al., 2016).

4.4.2 Atomic force microscopy

STM is capable of characterizing local topography and electronic structure of materials at the atomic-resolution level, and the research system is usually a conductor or semiconductor based on the tunneling current principle. AFM is another widely used form of SPM, which does not require an electrically conductive sample. But for a topography measurement, a rigid enough sample supported on a flat surface is needed. In 1986 Binnig et al. (1986) invented AFM, and they detected small forces between sample and tip by measuring the motion of a cantilever beam during the scanning of the sample. Since AFM can measure different kinds of forces, including magnetic forces, van der Waals forces, and electrostatic forces, it can be classified into subdisciplines such as MFM, Kelvin probe force microscopy, scanning electrostatic force microscopy, and friction force microscopy (Müller & Dufrêne, 2008).

4.4.2.1 Atomic imaging

AFM imaging can be categorized into static and dynamic operation modes. In static mode, AFM measures the force between tip and sample surface by the detection of the bending of the cantilever beam. While in dynamic mode, the cantilever is deliberately vibrated by an actuator. AFM measures the force by observing the changes in amplitude and frequency of the mechanically resonant cantilever beam (Giessibl, 2003). The operation and explanation of the static mode in the early years are simple, which directly reflected the distribution of surface force. Frequency-modulation AFM invented by Giessibl improves the microscope sensitivity to surface structure and imaging resolution. By using that, the (7 × 7) reconstruction on Si (1 1 1) surface can be imaged (Giessibl, 1995).

Under the same operation mode, Gross et al. from International Business Machines Corporation (IBM) Zurich Research Laboratory reported that the atomic molecular structure of pentacene can be precisely resolved by noncontact AFM with CO-terminated tip. The modification by functionalizing the AFM tip apex with atomically well-defined molecules has greatly improved the atomic resolution, which also exhibits great advantages over STM in resolving individual atoms within an adsorbed molecule, especially those on a metal substrate (Gross et al., 2009). There are different types of forces that can be measured by noncontact AFM between the tip and the adsorption molecules on sample surface. They are short-range Pauli repulsive force, van der Waals force, and the long-range electrostatic attractive forces. They contributed to image negative background, the dark halo surrounding the molecule, or atomic contrast of molecular structure. Thus AFM images are measured with CO molecule terminated tip, and their correlation with chemical structure of a molecule can be explained, for example, the intermolecular hydrogen bonding in 8-hydroxyquinoline was observed by the group of Qiu at the National Center for Nanoscience and Technology, China (Zhang, Chen, et al., 2013). Kawai et al. (2018) at the National Institute for Materials Science, Japan also used this technique to resolve the positions of boron and nitrogen dopant atoms in graphene (Fig. 4.11A).

Direct imaging of chemical bonds by AFM is helpful in understanding the chemical reaction and the synthesis mechanism of low-dimensional materials (Pavliček & Gross, 2017). Michael Crommie's group at the University of California, Berkeley observed the formation processes of covalent bonds between molecules (De Oteyza et al., 2013). Kaiser et al. from IBM Zurich Research Laboratory prepared high reactive cyclic C_{18} molecules by eliminating CO from the cyclocarbon oxide $C_{24}O_6$ molecule by AFM in situ manipulation. The isolated molecules were then characterized to be a polyynic structure with defined positions of alternating triple and single bond (Kaiser et al., 2019).

4.4.2.2 Thermal measurement

Thermal management is extremely important for the low-dimensional materials and devices. As the size of the transistor shrinks from

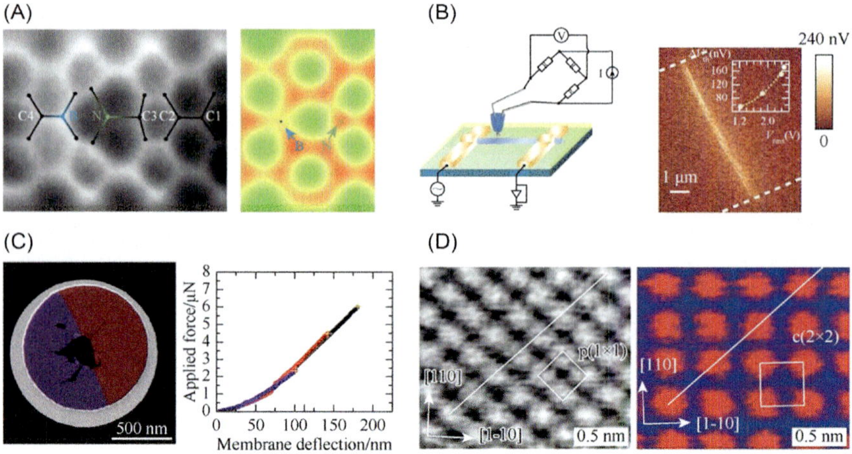

Figure 4.11 ***The characterization of structure and properties of low-dimensional materials by AFM.*** (A) Boron and nitrogen doped atoms in graphene nanoribbons (Kawai et al., 2018). (B) The temperature distribution of gallium nitride nanowires during on-current joule heating was characterized by scanning thermal microscopy (Soudi et al., 2011). (C) Fracture behavior and strength of grain boundaries of polycrystalline graphene (Rasool et al., 2013). (D) The atomic arrangement and spin structure of iron monolayer were mapped by magnetic exchange force microscopy (Schmidt et al., 2009). *AFM*, Atomic force microscopy.

micrometers to nanometers, the power density continues to increase, and the heat dissipation becomes one of the major challenges for high-performance nanotransistors. Thermal dissipation is also an important mechanism for introducing quantum state errors in quantum computers. Lowering the material dimension is an important strategy to improve the performance of thermoelectric materials and devices, as well as the efficiency of waste heat to power generation. Thermal measurements, meanwhile, are an important means of assessing the thermal properties of low-dimensional materials and understanding heat transport processes in low dimensions and small scales.

Temperature measurement, the basis of thermal measurement, includes thermometers, thermocouples, IR thermography (Teyssieux et al., 2007), electron spin resonance (ESR) spectroscopy (Neumann et al., 2013), Raman spectroscopy (Reparaz et al., 2014), and electron energy-loss spectroscopy (Mecklenburg et al., 2015). Each of these methods has its own characteristics and scope of applications. For instance, nanothermometry based on SPM can directly correlate thermal properties and local structures at nanometer resolution.

SThM is the SPM-based method for temperature or thermal property measurement using either Seebeck effect (Sadat et al., 2010) or resistance effect (Menges et al., 2016), etc., sensing mechanisms. It is based on a nanoscale thermocouple or resistor placed on an SPM head to obtain local thermal information by sensing the tip-sample thermal exchange processes (Zhang et al., n.d.). The thermal probe, including thermoelectric or resistive-based probe, is the key of an SThM system, which determines the range of operation and quality of measurement. For a Seebeck thermoelectric probe, conductive sample surface or coating is needed based on its working mode, either contact thermocouple or noncontact tunneling thermometry. Reddy's group at the University of Michigan used ultra-high vacuum SThM integrated with a nanoscale Au−Cr thermocouple to quantitatively measure the surface temperature fields of a platinum nanowire. It is demonstrated that the temperature resolution is up to 15 mK, and the spatial resolution is ~10 nm (Kim et al., 2012).

With a thermo-resistive probe, SThM measures the temperature-induced resistance changes by a Wheatstone bridge, where the sample or probe is heated, but there is no limitation to the conductivity of the sample. Gu et al. from Washington State University quantitatively studied the heat dissipation of a single GaN nanowire by spatially resolved Raman spectroscopy and SThM. By comparing the temperature distribution of the suspended and substrate-supported nanowires, it is found that the heat dissipated through the nanowire−substrate interface accounts for ~80%−93% of the total dissipated heat; accordingly, the heat dissipated at the nanowire−electrode interface plays a minor role. Due to phonon confinement and boundary scattering effects, the thermal conductivity of the nanowire is measured to ~40−60 W/m K, lower than the bulk GaN (Fig. 4.11B) (Soudi et al., 2011). Kwon et al. at Seoul University in South Korea applied the null point SThM to quantitatively characterize the temperature profile across a suspended graphene bridge at a ~50 nm resolution. The thermal conductivity of graphene at different temperatures was obtained by fitting the temperature profile, and the room temperature thermal conductivity of clean graphene is as high as 2000 W/m K (Yoon et al., 2014).

In consideration of the need for quantum devices to work at low temperatures, Halbertal et al. from the Weizmann Institute of Science in Israel have developed a nanoscale thermometer based on a superconducting quantum interference device (SQUID) for thermal measurements, at low temperatures and low heat dissipation conditions. The method has a spatial resolution of ~50 nm and an energy dissipation of ~40 fW at a working temperature of ~4.2 K. They can directly observe the changes in thermal dissipation due to single-electron charging of individual quantum dots in CNTs and clarify the origins of thermal dissipation (Halbertal et al., 2016).

4.4.2.3 Mechanical measurement

AFM is the measurement of forces between a sharp tip and a sample surface. The forces can be long-range van der Waals and electrostatic forces, or short-range chemical forces, etc. The latter force arises from the overlap of electron wave functions and from the repulsion of the ion cores. For homogeneous materials, it provides information of surface atom arrangement and roughness. For a material with different chemical compositions, a dynamic force microscopy with atomic resolution can be used to identify individual surface atoms, which is based the chemical force measurements due to the interaction between potential function of the tip and different surface atoms (Sugimoto et al., 2007).

AFM cantilever is usually manufactured by micromachining techniques, which is thin and is easily bent in the vertical direction, broader, and stiff in the horizontal direction. Consequently, it can only measure the force in the vertical direction, and the information in the horizontal direction is lost or reduced. Poggio et al. from the University of Basel modified AFM with a single nanowire as force sensors (Rossi et al., 2017). The nanowire has a small cross-sectional asymmetry, and its resonance mode can be split into doublets oscillating along vertical and horizontal directions, which is taken as a sensitive vectorial force sensor. The spatial tip-sample force derivatives in-plane can be obtained by monitoring the resonance frequency shift. They also mapped the vector plot of electrostatic force field on the nanometer scale, and this technique suggests a powerful capacity of mapping the size and direction of weak tip-sample forces.

AFM can measure the mechanical properties of individual low-dimensional nanostructures, including tension, compression, bending, and friction(Fig. 4.11C). Yu et al. measured the tensile strength of a multi-walled CNT mounted between two opposing AFM tips (force sensors) in a tensile-loading experiment. During the rigid AFM cantilever was driven upward, the soft AFM probe was bent upward and formed a deflection distance, together with the stretched length of a CNT imaged in SEM, the tensile strength was calculated to 11−63 GPa, which was attributed to the broken outermost layer of CNTs (Yu et al., 2000). Wu et al. applied AFM tip to laterally manipulate a single suspended gold nanowire and measured the mechanical properties. It is found that Young's modulus of nanowires is diameter-independent and comparable to that of bulk gold, while the yield strength shows a strong size effect and can be 100 times larger than bulk (Wu et al., 2005). Lee et al. measured the elastic properties and intrinsic breaking strength of suspended monolayer graphene membranes by AFM nanoindentation. It is found that the second- and third-order elastic stiffnesses for monolayer graphene are 340 and −690 N/m, respectively. The Young modulus of bulk graphite was then calculated to as high as 1 TPa and the breaking strength of 130 GPa (Lee et al., 2008). Kawai et al. (2016) found that the graphene nanoribbons can slide easily on a gold surface with an ultra-low friction force of only ~100 pN by dynamic AFM.

4.4.2.4 Magnetic measurement

Spin and magnetism are the basis of information storage, spintronics, quantum computer, brain neural activity, etc. There are many methods to measure magnetic properties, including SQUID (Wernsdorfer, 2009), Lorentz microscopy (McVitie et al., 1995), electron holography (Tonomura et al., 1982), MFM (Sáenz et al., 1987), Kerr microscopy (Argyle & McCord, 2000), STM (Verlhac et al., 2019; Wiesendanger et al., 1990), nuclear magnetic resonance (NMR) (Mamin et al., 2013), and electron paramagnetic resonance (EPR) spectroscopy-STM (Seifert et al., 2020). The magnetic resonance force microscopy (MRFM) and nano-SQUID in combination with SPM give the localized magnetic structure and properties of the material.

MRFM was invented by Martin and Wickramasinghe (1987) from IBM Watson Research Center. By measuring the magnetic force between the magnetized tip and the scanned surface of a sample, it is possible to analyze the magnetic fields and the magnetic interactions of low-dimensional structures or assemblies (Lu et al., 1997; Puntes et al., 2004; Sáenz et al., 1987). Rugar et al. showed that the improved MRFM can detect an individual electron spin below a sample surface with nanometer spatial resolution. The microwave magnetic field and the inhomogeneous field from the tip were taken as resonate slice, which was then swept back and forth through the sample with the vibration of the cantilever tip, until the appearance of an unpaired electron spin in silicon dioxide. It causes an alternating magnetic force on the cantilever and the shift in cantilever frequency (Rugar et al., 2004). The detection of individual spin state by MRFM represents the possibility to real-time readout quantum state and proceeds a variety of quantum measurements.

The Wiesendanger group of University of Hamburg realized the measurement of the arrangement of spins in real space and with atomic resolution by AFM-based magnetic exchange force microscopy. An external magnetic field is applied to align the magnetic polarization at the tip to optimize the interaction between tip and sample spins. The inter-spin interactions in the antiferromagnetic insulator, the (1 0 0) surface of NiO, can be investigated (Kaiser et al., 2007). Schmidt et al. characterized the atomic-scale antiferromagnetic structure of the iron monolayer on the W (0 0 1). Both the p (1 × 1) chemical surface unit cell and c (2 × 2) antiferromagnetic surface unit cell can be observed with magnetic insensitive and magnetic sensitive tip configurations (Fig. 4.11D) (Schmidt et al., 2009). Gai et al. from Oak Ridge National Laboratory characterized the magnetic structure of quasi2D crystal Fe_3GeTe_2 at different temperatures by MFM. They observed uniform contrast, no ordered magnetic structures, at room temperature, and branched and bubble structure at 204 K and below 150 K, respectively. They also determined the ferromagnetic transition at 214 K and the antiferromagnetic transition at 152 K (Yi et al., 2017).

The basic structure of SQUID is superconducting coils connected by two Josephson junctions. The coil current changes with the magnetic flux oscillation, $J_{\mathrm{max}} = 2J_0\left|\cos\left(\frac{q_e}{h}\phi\right)\right|$, and the current reaches the maximum when the magnetic flux in the coil is an integer multiple of the quantized magnetic flux $\frac{\pi h}{q}$. SQUID is extremely sensitive to weak magnetic fields, with the sensitivity at the level of 5×10^{-18} T, and the spatial resolution is normally limited by the effective size on the order of 1 μm. In 2013 Vasyukov et al. deposited two Nb or Pb superconducting leads connected to Au electrodes on the apex of a pulled quartz tube to fabricate a nanoscale SQUID probe, with the minimum size of 46 nm. With the improvement, the magnetic flux noise of the technique is as low as 50 $n\Phi_0\mathrm{Hz}^{-1/2}$, the spin sensitivity is 0.38 $\mu_B\mathrm{Hz}^{-1/2}$, and μ_B is the spin magnetic moment of a single electron (Vasyukov et al., 2013).

4.5 Spectroscopy

Characterization techniques, such as imaging and diffraction, are generally to probe the "structure or appearance" of a material, including size, shape, crystal structure, atomic occupation, and defects, in real and reciprocal spaces, respectively. The "composition or contents" of the material, such as atomic species, chemical bonding, valence, and energy bands, are usually characterized by various spectroscopic methods. Imaging, diffraction, and spectroscopy are techniques by collecting different signals from the similar interaction of a particle (wave) as a probe with matter. Taking electron signal as an example, TEM and SEM employ a beam of transmitted electrons and secondary electrons, respectively, while electron diffraction, EELS, and EDS utilize elastically scattered electrons, inelastically scattered electrons, and X-rays from the secondary excitation processes, respectively. According to theorems on energy and momentum conservation, the energy distribution, transfer, and transformation processes in low-dimensional materials and devices can be analyzed.

Spectroscopic techniques can be classified into several categories, according to the type of excitation "probe," they can include optical spectroscopy, X-ray spectroscopy, electron spectroscopy, and magnetic resonance spectroscopy. According to the type of energy, they can be divided into lattice vibration spectroscopy, electron energy spectroscopy, EPR spectroscopy, etc. Based on the process of interaction, they can be divided into absorption spectroscopy, emission spectroscopy, scattering spectroscopy, and so on. The commonly used spectroscopies in characterizing low-dimensional materials include ultraviolet—visible absorption spectroscopy, photoluminescence (PL) spectroscopy, and CL spectroscopy to characterize the optical properties; XPS, EELS, and EDS to characterize the chemical composition and state, IR absorption spectroscopy and Raman spectroscopy to characterize the vibration of the chemical bonding; and NMR and ESR spectroscopy to characterize the magnetic properties (Fig. 4.12).

4.5.1 Optical spectroscopy

4.5.1.1 Infrared absorption spectroscopy

IR absorption is the resonant absorption of IR light by molecules through changes in their intrinsic dipole moments. Molecular vibrations and lattice phonons can be considered as harmonic oscillators in quantum mechanics. The vibrational modes reflect the symmetry of molecules, and different modes have different characteristic vibrational energies, roughly on the order of 0.1 eV, which are in the range of IR range. Thus IR absorption spectroscopy can be used to detect the molecular vibrations and then characterize the chemical composition and type of bonding. Normally, the vibrations of polar molecules and the asymmetric vibrational modes of

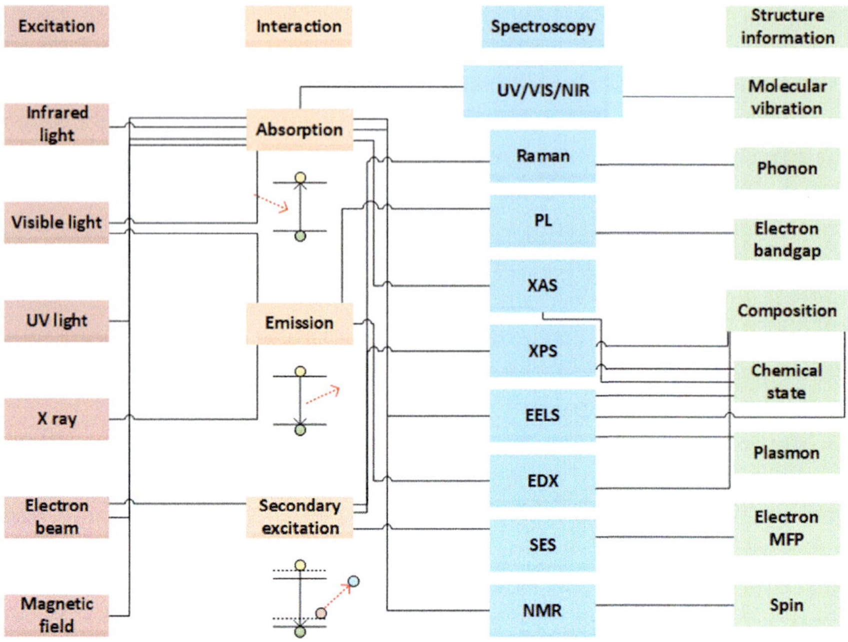

Figure 4.12 ***Fundamentals and classification of spectroscopic characterization techniques.*** According to the excitation signals and the process of interaction, they can be broadly classified into absorption, emission, and secondary excitation spectroscopies of light, X-rays, electrons, and magnets, the information obtained includes molecular vibrations, phonon structure, electron bandgap, chemical state, and spin structure.

nonpolar molecules have an IR response. In addition to molecular vibrations, IR absorption spectroscopy can also characterize low-lying energy excitation of a material. By combining with a magnetic field B, Jiang et al. from Columbia University measured the Landau level spectrum of graphene. They determined the linear relationship between the Landau levels of graphene and $\sqrt{B}$ by measuring the transition energies at different magnetic field strengths (B) (Jiang et al., 2007).

The catalysis of CO oxidation reaction at low temperatures is an important chemical reaction in the field of clean energy. Low-dimensional catalysts, including gold and platinum nanoparticles and single atoms, have been reported to have high catalytic activity, but the active sites and structures of the catalysts during the reaction are still controversial. Fourier-transform IR spectra and in situ IR spectra of adsorbed CO during reaction were collected to provide information about the dispersion and oxidation state of Pt and CO oxidation rate (Pedrero et al.,

2005; Qiao et al., 2011). Stair's group at Northwestern University, USA found that the IR spectroscopy can identify CO adsorbed on isolated Pt atoms or nanoparticles. They proposed that only Pt nanoparticles are catalytically active at lower temperatures, whereas Pt single atoms have reduced catalytic activity due to the strong binding of CO molecules (Ding et al., 2015).

4.5.1.2 Raman spectroscopy

Both Raman spectroscopy and IR absorption spectroscopy characterize the molecular and crystal vibrational modes. The difference is that Raman spectroscopy reflects the interaction between incident light and molecular dipoles. The electric dipole intensity μ of a molecule caused by electromagnetic waves is: $\mu = \alpha E_0 \cos(\omega_0 t)$, where α is the polarizability of the atoms, E is the electric field applied to a solid, and ω_0 is the optical frequency. The vibration mode of the molecule is expressed as: $Q = Q_0 \cos(\omega_R t)$, where ω_R is the vibration frequency. After the Taylor expansion of polarizability α to vibration intensity: $\mu = \alpha_0 E_0 \cos(\omega_0 t) + \frac{1}{2}\left(\frac{\partial \alpha}{\partial Q}\right)_{Q=0} Q_0 E_0 \cos[(\omega_0 - \omega_R)t] + \frac{1}{2}\left(\frac{\partial \alpha}{\partial Q}\right)_{Q=0} Q_0 E_0 \cos[(\omega_0 + \omega_R)t]$, where the first term represents Rayleigh elastic scattering, the second- and the third terms represent Raman scattering, Stokes lines with frequencies lower than the incident light and antiStokes line with frequencies higher than the incident light, respectively (Jorio et al., 2011).

Raman spectroscopy can give complete structural information for low-dimensional systems, and thereby sensitive to the microstructure of materials. For sp^2 carbon nanomaterials, information such as crystallite size, the presence of sp^3 hybridization and chemical impurities, doping, defect, elastic constants, and mass density can be obtained from Raman spectroscopy (Dresselhaus et al., 2010). Generally, the intensity of Raman spectrum is weak; thus surface-enhanced Raman scattering was applied to strongly increase Raman signals, so that even single molecule can be observed (Kneipp et al., 1997). SWCNT is a unique one-dimensional system, for instance, the electronic state density presents a sharp distribution of van Hove singularity. When the incident laser energy matches the van Hove singularity energy level E_{ii}, strong resonance absorption and Raman scattering occur. The signal intensity of resonance Raman scattering of an SWCNT with diameter of only about 1 nm is comparable to that of bulk silicon crystals; therefore it can be used to determine the structure of individual SWCNTs. One of the main features for SWCNTs is the radial breathing vibration mode (RBM), the vibrational motion in the radial direction of a nanotube. Dresselhaus's group at Massachusetts Institute of Technology used the RBM modes to characterize the chirality of a single SWCNT. The laser excitation energy limits the resonance energy window of SWCNTs; thereby, only the energy separation between van Hove singularities for a given chiral SWCNT within the window can be determined. The larger the SWCNT diameter, the more involved resonant SWCNTs and larger number of possible chiralities. They proposed that it is possible by resonant confocal micro-Raman spectroscopy to uniquely assign chiral indexes of SWCNTs, by decreasing nanotube diameters and measuring RBM peak position and using the theory of resonant transitions (Jorio et al., 2001).

4.5.1.3 Photoluminescence spectroscopy

PL spectroscopy probes the photon emission process that accompanies the transition of excited state electrons to the ground-state energy level when a material is excited by light. The luminescence process follows the law of conservation of momentum, and generally, only semiconductors with direct bandgap are able to detect PL signals. Due to the quantum confinement effect, the HOMO−LUMO gap of semiconductor quantum dots widens with decreasing crystallite size. Murray et al. reported the controlled synthesis of nearly monodisperse CdE (E = S, Se, Te) nanocrystallites and observed an inverse proportionality of the PL luminescence peak position with respect to crystallite size. The HOMO−LUMO gap increases from 1.7 eV of the bulk to 3 eV of the nanocrystallites, covering the entire visible light wavelength range (Murray et al., 1993). Peng's research group used PL to record the size change in situ during CdSe nanocrystal nucleation and growth. The average size of nanocrystals was determined from the offset of the peak position of the first excitation absorption, while the size distribution of nanocrystals was calculated from the width of the peaks, which elucidated the growth kinetics of nanocrystals (Qu et al., 2004). Qin et al. investigated the luminescence properties of single CdSe/CdS core/shell nanocrystals by using PL. They found that the PL decay dynamics of CdSe/CdS dots were closely related to the shell thickness, and the CdS shell thickness with 4−16 monolayers were all nonblinking within single-exciton regime, with "on" time being over 95% of the time (Qin et al., 2014). With the assist of PL characterization, the optoelectronic properties of semiconductor can be compared for a bulk and low-dimensional nanomaterial. Generally, the band structure of low-dimensional semiconducting materials is not only reflected in the bandgap associated with its size, but also the relative positions of the energy bands in different directions may change with the transformation from a 3D bulk to a low-dimensional structure. One of the typical examples is MoS_2, whose bulk crystal is an indirect bandgap semiconductor with a bandgap of about 1.3 eV, whereas a monolayer of MoS_2 with strong PL becomes a direct semiconductor with a bandgap of about 1.9 eV (Fig. 4.13) (Mak et al., 2010; Splendiani et al., 2010).

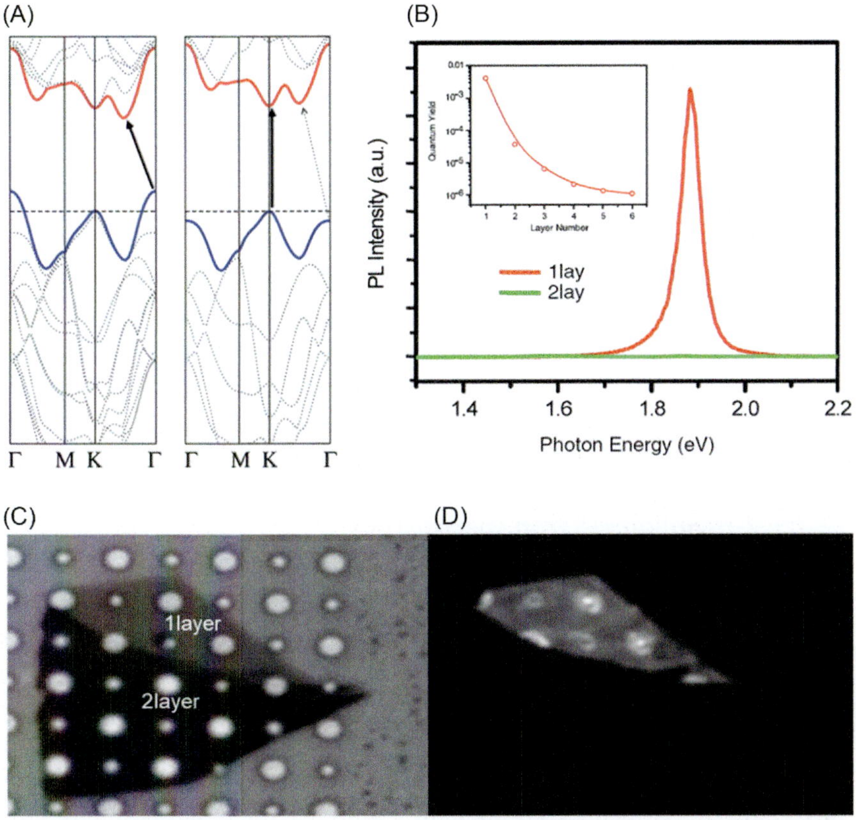

Figure 4.13 ***Electronic structure and PL spectra related to the number of MoS_2 layers.*** (A) Band structure of bilayer versus monolayer MoS_2, shifting from an indirect to a direct bandgap (Splendiani et al., 2010). (B) Layer-dependent PL spectra, with an exponential increase in PL intensity with decreasing number of layers. (C and D) An optical image and PL image of mono- and few-layer MoS_2 crystals on a silicon substrate (Mak et al., 2010). *PL*, Photoluminescence.

Optical absorption spectroscopy coupled with PL spectroscopy can be used to obtain both the absorption and luminescence characteristics of a material. For semiconducting SWCNTs, the absorption and excitation energies correspond to the E_{22} van Hove singularity transition, while the emission energy corresponds to the E_{11} van Hove singularity energy level difference, and thus the absorption—emission spectra can be used to determine the chiral distribution of semiconducting SWCNTs (Bachilo et al., 2002).

4.5.1.4 X-ray absorption spectroscopy

X-ray absorption spectroscopy, which also utilizes the interaction of photons with matter, is capable of detecting the transition of inner electrons to empty orbitals by using X-rays on the order of 10 keV. The absorption edge can determine the elemental species, the fine structure of the absorption spectrum can tell the coordination environment and chemical state. After the Fourier transform, the spatial information, such as the spacing between atoms and coordination number, can be obtained. In 2011 Zhang's group at the Dalian Institute of Chemical Physics, Chinese Academy of Sciences proposed the concept of single-atom catalysts and characterized the chemical state of the platinum catalysts by using extended X-ray absorption fine structure spectroscopy (EXAFS). In both normalized EXAFS and Fourier transform EXAFS, the peaks of Pt catalysts are located between the metal Pt foil and Pt oxide (PtO_2). They proposed that Pt single-atom catalysts are positively charged and would be oxidized by charge transfer from the oxide substrate (Qiao et al., 2011).

4.5.1.5 X-ray photoelectron spectroscopy

XPS is an important means of probing the chemical information and electronic structure of material surfaces using X-rays. In chemistry, the electronic structure of a material refers to the electronic distribution of inner and bonding electrons (valence electrons) in different energy orbitals, which is the basis for understanding structural stability, charge transfer, and reaction mechanisms. In physics, it refers mainly to the degrees of freedom of the electrons in the vicinity of the Fermi surface in terms of energy, momentum, spin, etc., which determines the nature of their transport properties and performances of electronic devices.

The basic principle of photoelectron spectroscopy is the photoelectric effect (Einstein, 1905), for which Albert Einstein was awarded the Nobel Prize in Physics in 1921. Swedish scientist Kai Siegbahn was awarded the Nobel Prize in Physics in 1981 for inventing high-resolution photoelectron spectroscopy (Siegbahn, 1982). When the incident photon energy is greater than the electronic work function of the material, the electron absorbs the photon energy and leaps to be emitted into the vacuum. Then the kinetic energy and momentum of the emitted electrons in different directions can be analyzed to obtain information about the electronic structure of the material. The photoelectron emission process is a transition process of electrons under electromagnetic field, and the photoelectron emission intensity (I) is expressed as $I \propto |<\phi_f|H_{int}|\phi_i>|^2\delta(E_f - E_i - \hbar\omega)F(T)$, in which the first term is the transition matrix of the action of the quantum state and photon perturbation (H_{int} is the interaction Hamiltonian of electromagnetic field) before and after the emission of electrons. The second term ensures energy conservation, where only photons with energy greater than the work function can excite photoelectrons, and the third term is the Fermi–Dirac distribution of electrons, reflecting the density of electronic states of the material.

XPS is capable of providing information on the chemical compositions and valence states of the surface layer of a material, which makes it suitable for studying the surface reaction mechanism of solid-state catalysts. In recent years, XPS can be used to characterize in situ reaction process under near atmospheric-pressure conditions, which constitutes a toolbox for in situ study of catalytic reactions together with high-temperature atmospheric-pressure STM, in situ XRD and ETEM (Lundgren et al., 2017). Lundgren's group at Lund University, Sweden investigated the CO oxidation reaction on Pd (1 0 0) crystalline surfaces by using the ambient XPS method. The XPS measurements contain information on all the involved molecules and surface structures during the reaction, such as adsorbed O, gas-phase CO, adsorbed CO and CO_2, metallic palladium, and CO-adsorbed palladium at different stages of the reaction with respect to the temperature and time. Thus information, such as catalyst state and reaction kinetics, are obtained in the reaction (Lundgren et al., 2017).

ARPES, based on the photoelectric effect, not only analyzes the energy of photoelectrons but also measures the direction of the emitted electrons and establishes a dispersion relation of the momentum and energy of electrons, which allows the reconfiguration of the energy band states. With the continuous improvement of light sources and equipment, ARPES can reach energy resolution of meV (Okazaki et al., 2012). ARPES possesses the capability to probe the fine electronic structure and the corresponding crystal structure, which is a key device for resolving the electronic structure of topological materials (Yang et al., 2018). It includes the Dirac cone linear dispersion relation for the surface states of topological insulators (König et al., 2007), the 3D Dirac cone for the surface Dirac semimetal structures (Liu, Jiang, et al., 2014; Liu, Zhou, et al., 2014; Wang et al., 2013). The use of polarized photons allows for the probing of spin-dependent electronic structures in materials, and the surface state spin chiral structure, that is, the electron spin–momentum locking relationship, is observed in 3D topological insulators (Chen, 2012; Hsieh et al., 2009).

4.5.2 Electron spectroscopy

The interaction of electron beam as a probe with materials generates a variety of signals. Elastic scattering makes electron diffraction and high-resolution imaging. Inelastic scattering is the basis of chemical analysis, for example, chemistry and electronic structure of the specimen, which include energy-loss electrons, characteristic X-rays, secondary electrons and CL. Electron spectroscopy can be classified into secondary electron energy spectroscopy (SEES), Auger electron spectroscopy, EELS, and EDS.

4.5.2.1 Electron energy–loss spectroscopy

The interaction of high energy electrons and specimens can excite the electrons from materials and make them transition from high energy state to vacuum, then the incident electrons lost parts of energies. When the transmitted electrons pass through magnetic lens, the spread of different electron energies can be measured by photomultiplier tube and CCD, that is, EELS can be obtained. It includes a variety of chemical, electron, optical, and photon structure information. EELS can detect all elements, including hydrogen, the higher the element number, the harder the excitation of inner electrons. Thus the lighter element can be easier to be detected, excellently complementary to EDS that characterizes mainly heavier elements.

The spatial resolution of EELS mainly depends on the electron beam (probe) size. In the spherical aberration-corrected STEM system, EELS can reach single-atom level resolution. The energy resolution of EELS is related to the electron gun, 0.2–0.3 eV can be reached for a cold field emission electron gun, and the energy spread of field emission electron gun together with monochromator can reach 0.01 eV, so that EELS can obtain information on fine electronic and phonon structure of materials.

EELS can be divided into three parts: (1) zero-loss peak with elastic electrons, (2) low-loss region (below 50 eV), corresponding the plasmon excitation, interband transition, and phonon information, (3) high-energy loss region, corresponding to the inner shell electrons, including material composition and bonding information.

Due to the high spatial and energy resolution and high signal to noise of STEM−EELS, chemical analysis can be performed at atomic resolution (Kimoto et al., 2007; Muller et al., 2008, 2009). Pennycook et al. at Oak Ridge National Laboratory, USA used the STEM−EELS technique to resolve impurity atoms such as individual oxygen and carbon atoms in the monolayer of boron nitride (Krivanek et al., 2010). Suenaga's group at AIST, Japan used spherical aberration-corrected STEM imaging with elemental characteristic absorption edge analysis by EELS to resolve individual gadolinium atoms inside a single chain of metallofullerene molecules within SWCNTs. EELS analysis of graphene at atom resolution revealed the electronic and bonding structures of the edge atoms different from the bulk atoms and achieved direct measurement of phonon dispersions in monolayer graphene (Senga et al., 2019; Suenaga & Koshino, 2010; Suenaga et al., 2000).

The surface plasmon oscillations of nanostructures are mainly distributed over the several eV energy range, which have many applications in the fields of catalysis and surface-enhanced Raman spectroscopy. As shown in Fig. 4.14, the combination of EELS with ADF images and CL maps in the STEM system can directly correlate the topographical features and electronic structure of low-dimensional structures. For instance, the nanocrystals are shown with the spatial distribution of the characteristic energy of the plasma oscillations, which can provide a powerful guide for the design and application of surface plasmon oscillation effect (Colliex et al., 2016).

4.5.2.2 Energy-dispersive of X-ray spectroscopy

Generally, many effects will occur during the interaction between the microscope's high energy electrons and the atoms of a sample, one is the formation of characteristic X-rays. That is, the ejection of electrons from the inner shell of an atom creates an electron vacancy, which is then occupied by an outer shell electron. As the higher energy electron drops to a lower level shell, it loses energy and leads to the emission of X-rays. Because the excess energy for each element is unique, the EDS for detecting and measuring characteristic X-rays can identify elements in a sample.

EDS and STEM imaging are combined to obtain both EDS spectra and spatial distributions of components. In recent years, subangstrom high-brightness probe size can be produced by spherical aberration-corrected electron microscopy, which is also possible to possess larger spaced pole pieces than conventional electron microscopy and new silicon

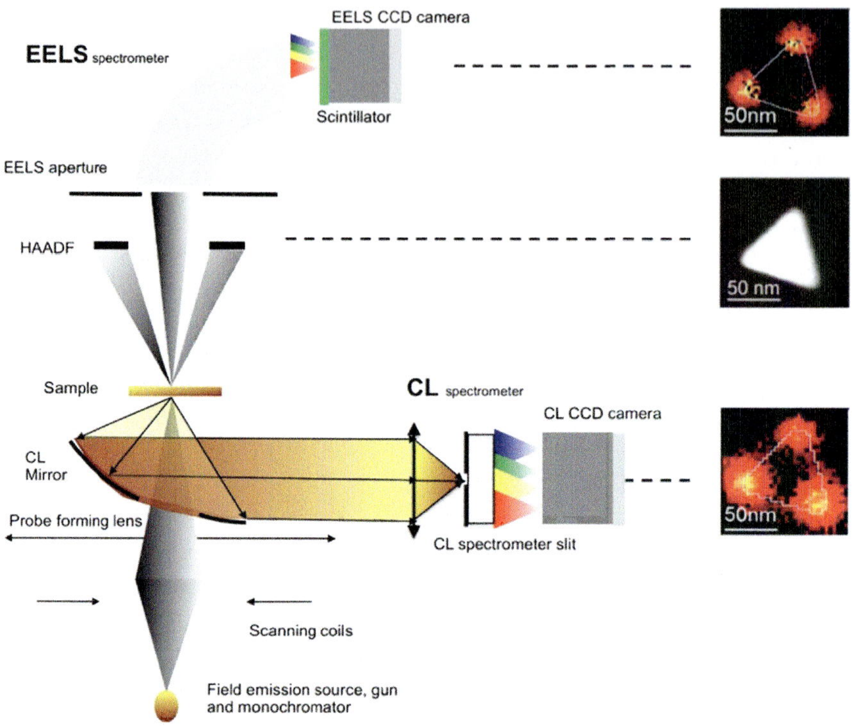

Figure 4.14 ***STEM-EELS characterization of surface plasma oscillations in low-dimensional structures.*** It is associated with STEM-ADF images and CL maps (Colliex et al., 2016). *ADF*, Annular dark field; *CL*, cathodoluminescence; *EELS*, electron energy loss spectroscopy; *STEM*, scanning TEM.

drift detectors with larger areas and acquisition angles of more than 0.9 sr. Therefore it is feasible to obtain nanoscale or atom-resolved compositional distributions in a few minutes.

Heavier elements, such as metals, have stronger EDS signals, low spectral backgrounds, and high signal-to-noise ratios, so EDS is suitable for characterizing the elemental distribution of metals in alloy catalyst nanoparticles. For example, it can be used to determine whether they are solid solution or phase-separated core/shell alloys. This technique can help to understand the relationships among the structure, catalytic activity, and thermal stability of catalysts (Wu et al., 2011; Zhang et al., 2011).

4.5.2.3 Secondary electron energy spectroscopy

Secondary electrons are emitted when a material is excited by incident electrons. Analyzing the secondary electron energy distribution, known as SEES, allows a rapid analysis of the surface chemical of materials. However, 2D materials with atomic-level thickness are usually grown or placed on a substrate; in the majorities of cases, the structural characterization of 2D materials inevitably includes signals from the substrate as well as from their interactions. Thus how to extract intrinsic properties of 2D materials from an energy spectrum with high background and low signal intensity is a major challenge.

SEES is capable of obtaining information on the number of layers, composition, and cleanliness of 2D materials (Hiura et al., 2010). Traditional SEES analyses are based on physical modeling methods, which take into account the excitation modes of the electrons in the inner shell layer, and the energy spectrum is processed by splitting the characteristic peaks (Lander, 1953; Von Koch, 1970). Da and Ding et al. proposed the "white electron" spectrum, mainly including secondary electrons, which took the spectrum as a whole and can fully explore the structural information in it. In electrical property measurements, the four-probe method can eliminate the effect of contact resistance and obtain the intrinsic conductive properties of materials. Inspired by the four-point probe method for resistance measurement, Da et al. (2017) proposed the "virtual substrate method" to obtain the intrinsic electron−electron interactions of 2D materials without the influence of underlying substrate signals from four interrelated measurements. Using this method, they measured the electrons in transmitted spectra associated with the number of layers of 2D materials such as graphene and MoS_2 on different substrates and quantitatively calculated the inelastic mean free path.

4.5.3 Magnetic resonance spectroscopy

Both the nuclear and electron spin undergo Zeeman energy level splitting as well as electromagnetic wave resonance absorption and emission in magnetic fields. These form the basis of NMR spectroscopy and EPR spectroscopy, respectively, which characterize the magnetic properties of materials. Consider the Hamiltonian in a magnetic field B_0 for a system consisting of one electron spin (S) and l nuclear spins (I):

$$H_0 = H_{EZ} + H_{NZ} + H_{HF} + H_{NQ} + H_{NN} + H_{ZFS}$$

$$H_0 = \mu_e B_0 g S/h - \mu_n \sum_{k=1}^{l} g_{n,k} B_0 I_k/h + \sum_{k=1}^{l} S A_k I_k + \Sigma_{I_k > \frac{1}{2}} I_k Q_k I_k$$

$$+ \Sigma_{i \neq k} I_i d_{ik} I_k + SDS$$

The six terms are the electron Zeeman, nuclear Zeeman, hyperfine, nuclear quadrupole, nuclear–nuclear, and the zero-field splitting interactions (Roessler & Salvadori, 2018). It can be seen that EPR and NMR correspond to the first term in the Hamiltonian quantities for the electron and the second term for the quantized energy levels of the atomic nucleus spin in the magnetic field, respectively. In addition, the third-term hyperfine interaction coupling electron and nuclear spins is the basis for quantum computers to measure, read, and store spin states. Due to the Pauli incompatibility principle, nuclei and electrons as fermions tend to have opposite combinations of spins, and thus only nuclei with an odd number of nucleons and unpaired electrons are NMR and EPR active, respectively.

Both electron and nuclear Zeeman interactions have a proportionality constant, the *g*-factor, which is affected by environment, leading to shifts in resonance energies and frequencies. This is the basic principle of NMR and EPR to characterize the chemical environment. The electron Bohr magneton $\left(\mu_{\mathrm{B}} = \frac{eh}{2m_{\mathrm{e}}}\right)$ is larger than the nuclear magneton $\left(\mu_{\mathrm{n}} = \frac{eh}{2m_{\mathrm{p}}}\right)$ because the electron mass is much lighter than the nucleus. The Zeeman splitting energy of the electron spin is higher than that of the nucleus in a magnetic field of the same strength; thus EPR uses a weaker magnetic field (≈ 0.1 T), a higher resonance frequency ($\approx$ GHz), a higher signal strength, and a higher detection sensitivity (μM).

NMR and EPR show a great advantage over optical and electron spectroscopies, that is, the magnetic field has a greater depth of penetration than light and electrons. While optical and electron spectroscopies provide more information on the surface structure of materials, magnetic spectroscopy can provide information on the bulk, which can nondestructively characterize electrochemical energy storage and energy conversion devices, such as batteries, capacitors, and fuel cells, under actual working conditions. This is important for understanding the working and destruction mechanisms of the devices (Blanc et al., 2013).

Grey's group at the University of Cambridge performed in situ studies of secondary metal-ion batteries such as silicon anodes (Key et al., 2009), lithium metal electrodes (Bhattacharyya et al., 2010), lithium–sulfur batteries (See et al., 2014), and sodium-ion batteries (Stratford et al., 2017) by using NMR. They observed the changes in the ^{7}Li peak position of silicon electrodes in an actual working lithium-ion battery, suggesting that the silicon clusters and isolated silicon atoms were embedded in a lithium matrix during the first discharge, and the Si clusters were broken apart at the end of the discharge. It was found that spontaneous reaction of lithium silicide with the electrolyte resulted in self-discharge and potential capacity loss (Key et al., 2009).

Metals have a skin effect on the radio frequency signals used in the NMR techniques, and the skin depth is related to the density of metal, the magnetic permeability, and the frequency of electromagnetic wave. This effect can be used to distinguish between the bulk lithium-metal electrodes and the tiny dendritic structures of lithium formed during charging and discharging. Such dendrites are one of the main dangers to the safety of lithium metal batteries and can lead to short circuits or even fires. The intensity of the ^{7}Li peak of bulk metal is proportional to its area, while the ^{7}Li signal of lithium dendrites or mossy structures is proportional to their volume or mass. Through the changes of ^{7}Li signals of lithium dendrites in two kinds of ionic liquids with a number of times of charge—discharge cycles, Bhattacharyya et al. (2010) found that the ionic liquid electrolyte can suppress the growth of lithium dendrites.

4.6 Summary

Nowadays, transistors can be sized down to 5 nm, while catalysts can also be engineered to the single-atom level. As low-dimensional materials and devices are used in an even wider range of applications and ever smaller scales, it is critical to understand their structure—property relationships at the atomic scale. Just as TEM has the capability to see individual atoms, STM and AFM can touch and move individual atoms, respectively. Spectroscopic analyses are also able to provide information on the energy of lattice vibrations from the nuclear spin to the electron spin, and from inner-electron binding energies to valence-electron bandgaps and energy distributions. Notably, modern characterization techniques are complemented by the development of low-dimensional materials and devices.

Regarding prospective approaches to structural characterization of low-dimensional material structures, we believe that it is necessary to systematically characterize low-dimensional devices with suitable characterization means, to establish processing—structure—property relationships and go deeper into the quantum information level.

First of all, it is advised to choose appropriate means of characterization according to the purpose. TEM and SPM can analyze individual atoms one by one, but at the same time the sampling rate is inevitably limited, EELS can provide a wealth of information from the phonon to the photon to the inner layer of the electron binding energy, but high-energy electron beam will inevitably cause irradiation damage. In general, the higher the spatial resolution and the more focused the energy, the more severe the damage to the sample and the higher the "cleanliness" of the sample. Therefore we need to consider what the purpose of the characterization is then to choose the appropriate means of characterization accordingly.

Second, low-dimensional materials and devices should be deliberately characterized in a systematic way. Taking lithium-ion batteries as an example, it involves several processes and evolutions. Spatially, they include the atomic scale of the intercalation and alloying reaction, the electrode material—electrolyte interfacial reaction, the volume expansion and stress at the particle level, the lithium-ion consumption at the pack-level, and the heat dissipation management of the macroscopic battery modules. Temporally, they include the micro-processes of the electron transport, ion diffusion, and interfacial reaction, the charging and discharging cycle, and battery cycle life in several orders of magnitude. This requires the establishment of a long-term monitoring and characterization system that links coupled multiple space and time scales.

Once again, it is necessary to understand a processing—structure—property relationship. By combining various characterization techniques to form a multidimensional data association space, we might avoid bias from "blind men feeling an elephant." By developing in situ characterization techniques to obtain real-time structural changes in the process of material growth, work, and failure, we might establish a direct relationship between them. With the association of multidimensional big data, we might fully explore the implicit relationships by making use of big data analysis and machine learning techniques.

Finally, it is urgently needed to develop characterization methods for quantum materials. Currently, the structural characterization of low-dimensional materials is more of a "deterministic" information measurement of atomic positions and electron energies. Quantum materials can be regarded as new materials with even lower dimensions than low-dimensional materials. Better tools make agood job. The structural characterization of quantum materials and the analysis and manipulation of quantum information will be the important directions for future development.

References

Argyle, B. E., & McCord, J. G. (2000). New laser illumination method for Kerr microscopy. *Journal of Applied Physics*, *87*(9), 6487–6489. Available from https://doi.org/10.1063/1.372746, http://scitation.aip.org/content/aip/journal/jap.

Bachilo, S. M., Strano, M. S., Kittrell, C., Hauge, R. H., Smalley, R. E., & Weisman, R. B. (2002). Structure-assigned optical spectra of single-walled carbon nanotubes. *Science (New York, N.Y.)*, *298*(5602), 2361–2366. Available from https://doi.org/10.1126/science.1078727.

Bartels, L., Meyer, G., & Rieder, K.-H. (1997). Controlled vertical manipulation of single CO molecules with the scanning tunneling microscope: A route to chemical contrast. *Applied Physics Letters*, *71*(2), 213–215. Available from https://doi.org/10.1063/1.119503.

Barwick, B., Park, H. S., Kwon, O.-H., Baskin, J. S., & Zewail, A. H. (2008). 4D imaging of transient structures and morphologies in ultrafast electron microscopy. *Science (New York, N.Y.)*, *322*(5905), 1227–1231. Available from https://doi.org/10.1126/science.1164000.

Bhattacharyya, R., Key, B., Chen, H., Best, A. S., Hollenkamp, A. F., & Grey, C. P. (2010). In situ NMR observation of the formation of metallic lithium microstructures in lithium batteries. *Nature Materials*, *9*(6), 504–510. Available from https://doi.org/10.1038/nmat2764.

Binnig, G., Quate, C. F., & Gerber, C. (1986). Atomic force microscope. *Physical Review Letters*, *56*(9), 930–933. Available from https://doi.org/10.1103/physrevlett.56.930.

Binnig, G., & Rohrer, H. (1987). Scanning tunneling microscopy—From birth to adolescence. *Reviews of Modern Physics*, *59*(3), 615–625. Available from https://doi.org/10.1103/revmodphys.59.615.

Binnig, G., Rohrer, H., Gerber, C., & Weibel, E. (1982). Surface studies by scanning tunneling microscopy. *Physical Review Letters*, *49*(1), 57–61. Available from https://doi.org/10.1103/physrevlett.49.57.

Blanc, F., Leskes, M., & Grey, C. P. (2013). In situ solid-state NMR spectroscopy of electrochemical cells: Batteries, supercapacitors, and fuel cells. *Accounts of Chemical Research*, *46*(9), 1952–1963. Available from https://doi.org/10.1021/ar400022u.

Böller, B., Durner, K. M., & Wintterlin, J. (2019). The active sites of a working Fischer–Tropsch catalyst revealed by operando scanning tunnelling microscopy. *Nature Catalysis.*, *2*(11), 1027–1034. Available from https://doi.org/10.1038/s41929-019-0360-1, https://www.nature.com/natcatal/.

Bonnell, D. A., Basov, D. N., Bode, M., Diebold, U., Kalinin, S. V., Madhavan, V., Novotny, L., Salmeron, M., Schwarz, U. D., & Weiss, P. S. (2012). Imaging physical phenomena with local probes: From electrons to photons. *Reviews of Modern Physics*, *84*(3), 1343–1381. Available from https://doi.org/10.1103/RevModPhys.84.1343, http://oai.aps.org/filefetch?identifier = 10.1103/RevModPhys.84.1343&component = fulltext&description = markup&format = xml, United States.

Chen, C. C., Zhu, C., White, E. R., Chiu, C. Y., Scott, M. C., Regan, B. C., Marks, L. D., Huang, Y., & Miao, J. (2013). Three-dimensional imaging of dislocations in a nanoparticle at atomic resolution. *Nature*, *496*(7443), 74–77. Available from https://doi.org/10.1038/nature12009.

Chen, Y. (2012). Studies on the electronic structures of three-dimensional topological insulators by angle resolved photoemission spectroscopy. *Frontiers of Physics.*, 7(2), 175–192. Available from https://doi.org/10.1007/s11467-011-0197-9.

Cheng, L., Yang, L., Zeng, R., Ming, Y., Da, B., & Ding, Z. (2021). Theoretical perspective of atomic resolution secondary electron imaging. *The Journal of Physical Chemistry C.*, *125*(19), 10458–10472. Available from https://doi.org/10.1021/acs.jpcc.1c01382.

Chung, H. T., Cullen, D. A., Higgins, D., Sneed, B. T., Holby, E. F., More, K. L., & Zelenay, P. (2017). Direct atomic-level insight into the active sites of a high-performance PGM-free ORR catalyst. *Science (New York, N.Y.)*, *357*(6350), 479–484. Available from https://doi.org/10.1126/science.aan2255, http://science.sciencemag.org/content/sci/357/6350/479.full.pdf.

Colliex, C., Kociak, M., & Stéphan, O. (2016). Electron energy loss spectroscopy imaging of surface plasmons at the nanometer scale. *Ultramicroscopy*, *162*, A1–A24. Available from https://doi.org/10.1016/j.ultramic.2015.11.012.

Cowley, J. M., & Iijima, S. (1976). *The direct imaging of crystal structures* (pp. 123–136). Springer Science and Business Media LLC. Available from 10.1007/978-3-642-66196-9_6.

Cumings, J., Zettl, A., McCartney, M. R., & Spence, J. C. H. (2002). Electron holography of field-emitting carbon nanotubes. *Physical Review Letters*, *88*(5). Available from https://doi.org/10.1103/physrevlett.88.056804.

Da, B., Liu, J., Yamamoto, M., Ueda, Y., Watanabe, K., Cuong, N. T., Li, S., Tsukagoshi, K., Yoshikawa, H., Iwai, H., Tanuma, S., Guo, H., Gao, Z., Sun, X., & Ding, Z. (2017). Virtual substrate method for nanomaterials characterization. *Nature Communications.*, *8*(1). Available from https://doi.org/10.1038/ncomms15629.

Danz, T., Domröse, T., & Ropers, C. (2021). Ultrafast nanoimaging of the order parameter in a structural phase transition. *Science (New York, N.Y.)*, *371*(6527), 371–374. Available from https://doi.org/10.1126/science.abd2774, https://science.sciencemag.org/content/sci/371/6527/371.full.pdf.

De Jonge, N., & Ross, F. M. (2011). Electron microscopy of specimens in liquid. *Nature Nanotechnology*, *6*(11), 695–704. Available from https://doi.org/10.1038/nnano.2011.161, http://www.nature.com/nnano/index.html.

De Oteyza, D. G., Gorman, P., Chen, Y. C., Wickenburg, S., Riss, A., Mowbray, D. J., Etkin, G., Pedramrazi, Z., Tsai, H. Z., Rubio, A., Crommie, M. F., & Fischer, F. R. (2013). Direct imaging of covalent bond structure in single-molecule chemical reactions. *Science (New York, N.Y.)*, *340*(6139), 1434–1437. Available from https://doi.org/10.1126/science.1238187, http://www.sciencemag.org/content/340/6139/1434.full.pdf.

Dean, C. R., Young, A. F., Meric, I., Lee, C., Wang, L., Sorgenfrei, S., Watanabe, K., Taniguchi, T., Kim, P., Shepard, K. L., & Hone, J. (2010). Boron nitride substrates for high-quality graphene electronics. *Nature Nanotechnology.*, *5*(10), 722–726. Available from https://doi.org/10.1038/nnano.2010.172.

DeRita, L., Resasco, J., Dai, S., Boubnov, A., Thang, H. V., Hoffman, A. S., Ro, I., Graham, G. W., Bare, S. R., Pacchioni, G., Pan, X., & Christopher, P. (2019). Structural evolution of atomically dispersed Pt catalysts dictates reactivity. *Nature Materials*, *18*(7), 746–751. Available from https://doi.org/10.1038/s41563-019-0349-9.

Ding, K., Gulec, A., Johnson, A. M., Schweitzer, N. M., Stucky, G. D., Marks, L. D., & Stair, P. C. (2015). Identification of active sites in CO oxidation and water-gas shift over supported Pt catalysts. *Science (New York, N.Y.)*, *350*(6257), 189–192. Available from https://doi.org/10.1126/science.aac6368, http://www.sciencemag.org/content/350/6257/189.full.pdf.

Dresselhaus, M. S., Jorio, A., Hofmann, M., Dresselhaus, G., & Saito, R. (2010). Perspectives on carbon nanotubes and graphene Raman spectroscopy. *Nano Letters*, *10* (3), 751–758. Available from https://doi.org/10.1021/nl904286r.

Dunin-Borkowski, R. E., Kasama, T., Wei, A., Tripp, S. L., Hytch, M. J., Snoeck, E., Harrison, R. J., & Putnis, A. (2004). Off-axis electron holography of magnetic nanowires and chains, rings, and planar arrays of magnetic nanoparticles. *Microscopy Research and Technique*, *64*(5–6), 390–402. Available from https://doi.org/10.1002/jemt.20098, http://onlinelibrary.wiley.com/journal/10.1002/(ISSN)1097-0029.

Eigler, D. M., & Schweizer, E. K. (1990). Positioning single atoms with a scanning tunnelling microscope. *Nature, 344*(6266), 524–526. Available from https://doi.org/10.1038/344524a0.

Fang, S., Wen, Y., Allen, C. S., Ophus, C., Han, G. G. D., Kirkland, A. I., Kaxiras, E., & Warner, J. H. (2019). Atomic electrostatic maps of 1D channels in 2D semiconductors using 4D scanning transmission electron microscopy. *Nature Communications., 10*(1). Available from https://doi.org/10.1038/s41467-019-08904-9, http://www.nature.com/ncomms/index.html.

Einstein, A. (1905). Concerning an heuristic point of view towards the emission and transformation of light. *Annalen der Physik, 8,* 132–148.

Fang, Z., Li, X., Shi, W., Li, Z., Guo, Y., Chen, Q., Peng, L., & Wei, X. (2020). Interlayer binding energy of hexagonal MoS_2 as determined by an in situ peeling-to-fracture method. *The Journal of Physical Chemistry C., 124*(42), 23419–23425. Available from https://doi.org/10.1021/acs.jpcc.0c06828.

Feynman, R. (1960). There's plenty of room at the bottom. *Engineering and Science, 23*(5), 22–36. Available from https://resolver.caltech.edu/CaltechES:23.5.1960Bottom.

Findlay, S. D., Shibata, N., Sawada, H., Okunishi, E., Kondo, Y., & Ikuhara, Y. (2010). Dynamics of annular bright field imaging in scanning transmission electron microscopy. *Ultramicroscopy, 110*(7), 903–923. Available from https://doi.org/10.1016/j.ultramic.2010.04.004.

Friedrich, H., De Jongh, P. E., Verkleij, A. J., & De Jong, K. P. (2009). Electron tomography for heterogeneous catalysts and related nanostructured materials. *Chemical Reviews, 109*(5), 1613–1629. Available from https://doi.org/10.1021/cr800434tNetherlands, http://pubs.acs.org/doi/pdfplus/10.1021/cr800434t.

Gao, W., Addiego, C., Wang, H., Yan, X., Hou, Y., Ji, D., Heikes, C., Zhang, Y., Li, L., Huyan, H., Blum, T., Aoki, T., Nie, Y., Schlom, D. G., Wu, R., & Pan, X. (2019). Real-space charge-density imaging with sub-ångström resolution by four-dimensional electron microscopy. *Nature, 575*(7783), 480–484. Available from https://doi.org/10.1038/s41586-019-1649-6.

Gault, B., Moody, M. P., Cairney, J. M., & Ringer, S. P. (2012). Atom probe crystallography. *Materials Today, 15*(9), 378–386. Available from https://doi.org/10.1016/S1369-7021(12)70164-5.

Giessibl, F. J. (1995). Atomic resolution of the silicon (111)-(7 × 7) surface by atomic force microscopy. *Science (New York, N. Y.), 267*(5194), 68–71. Available from https://doi.org/10.1126/science.267.5194.68.

Giessibl, F. J. (2003). Advances in atomic force microscopy. *Reviews of Modern Physics., 75*(3), 949–983. Available from https://doi.org/10.1103/RevModPhys.75.949.

González, J. C., Hernández, J. C., López-Haro, M., del Río, E., Delgado, J. J., Hungría, A. B., Trasobares, S., Bernal, S., Midgley, P. A., & Calvino, J. J. (2009). 3D characterization of gold nanoparticles supported on heavy metal oxide catalysts by HAADF-STEM electron tomography. *Angewandte Chemie International Edition, 48*(29), 5313–5315. Available from https://doi.org/10.1002/anie.200901308.

Gross, L., Mohn, F., Moll, N., Liljeroth, P., & Meyer, G. (2009). The chemical structure of a molecule resolved by atomic force microscopy. *Science (New York, N.Y.), 325*(5944), 1110–1114. Available from https://doi.org/10.1126/science.1176210.

Gu, L., Zhu, C., Li, H., Yu, Y., Li, C., Tsukimoto, S., Maier, J., & Ikuhara, Y. (2011). Direct observation of lithium staging in partially delithiated $LiFePO_4$ at atomic resolution. *Journal of the American Chemical Society, 133*(13), 4661–4663. Available from https://doi.org/10.1021/ja109412x.

Haider, M., Rose, H., Uhlemann, S., Schwan, E., Kabius, B., & Urban, K. (1998). A spherical-aberration-corrected 200kV transmission electron microscope. *Ultramicroscopy, 75*(1), 53–60. Available from https://doi.org/10.1016/s0304-3991(98)00048-5.

Halbertal, D., Cuppens, J., Shalom, M. B., Embon, L., Shadmi, N., Anahory, Y., Naren, H. R., Sarkar, J., Uri, A., Ronen, Y., Myasoedov, Y., Levitov, L. S., Joselevich, E., Geim, A. K., & Zeldov, E. (2016). Nanoscale thermal imaging of dissipation in quantum systems. *Nature*, *539*(7629), 407–410. Available from https://doi.org/10.1038/nature19843, http://www.nature.com/nature/index.html.

Hansen, T. W., Wagner, J. B., Hansen, P. L., Dahl, S., Topsøe, H., & Jacobsen, C. J. H. (2001). Atomic-resolution in situ transmission electron microscopy of a promoter of a heterogeneous catalyst. *Science (New York, N.Y.)*, *294*(5546), 1508–1510. Available from https://doi.org/10.1126/science.1064399.

Harada, K., Matsuda, T., Bonevich, J., Igarashi, M., Kondo, S., Pozzi, G., Kawabe, U., & Tonomura, A. (1992). Real-time observation of vortex lattices in a superconductor by electron microscopy. *Nature*, *360*(6399), 51–53. Available from https://doi.org/10.1038/360051a0.

Haruta, M., Yamada, N., Kobayashi, T., & Iijima, S. (1989). Gold catalysts prepared by coprecipitation for low-temperature oxidation of hydrogen and of carbon monoxide. *Journal of Catalysis.*, *115*(2), 301–309. Available from https://doi.org/10.1016/0021-9517(89)90034-1.

Helveg, S., López-Cartes, C., Sehested, J., Hansen, P. L., Clausen, B. S., Rostrup-Nielsen, J. R., Abild-Pedersen, F., & Nørskov, J. K. (2004). Atomic-scale imaging of carbon nanofibre growth. *Nature*, *427*(6973), 426–429. Available from https://doi.org/10.1038/nature02278.

Hill, H. M., Rigosi, A. F., Rim, K. T., Flynn, G. W., & Heinz, T. F. (2016). Band alignment in MoS_2/WS_2 transition metal dichalcogenide heterostructures probed by scanning tunneling microscopy and spectroscopy. *Nano Letters*, *16*(8), 4831–4837. Available from https://doi.org/10.1021/acs.nanolett.6b01007, http://pubs.acs.org/journal/nalefd.

Hiura, H., Miyazaki, H., & Tsukagoshi, K. (2010). Determination of the number of graphene layers: Discrete distribution of the secondary electron intensity stemming from individual graphene layers. *Applied Physics Express*, *3*(9). Available from https://doi.org/10.1143/apex.3.095101.

Hla, S. W., Bartels, L., Meyer, G., & Rieder, K. H. (2000). Inducing all steps of a chemical reaction with the scanning tunneling microscope tip: Towards single molecule engineering. *Physical Review Letters*, *85*(13), 2777–2780. Available from https://doi.org/10.1103/PhysRevLett.85.2777.

Hofmann, S., Sharma, R., Wirth, C. T., Cervantes-Sodi, F., Ducati, C., Kasama, T., Dunin-Borkowski, R. E., Drucker, J., Bennett, P., & Robertson, J. (2008). Ledge-flow-controlled catalyst interface dynamics during Si nanowire growth. *Nature Materials*, 7(5), 372–375. Available from https://doi.org/10.1038/nmat2140, http://www.nature.com/nmat/.

Holler, M., Guizar-Sicairos, M., Tsai, E. H. R., Dinapoli, R., Müller, E., Bunk, O., Raabe, J., & Aeppli, G. (2017). High-resolution non-destructive three-dimensional imaging of integrated circuits. *Nature*, *543*(7645), 402–406. Available from https://doi.org/10.1038/nature21698, http://www.nature.com/nature/index.html.

Hong, Y. L., Liu, Z., Wang, L., Zhou, T., Ma, W., Xu, C., Feng, S., Chen, L., Chen, M. L., Sun, D. M., Chen, X. Q., Cheng, H. M., & Ren, W. (2020). Chemical vapor deposition of layered two-dimensional $MoSi_2N_4$ materials. *Science (New York, N.Y.)*, *369*(6504), 670–674. Available from https://doi.org/10.1126/science.abb7023, https://science.sciencemag.org/content/369/6504/670/tab-pdf.

Hsia, F. C., Tang, D. M., Jevasuwan, W., Fukata, N., Zhou, X., Mitome, M., Bando, Y., Nordling, T. E. M., & Golberg, D. (2019). Realization and direct observation of five normal and parametric modes in silicon nanowire resonators by: In situ transmission electron microscopy. *Nanoscale Advances, 1*(5), 1784−1790. Available from https://doi.org/10.1039/c8na00373d, pubs.rsc.org/en/journals/journalissues/na?_ga2.190536939.1555337663.1552312502-1364180372.1550481316#!issueidna001002&typecurrent&issnonline2516-0230.

Hsieh, D., Xia, Y., Qian, D., Wray, L., Dil, J. H., Meier, F., Osterwalder, J., Patthey, L., Checkelsky, J. G., Ong, N. P., Fedorov, A. V., Lin, H., Bansil, A., Grauer, D., Hor, Y. S., Cava, R. J., & Hasan, M. Z. (2009). A tunable topological insulator in the spin helical Dirac transport regime. *Nature, 460*(7259), 1101−1105. Available from https://doi.org/10.1038/nature08234.

Hulva, J., Meier, M., Bliem, R., Jakub, Z., Kraushofer, F., Schmid, M., Diebold, U., Franchini, C., & Parkinson, G. S. (2021). Unraveling CO adsorption on model single-atom catalysts. *Science (New York, N.Y.), 371*(6527), 375−379. Available from https://doi.org/10.1126/science.abe5757.

Iijima, S. (1971). High-resolution electron microscopy of crystal lattice of titanium-niobium oxide. *Journal of Applied Physics, 42*(13), 5891−5893. Available from https://doi.org/10.1063/1.1660042.

Iijima, S. (1991). Helical microtubules of graphitic carbon. *Nature, 354*(6348), 56−58. Available from https://doi.org/10.1038/354056a0.

Iijima, S., & Ichihashi, T. (1993). Single-shell carbon nanotubes of 1-nm diameter. *Nature, 363*(6430), 603−605. Available from https://doi.org/10.1038/363603a0.

Inada, H., Su, D., Egerton, R. F., Konno, M., Wu, L., Ciston, J., Wall, J., & Zhu, Y. (2011). Atomic imaging using secondary electrons in a scanning transmission electron microscope: Experimental observations and possible mechanisms. *Ultramicroscopy, 111* (7), 865−876. Available from https://doi.org/10.1016/j.ultramic.2010.10.002.

Ishikawa, R., Findlay, S. D., Seki, T., Sánchez-Santolino, G., Kohno, Y., Ikuhara, Y., & Shibata, N. (2018). Direct electric field imaging of graphene defects. *Nature Communications., 9*(1). Available from https://doi.org/10.1038/s41467-018-06387-8, http://www.nature.com/ncomms/index.html.

Ishikawa, R., Okunishi, E., Sawada, H., Kondo, Y., Hosokawa, F., & Abe, E. (2011). Direct imaging of hydrogen-atom columns in a crystal by annular bright-field electron microscopy. *Nature Materials, 10*(4), 278−281. Available from https://doi.org/10.1038/nmat2957.

Jiang, Z., Henriksen, E. A., Tung, L. C., Wang, Y.-J., Schwartz, M. E., Han, M. Y., Kim, P., & Stormer, H. L. (2007). Infrared spectroscopy of Landau levels of graphene. *Physical Review Letters, 98*(19). Available from https://doi.org/10.1103/physrevlett.98.197403.

Jorio, A., Saito, R., Dresselhaus, G., & Dresselhaus, M. S. (2011). *Raman spectroscopy: From graphite to sp2 nanocarbons. Raman spectroscopy in graphene related systems* (pp. 73−101). Wiley. Available from 10.1002/9783527632695.ch4.

Jorio, A., Saito, R., Hafner, J. H., Lieber, C. M., Hunter, M., McClure, T., Dresselhaus, G., & Dresselhaus, M. S. (2001). Structural (n, m) determination of isolated single-wall carbon nanotubes by resonant Raman scattering. *Physical Review Letters, 86*(6), 1118−1121. Available from https://doi.org/10.1103/physrevlett.86.1118.

Kaiser, K., Scriven, L. M., Schulz, F., Gawel, P., Gross, L., & Anderson, H. L. (2019). An sp-hybridized molecular carbon allotrope, cyclo[18]carbon. *Science (New York, N.Y.), 365*(6459), 1299−1301. Available from https://doi.org/10.1126/science.aay1914, https://science.sciencemag.org/content/365/6459/1299/tab-pdf.

Kaiser, U., Schwarz, A., & Wiesendanger, R. (2007). Magnetic exchange force microscopy with atomic resolution. *Nature, 446*(7135), 522−525. Available from https://doi.org/10.1038/nature05617.

Kawai, S., Benassi, A., Gnecco, E., Söde, H., Pawlak, R., Feng, X., Müllen, K., Passerone, D., Pignedoli, C. A., Ruffieux, P., Fasel, R., & Meyer, E. (2016).

Superlubricity of graphene nanoribbons on gold surfaces. *Science (New York, N.Y.), 351*(6276), 957–961. Available from https://doi.org/10.1126/science.aad3569.

Kawai, S., Nakatsuka, S., Hatakeyama, T., Pawlak, R., Meier, T., Tracey, J., Meyer, E., & Foster, A. S. (2018). Multiple heteroatom substitution to graphene nanoribbon. *Science Advances., 4*(4). Available from https://doi.org/10.1126/sciadv.aar7181.

Key, B., Bhattacharyya, R., Morcrette, M., Seznéc, V., Tarascon, J.-M., & Grey, C. P. (2009). Real-time NMR investigations of structural changes in silicon electrodes for lithium-ion batteries. *Journal of the American Chemical Society, 131*(26), 9239–9249. Available from https://doi.org/10.1021/ja8086278.

Kim, K., Jeong, W., Lee, W., & Reddy, P. (2012). Ultra-high vacuum scanning thermal microscopy for nanometer resolution quantitative thermometry. *ACS Nano., 6*(5), 4248–4257. Available from https://doi.org/10.1021/nn300774n.

Kimoto, K., Asaka, T., Nagai, T., Saito, M., Matsui, Y., & Ishizuka, K. (2007). Element-selective imaging of atomic columns in a crystal using STEM and EELS. *Nature, 450* (7170), 702–704. Available from https://doi.org/10.1038/nature06352.

Kimouche, A., Ervasti, M. M., Drost, R., Halonen, S., Harju, A., Joensuu, P. M., Sainio, J., & Liljeroth, P. (2015). Ultra-narrow metallic armchair graphene nanoribbons. *Nature Communications., 6*. Available from https://doi.org/10.1038/ncomms10177, http://www.nature.com/ncomms/index.html.

Kneipp, K., Wang, Y., Kneipp, H., Perelman, L. T., Itzkan, I., Dasari, R. R., & Feld, M. S. (1997). Single molecule detection using surface-enhanced Raman scattering (SERS). *Physical Review Letters, 78*(9), 1667–1670. Available from https://doi.org/10.1103/physrevlett.78.1667.

König, M., Wiedmann, S., Brüne, C., Roth, A., Buhmann, H., Molenkamp, L. W., Qi, X.-L., & Zhang, S.-C. (2007). Quantum spin Hall insulator state in HgTe quantum wells. *Science (New York, N.Y.), 318*(5851), 766–770. Available from https://doi.org/10.1126/science.1148047.

Krivanek, O. L., Chisholm, M. F., Nicolosi, V., Pennycook, T. J., Corbin, G. J., Dellby, N., Murfitt, M. F., Own, C. S., Szilagyi, Z. S., Oxley, M. P., Pantelides, S. T., & Pennycook, S. J. (2010). Atom-by-atom structural and chemical analysis by annular dark-field electron microscopy. *Nature, 464*(7288), 571–574. Available from https://doi.org/10.1038/nature08879.

Krumeich, F., Müller, E., Wepf, R. A., & Nesper, R. (2011). Characterization of catalysts in an aberration-corrected scanning transmission electron microscope. *Journal of Physical Chemistry C, 115*(4), 1080–1083. Available from https://doi.org/10.1021/jp105997h.

Lander, J. J. (1953). Auger peaks in the energy spectra of secondary electrons from various materials. *Physical Review., 91*(6), 1382–1387. Available from https://doi.org/10.1103/physrev.91.1382.

Lee, C., Wei, X., Kysar, J. W., & Hone, J. (2008). Measurement of the elastic properties and intrinsic strength of monolayer graphene. *Science (New York, N.Y.), 321*(5887), 385–388. Available from https://doi.org/10.1126/science.1157996.

Lee, H. J., & Ho, W. (1999). Single-bond formation and characterization with a scanning tunneling microscope. *Science (New York, N.Y.), 286*(5445), 1719–1722. Available from https://doi.org/10.1126/science.286.5445.1719.

Li, H., Wang, J., Gao, S., Chen, Q., Peng, L., Liu, K., & Wei, X. (2017). Superlubricity between MoS_2 monolayers. *Advanced Materials., 29*(27). Available from https://doi.org/10.1002/adma.201701474, http://www3.interscience.wiley.com/journal/119030556/issue.

Li, H., Wei, X., Wu, G., Gao, S., Chen, Q., & Peng, L.-M. (2018). Interlayer electrical resistivity of rotated graphene layers studied by in-situ scanning electron microscopy. *Ultramicroscopy, 193*, 90–96. Available from https://doi.org/10.1016/j.ultramic.2018.06.015.

Li, J., He, Y., Han, Y., Liu, K., Wang, J., Li, Q., Fan, S., & Jiang, K. (2012). Direct identification of metallic and semiconducting single-walled carbon nanotubes in scanning electron microscopy. *Nano Letters, 12*(8), 4095–4101. Available from https://doi.org/10.1021/nl301561f.

Li, Z., Sun, S., Li, Z. A., Zhang, M., Cao, G., Tian, H., Yang, H., & Li, J. (2017). Ultrafast structural dynamics of boron nitride nanotubes studied using transmitted electrons. *Nanoscale., 9*(35), 13313–13319. Available from https://doi.org/10.1039/c7nr04162d, http://www.rsc.org/publishing/journals/NR/Index.asp.

Liao, H. G., Cui, L., Whitelam, S., & Zheng, H. (2012). Real-time imaging of Pt_3Fe nanorod growth in solution. *Science (New York, N.Y.), 336*(6084), 1011–1014. Available from https://doi.org/10.1126/science.1219185, http://www.sciencemag.org/content/336/6084/1011.full.pdf.

Liao, M., Jiang, S., Hu, C., Zhang, R., Kuang, Y., Zhu, J., Zhang, Y., & Dong, Z. (2016). Tip-enhanced Raman spectroscopic imaging of individual carbon nanotubes with subnanometer resolution. *Nano Letters, 16*(7), 4040–4046. Available from https://doi.org/10.1021/acs.nanolett.6b00533.

Lin, P. A., Gomez-Ballesteros, J. L., Burgos, J. C., Balbuena, P. B., Natarajan, B., & Sharma, R. (2017). Direct evidence of atomic-scale structural fluctuations in catalyst nanoparticles. *Journal of Catalysis., 349*, 149–155. Available from https://doi.org/10.1016/j.jcat.2017.03.009, http://www.elsevier.com/inca/publications/store/6/2/2/8/5/8/index.htt.

Linck, M., Hartel, P., Uhlemann, S., Kahl, F., Müller, H., Zach, J., Biskupek, J., Niestadt, M., Kaiser, U., & Haider, M. (2016). Performance of the SALVE-microscope: Atomic-resolution TEM imaging at 20 kV. *Microscopy and Microanalysis., 22*(S3), 878–879. Available from https://doi.org/10.1017/S1431927616005237.

Liu, Z. K., Jiang, J., Zhou, B., Wang, Z. J., Zhang, Y., Weng, H. M., Prabhakaran, D., Mo, S.-K., Peng, H., Dudin, P., Kim, T., Hoesch, M., Fang, Z., Dai, X., Shen, Z. X., Feng, D. L., Hussain, Z., & Chen, Y. L. (2014). A stable three-dimensional topological Dirac semimetal Cd_3As_2. *Nature Materials, 13*(7), 677–681. Available from https://doi.org/10.1038/nmat3990.

Liu, Z. K., Zhou, B., Zhang, Y., Wang, Z. J., Weng, H. M., Prabhakaran, D., Mo, S.-K., Shen, Z. X., Fang, Z., Dai, X., Hussain, Z., & Chen, Y. L. (2014). Discovery of a three-dimensional topological Dirac semimetal, Na_3Bi. *Science (New York, N.Y.), 343*(6173), 864–867. Available from https://doi.org/10.1126/science.1245085.

Lorenz, U. J., & Zewail, A. H. (2014). Observing liquid flow in nanotubes by 4D electron microscopy. *Science (New York, N.Y.), 344*(6191), 1496–1500. Available from https://doi.org/10.1126/science.1253618, http://www.sciencemag.org/content/344/6191/1496.full.pdf.

Lu, Q., Chen, C. C., & De Lozanne, A. (1997). Observation of magnetic domain behavior in colossal magnetoresistive materials with a magnetic force microscope. *Science (New York, N.Y.), 276*(5321), 2006–2008. Available from https://doi.org/10.1126/science.276.5321.2006.

Lu, X., Grobis, M., Khoo, K. H., Louie, S. G., & Crommie, M. F. (2003). Spatially mapping the spectral density of a single C_{60} molecule. *Physical Review Letters, 90*(9). Available from https://doi.org/10.1103/physrevlett.90.096802.

Lundgren, E., Zhang, C., Merte, L. R., Shipilin, M., Blomberg, S., Hejral, U., Zhou, J., Zetterberg, J., & Gustafson, J. (2017). Novel in situ techniques for studies of model catalysts. *Accounts of Chemical Research, 50*(9), 2326–2333. Available from https://doi.org/10.1021/acs.accounts.7b00281, http://pubs.acs.org/journal/achre4.

Lyeo, H. K., Khajetoorians, A. A., Shi, L., Pipe, K. P., Ram, R. J., Shakouri, A., & Shih, C. K. (2004). Profiling the thermoelectric power of semiconductor junctions with nanometer resolution. *Science (New York, N.Y.), 303*(5659), 816–818. Available from https://doi.org/10.1126/science.1091600.

Mak, K. F., Lee, C., Hone, J., Shan, J., & Heinz, T. F. (2010). Atomically thin MoS_2: A new direct-gap semiconductor. *Physical Review Letters*, *105*(13). Available from https://doi.org/10.1103/PhysRevLett.105.136805, http://oai.aps.org/oai?verb = GetRecord&Identifier = oai:aps.org:PhysRevLett.105.136805&metadataPrefix = oai_apsmeta_2, United States.

Mamin, H. J., Kim, M., Sherwood, M. H., Rettner, C. T., Ohno, K., Awschalom, D. D., & Rugar, D. (2013). Nanoscale nuclear magnetic resonance with a nitrogen-vacancy spin sensor. *Science (New York, N.Y.)*, *339*(6119), 557–560. Available from https://doi.org/10.1126/science.1231540.

Martin, Y., & Wickramasinghe, H. K. (1987). Magnetic imaging by "force microscopy" with 1000 Å resolution. *Applied Physics Letters*, *50*(20), 1455–1457. Available from https://doi.org/10.1063/1.97800.

Martinez-Castro, J., Mauro, D., Pásztor, Á., Gutiérrez-Lezama, I., Scarfato, A., Morpurgo, A. F., & Renner, C. (2018). Scanning tunneling microscopy of an air sensitive dichalcogenide through an encapsulating layer. *Nano Letters*, *18*(11), 6696–6702. Available from https://doi.org/10.1021/acs.nanolett.8b01978.

McVitie, S., Chapman, J. N., Zhou, L., Heyderman, L. J., & Nicholson, W. A. P. (1995). In-situ magnetising experiments using coherent magnetic imaging in TEM. *Journal of Magnetism and Magnetic Materials*, *148*(1–2), 232–236. Available from https://doi.org/10.1016/0304-8853(95)00219-7.

Mecklenburg, M., Hubbard, W. A., White, E. R., Dhall, R., Cronin, S. B., Aloni, S., & Regan, B. C. (2015). Nanoscale temperature mapping in operating microelectronic devices. *Science (New York, N.Y.)*, *347*(6222), 629–632. Available from https://doi.org/10.1126/science.aaa2433, http://www.sciencemag.org/content/347/6222/629.full.pdf.

Meier, F., Zhou, L., Wiebe, J., & Wiesendanger, R. (2008). Revealing magnetic interactions from single-atom magnetization curves. *Science (New York, N.Y.)*, *320*(5872), 82–86. Available from https://doi.org/10.1126/science.1154415.

Menges, F., Mensch, P., Schmid, H., Riel, H., Stemmer, A., & Gotsmann, B. (2016). Temperature mapping of operating nanoscale devices by scanning probe thermometry. *Nature Communications.*, 7(1). Available from https://doi.org/10.1038/ncomms10874.

Mermin, N. D., & Wagner, H. (1966). Absence of ferromagnetism or antiferromagnetism in one- or two-dimensional isotropic Heisenberg models. *Physical Review Letters*, *17*(22), 1133–1136. Available from https://doi.org/10.1103/physrevlett.17.1133.

Meyer, J. C., Geim, A. K., Katsnelson, M. I., Novoselov, K. S., Booth, T. J., & Roth, S. (2007). The structure of suspended graphene sheets. *Nature*, *446*(7131), 60–63. Available from https://doi.org/10.1038/nature05545.

Miao, J., Ercius, P., & Billinge, S. J. L. (2016). Atomic electron tomography: 3D structures without crystals. *Science (New York, N.Y.)*, *353*(6306). Available from https://doi.org/10.1126/science.aaf2157, http://science.sciencemag.org/content/sci/353/6306/aaf2157.full.pdf.

Motta, A., Berning, M., Boergens, K. M., Staffler, B., Beining, M., Loomba, S., Hennig, P., Wissler, H., & Helmstaedter, M. (2019). Dense connectomic reconstruction in layer 4 of the somatosensory cortex. *Science (New York, N.Y.)*, *366*(6469). Available from https://doi.org/10.1126/science.aay3134, https://science.sciencemag.org/content/366/6469/eaay3134/tab-pdf.

Muller, D. A. (2009). Structure and bonding at the atomic scale by scanning transmission electron microscopy. *Nature Materials*, *8*(4), 263–270. Available from https://doi.org/10.1038/nmat2380, http://www.nature.com/nmat/.

Muller, D. A., Fitting Kourkoutis, L., Murfitt, M., Song, J. H., Hwang, H. Y., Silcox, J., Dellby, N., & Krivanek, O. L. (2008). Atomic-scale chemical imaging of composition and bonding by aberration-corrected microscopy. *Science (New York, N.Y.), 319* (5866), 1073–1076. Available from https://doi.org/10.1126/science.1148820.

Müller, D. J., & Dufrêne, Y. F. (2008). Atomic force microscopy as a multifunctional molecular toolbox in nanobiotechnology. *Nature Nanotechnology., 3*(5), 261–269. Available from https://doi.org/10.1038/nnano.2008.100, http://www.nature.com/nnano/index.html.

Murray, C. B., Norris, D. J., & Bawendi, M. G. (1993). Synthesis and characterization of nearly monodisperse CdE (E = sulfur, selenium, tellurium) semiconductor nanocrystallites. *Journal of the American Chemical Society, 115*(19), 8706–8715. Available from https://doi.org/10.1021/ja00072a025.

Néel, N., Kröger, J., Limot, L., Frederiksen, T., Brandbyge, M., & Berndt, R. (2007). Controlled contact to a C_{60} molecule. *Physical Review Letters, 98*(6). Available from https://doi.org/10.1103/physrevlett.98.065502.

Neumann, P., Jakobi, I., Dolde, F., Burk, C., Reuter, R., Waldherr, G., Honert, J., Wolf, T., Brunner, A., Shim, J. H., Suter, D., Sumiya, H., Isoya, J., & Wrachtrup, J. (2013). High-precision nanoscale temperature sensing using single defects in diamond. *Nano Letters, 13*(6), 2738–2742. Available from https://doi.org/10.1021/nl401216y.

Novoselov, K. S., Geim, A. K., Morozov, S. V., Jiang, D., Zhang, Y., Dubonos, S. V., Grigorieva, I. V., & Firsov, A. A. (2004). Electric field effect in atomically thin carbon films. *Science (New York, N.Y.), 306*(5696), 666–669. Available from https://doi.org/10.1126/science.1102896.

Odom, T. W., Huang, J. L., Kim, P., & Lieber, C. M. (1998). Atomic structure and electronic properties of single-walled carbon nanotubes. *Nature, 391*(6662), 62–64. Available from https://doi.org/10.1038/34145.

Okazaki, K., Ota, Y., Kotani, Y., Malaeb, W., Ishida, Y., Shimojima, T., Kiss, T., Watanabe, S., Chen, C.-T., Kihou, K., Lee, C. H., Iyo, A., Eisaki, H., Saito, T., Fukazawa, H., Kohori, Y., Hashimoto, K., Shibauchi, T., Matsuda, Y., ... Shin, S. (2012). Octet-line node structure of superconducting order parameter in KFe_2As_2. *Science (New York, N.Y.), 337*(6100), 1314–1317. Available from https://doi.org/10.1126/science.1222793.

Okunishi, E., Ishikawa, I., Sawada, H., Hosokawa, F., Hori, M., & Kondo, Y. (2009). Visualization of light elements at ultrahigh resolution by STEM annular bright field microscopy. *Microscopy and Microanalysis., 15*(S2), 164–165. Available from https://doi.org/10.1017/s1431927609093891.

Ouyang, M., Huang, J. L., Cheung, C. L., & Lieber, C. M. (2001). Atomically resolved single-walled carbon nanotube intramolecular junctions. *Science (New York, N.Y.), 291* (5501), 97–100. Available from https://doi.org/10.1126/science.291.5501.97.

Park, J., Elmlund, H., Ercius, P., Yuk, J. M., Limmer, D. T., Chen, Q., Kim, K., Han, S. H., Weitz, D. A., Zettl, A., & Paul Alivisatos, A. (2015). 3D structure of individual nanocrystals in solution by electron microscopy. *Science (New York, N.Y.), 349*(6245), 290–295. Available from https://doi.org/10.1126/science.aab1343.

Park, J. Y., Maier, S., Hendriksen, B., & Salmeron, M. (2010). Sensing current and forces with SPM. *Materials Today., 13*(10), 38–45. Available from https://doi.org/10.1016/S1369-7021(10)70185-1.

Pascual, J. I., Méndez, J., Gómez-Herrero, J., Baró, A. M., García, N., & Vu Thien Binh. (1993). Quantum contact in gold nanostructures by scanning tunneling microscopy. *Physical Review Letters, 71*(12), 1852–1855. Available from https://doi.org/10.1103/physrevlett.71.1852.

Pavliček, N., & Gross, L. (2017). Generation, manipulation and characterization of molecules by atomic force microscopy. *Nature Reviews Chemistry., 1*(1). Available from https://doi.org/10.1038/s41570-016-0005.

Pedrero, C., Waku, T., & Iglesia, E. (2005). Oxidation of CO in H_2−CO mixtures catalyzed by platinum: alkali effects on rates and selectivity. *Journal of Catalysis.*, *233*(1), 242−255. Available from https://doi.org/10.1016/j.jcat.2005.04.005.

Pennycook, S. J., & Howie, A. (2006). Study of single-electron excitations by electron microscopy II. Cathodoluminescence image contrast from localized energy transfers. *Philosophical Magazine A.*, *41*(6), 809−827. Available from https://doi.org/10.1080/01418618008243890.

Picher, M., Lin, P. A., Gomez-Ballesteros, J. L., Balbuena, P. B., & Sharma, R. (2014). Nucleation of graphene and its conversion to single-walled carbon nanotubes. *Nano Letters*, *14*(11), 6104−6108. Available from https://doi.org/10.1021/nl501977b, http://pubs.acs.org/journal/nalefd.

Puntes, V. F., Gorostiza, P., Aruguete, D. M., Bastus, N. G., & Alivisatos, A. P. (2004). Collective behaviour in two-dimensional cobalt nanoparticle assemblies observed by magnetic force microscopy. *Nature Materials*, *3*(4), 263−268. Available from https://doi.org/10.1038/nmat1094, http://www.nature.com/nmat/.

Qiao, B., Wang, A., Yang, X., Allard, L. F., Jiang, Z., Cui, Y., Liu, J., Li, J., & Zhang, T. (2011). Single-atom catalysis of CO oxidation using Pt1/FeO_x. *Nature Chemistry.*, *3*(8), 634−641. Available from https://doi.org/10.1038/nchem.1095.

Qin, H., Niu, Y., Meng, R., Lin, X., Lai, R., Fang, W., & Peng, X. (2014). Single-dot spectroscopy of zinc-blende CdSe/CdS core/shell nanocrystals: Nonblinking and correlation with ensemble measurements. *Journal of the American Chemical Society*, *136*(1), 179−187. Available from https://doi.org/10.1021/ja4078528.

Qin, R., Liu, K., Wu, Q., & Zheng, N. (2020). Surface coordination chemistry of atomically dispersed metal catalysts. *Chemical Reviews*, *120*(21), 11810−11899. Available from https://doi.org/10.1021/acs.chemrev.0c00094.

Qu, L., Yu, W. W., & Peng, X. (2004). In situ observation of the nucleation and growth of CdSe nanocrystals. *Nano Letters*, *4*(3), 465−469. Available from https://doi.org/10.1021/nl035211r.

Rasool, H. I., Ophus, C., Klug, W. S., Zettl, A., & Gimzewski, J. K. (2013). Measurement of the intrinsic strength of crystalline and polycrystalline graphene. *Nature Communications.*, *4*. Available from https://doi.org/10.1038/ncomms3811, http://www.nature.com/ncomms/index.html.

Reparaz, J. S., Chavez-Angel, E., Wagner, M. R., Graczykowski, B., Gomis-Bresco, J., Alzina, F., & Sotomayor Torres, C. M. (2014). A novel contactless technique for thermal field mapping and thermal conductivity determination: Two-laser Raman thermometry. *Review of Scientific Instruments.*, *85*(3). Available from https://doi.org/10.1063/1.4867166.

Repp, J., Meyer, G., Stojković, S. M., Gourdon, A., & Joachim, C. (2005). Molecules on insulating films: Scanning-tunneling microscopy imaging of individual molecular orbitals. *Physical Review Letters*, *94*(2). Available from https://doi.org/10.1103/PhysRevLett.94.026803Switzerland, http://scitation.aip.org/getpdf/servlet/GetPDFServlet?filetype = pdf&id = PRLTAO000094000002026803000001&idtype = cvips.

Roessler, M. M., & Salvadori, E. (2018). Principles and applications of EPR spectroscopy in the chemical sciences. *Chemical Society Reviews*, *47*(8), 2534−2553. Available from https://doi.org/10.1039/c6cs00565a, http://pubs.rsc.org/en/journals/journal/cs.

Rossi, N., Braakman, F. R., Cadeddu, D., Vasyukov, D., Tütüncüoglu, G., Fontcuberta, I. M. A., & Poggio, M. (2017). Vectorial scanning force microscopy using a nanowire sensor. *Nature Nanotechnology.*, *12*(2), 150−155. Available from https://doi.org/10.1038/nnano.2016.189, http://www.nature.com/nnano/index.html.

Rugar, D., Budakian, R., Mamin, H. J., & Chui, B. W. (2004). Single spin detection by magnetic resonance force microscopy. *Nature*, *430*(6997), 329−332. Available from https://doi.org/10.1038/nature02658.

Sadat, S., Tan, A., Chua, Y. J., & Reddy, P. (2010). Nanoscale thermometry using point contact thermocouples. *Nano Letters, 10*(7), 2613–2617. Available from https://doi.org/10.1021/nl101354e.

Sáenz, J. J., García, N., Grütter, P., Meyer, E., Heinzelmann, H., Wiesendanger, R., Rosenthaler, L., Hidber, H. R., & Güntherodt, H.-J. (1987). Observation of magnetic forces by the atomic force microscope. *Journal of Applied Physics, 62*(10), 4293–4295. Available from https://doi.org/10.1063/1.339105.

Sawada, H., Sasaki, T., Hosokawa, F., & Suenaga, K. (2015). Atomic-resolution STEM imaging of graphene at low voltage of 30kV with resolution enhancement by using large convergence angle. *Physical Review Letters, 114*(16). Available from https://doi.org/10.1103/physrevlett.114.166102.

Scherzer, O. (1947). Spharische und chromatische Korrektur von Elektronen-Linsen. *Optik., 2*, 114–132.

Schmidt, R., Lazo, C., Hölscher, H., Pi, U. H., Caciuc, V., Schwarz, A., Wiesendanger, R., & Heinze, S. (2009). Probing the magnetic exchange forces of iron on the atomic scale. *Nano Letters, 9*(1), 200–204. Available from https://doi.org/10.1021/nl802770x.

Scott, M. C., Chen, C. C., Mecklenburg, M., Zhu, C., Xu, R., Ercius, P., Dahmen, U., Regan, B. C., & Miao, J. (2012). Electron tomography at 2.4-ångström resolution. *Nature, 483*(7390), 444–447. Available from https://doi.org/10.1038/nature10934.

See, K. A., Leskes, M., Griffin, J. M., Britto, S., Matthews, P. D., Emly, A., Van Der Ven, A., Wright, D. S., Morris, A. J., Grey, C. P., & Seshadri, R. (2014). Ab initio structure search and in situ^{7}Li NMR studies of discharge products in the Li-S battery system. *Journal of the American Chemical Society, 136*(46), 16368–16377. Available from https://doi.org/10.1021/ja508982p, http://pubs.acs.org/journal/jacsat.

Seifert, T. S., Kovarik, S., Juraschek, D. M., Spaldin, N. A., Gambardella, P., & Stepanow, S. (2020). Longitudinal and transverse electron paramagnetic resonance in a scanning tunneling microscope. *Science Advances., 6*(40). Available from https://doi.org/10.1126/sciadv.abc5511, https://advances.sciencemag.org/content/6/40/eabc5511/tab-pdf.

Senga, R., Suenaga, K., Barone, P., Morishita, S., Mauri, F., & Pichler, T. (2019). Position and momentum mapping of vibrations in graphene nanostructures. *Nature, 573*(7773), 247–250. Available from https://doi.org/10.1038/s41586-019-1477-8.

Shibata, N., Findlay, S. D., Kohno, Y., Sawada, H., Kondo, Y., & Ikuhara, Y. (2012). Differential phase-contrast microscopy at atomic resolution. *Nature Physics., 8*(8), 611–615. Available from https://doi.org/10.1038/nphys2337.

Siegbahn, K. (1982). Electron spectroscopy for atoms, molecules, and condensed matter. *Reviews of Modern Physics., 54*(3), 709–728. Available from https://doi.org/10.1103/revmodphys.54.709.

Soudi, A., Dawson, R. D., & Gu, Y. (2011). Quantitative heat dissipation characteristics in current-carrying GaN nanowires probed by combining scanning thermal microscopy and spatially resolved Raman spectroscopy. *ACS Nano., 5*(1), 255–262. Available from https://doi.org/10.1021/nn102818s.

Splendiani, A., Sun, L., Zhang, Y., Li, T., Kim, J., Chim, C.-Y., Galli, G., & Wang, F. (2010). Emerging photoluminescence in monolayer MoS_2. *Nano Letters, 10*(4), 1271–1275. Available from https://doi.org/10.1021/nl903868w.

Stratford, J. M., Mayo, M., Allan, P. K., Pecher, O., Borkiewicz, O. J., Wiaderek, K. M., Chapman, K. W., Pickard, C. J., Morris, A. J., & Grey, C. P. (2017). Investigating sodium storage mechanisms in tin anodes: A combined pair distribution function analysis, density functional theory, and solid-state NMR approach. *Journal of the American Chemical Society, 139*(21), 7273–7286. Available from https://doi.org/10.1021/jacs.7b01398, http://pubs.acs.org/journal/jacsat.

Suenaga, K., & Koshino, M. (2010). Atom-by-atom spectroscopy at graphene edge. *Nature*, *468*(7327), 1088–1090. Available from https://doi.org/10.1038/nature09664.

Suenaga, K., Tencé, M., Mory, C., Colliex, C., Kato, H., Okazaki, T., Shinohara, H., Hirahara, K., Bandow, S., & Iijima, S. (2000). Element-selective single atom imaging. *Science (New York, N.Y.)*, *290*(5500), 2280–2282. Available from https://doi.org/10.1126/science.290.5500.2280.

Sugimoto, Y., Pou, P., Abe, M., Jelinek, P., Pérez, R., Morita, S., & Custance, Ó. (2007). Chemical identification of individual surface atoms by atomic force microscopy. *Nature*, *446*(7131), 64–67. Available from https://doi.org/10.1038/nature05530.

Tang, D. M., Kvashnin, D. G., Najmaei, S., Bando, Y., Kimoto, K., Koskinen, P., Ajayan, P. M., Yakobson, B. I., Sorokin, P. B., Lou, J., & Golberg, D. (2014). Nanomechanical cleavage of molybdenum disulphide atomic layers. *Nature Communications.*, *5*. Available from https://doi.org/10.1038/ncomms4631, http://www.nature.com/ncomms/index.html.

Tang, D. M., Ren, C. L., Wang, M. S., Wei, X., Kawamoto, N., Liu, C., Bando, Y., Mitome, M., Fukata, N., & Golberg, D. (2012). Mechanical properties of Si nanowires as revealed by in situ transmission electron microscopy and molecular dynamics simulations. *Nano Letters*, *12*(4), 1898–1904. Available from https://doi.org/10.1021/nl204282y.

Tang, D. M., Yin, L. C., Li, F., Liu, C., Yu, W. J., Hou, P. X., Wu, B., Lee, Y. H., Ma, X. L., & Cheng, H. M. (2010). Carbon nanotube-clamped metal atomic chain. *Proceedings of the National Academy of Sciences of the United States of America.*, *107*(20), 9055–9059. Available from https://doi.org/10.1073/pnas.0914970107China, http://www.pnas.org/content/107/20/9055.full.pdf.

Tang, D.-M., Erohin, S. V., Kvashnin, D. G., Demin, V. A., Cretu, O., Jiang, S., Zhang, L., Hou, P.-X., Chen, G., Futaba, D. N., Zheng, Y., Xiang, R., Zhou, X., Hsia, F.-C., Kawamoto, N., Mitome, M., Nemoto, Y., Uesugi, F., Takeguchi, M., ... Golberg, D. (2021). Semiconductor nanochannels in metallic carbon nanotubes by thermomechanical chirality alteration. *Science (New York, N.Y.)*, *374*(6575), 1616–1620. Available from https://doi.org/10.1126/science.abi8884.

Tang, Z., Li, X., Wu, G., Gao, S., Chen, Q., Peng, L., & Wei, X. (2016). Whole-journey nanomaterial research in an electron microscope: From material synthesis, composition characterization, property measurements to device construction and tests. *Nanotechnology*, *27*(48). Available from https://doi.org/10.1088/0957-4484/27/48/485710.

Tanji, T., Ru, Q., & Tonomura, A. (1996). Differential microscopy by conventional electron off-axis holography. *Applied Physics Letters*, *69*(18), 2623–2625. Available from https://doi.org/10.1063/1.117555.

Textor, M., & de Jonge, N. (2018). Strategies for preparing graphene liquid cells for transmission electron microscopy. *Nano Letters*, *18*(6), 3313–3321. Available from https://doi.org/10.1021/acs.nanolett.8b01366.

Teyssieux, D., Thiery, L., & Cretin, B. (2007). Near-infrared thermography using a charge-coupled device camera: Application to microsystems. *Review of Scientific Instruments.*, *78*(3). Available from https://doi.org/10.1063/1.2714040.

Therrien, A. J., Hensley, A. J. R., Marcinkowski, M. D., Zhang, R., Lucci, F. R., Coughlin, B., Schilling, A. C., McEwen, J. S., & Sykes, E. C. H. (2018). An atomic-scale view of single-site Pt catalysis for low-temperature CO oxidation. *Nature Catalysis.*, *1*(3), 192–198. Available from https://doi.org/10.1038/s41929-018-0028-2, https://www.nature.com/natcatal/.

Tison, Y., Lin, H., Lagoute, J., Repain, V., Chacon, C., Girard, Y., Rousset, S., Henrard, L., Zheng, B., Susi, T., Kauppinen, E. I., Ducastelle, F., & Loiseau, A. (2013).

Identification of nitrogen dopants in single-walled carbon nanotubes by scanning tunneling microscopy. *ACS Nano.*, 7(8), 7219–7226. Available from https://doi.org/10.1021/nn4026146.

Tonomura, A., Matsuda, T., Endo, J., Arii, T., & Mihama, K. (1980). Direct observation of fine structure of magnetic domain walls by electron holography. *Physical Review Letters*, *44*(21), 1430–1433. Available from https://doi.org/10.1103/physrevlett.44.1430.

Tonomura, A., Matsuda, T., Tanabe, H., Osakabe, N., Endo, J., Fukuhara, A., Shinagawa, K., & Fujiwara, H. (1982). Electron holography technique for investigating thin ferromagnetic films. *Physical Review B.*, *25*(11), 6799–6804. Available from https://doi.org/10.1103/physrevb.25.6799.

Tuxen, A., Kibsgaard, J., Gøbel, H., Lægsgaard, E., Topsøe, H., Lauritsen, J. V., & Besenbacher, F. (2010). Size threshold in the dibenzothiophene adsorption on MoS_2 nanoclusters. *ACS Nano.*, *4*(8), 4677–4682. Available from https://doi.org/10.1021/nn1011013.

Vasyukov, D., Anahory, Y., Embon, L., Halbertal, D., Cuppens, J., Neeman, L., Finkler, A., Segev, Y., Myasoedov, Y., Rappaport, M. L., Huber, M. E., & Zeldov, E. (2013). A scanning superconducting quantum interference device with single electron spin sensitivity. *Nature Nanotechnology.*, *8*(9), 639–644. Available from https://doi.org/10.1038/nnano.2013.169.

Verlhac, B., Bachellier, N., Garnier, L., Ormaza, M., Abufager, P., Robles, R., Bocquet, M.-L., Ternes, M., Lorente, N., & Limot, L. (2019). Atomic-scale spin sensing with a single molecule at the apex of a scanning tunneling microscope. *Science (New York, N. Y.)*, *366*(6465), 623–627. Available from https://doi.org/10.1126/science.aax8222.

Voiry, D., Salehi, M., Silva, R., Fujita, T., Chen, M., Asefa, T., Shenoy, V. B., Eda, G., & Chhowalla, M. (2013). Conducting MoS_2 nanosheets as catalysts for hydrogen evolution reaction. *Nano Letters*, *13*(12), 6222–6227. Available from https://doi.org/10.1021/nl403661s.

Von Koch, C. V. (1970). Evidence of plasmons in secondary electron emission spectra. *Physical Review Letters*, *25*(12), 792–794. Available from https://doi.org/10.1103/PhysRevLett.25.792.

Wang, J., Jin, X., Liu, Z., Yu, G., Ji, Q., Wei, H., Zhang, J., Zhang, K., Li, D., Yuan, Z., Li, J., Liu, P., Wu, Y., Wei, Y., Wang, J., Li, Q., Zhang, L., Kong, J., Fan, S., & Jiang, K. (2018). Growing highly pure semiconducting carbon nanotubes by electrotwisting the helicity. *Nature Catalysis.*, *1*(5), 326–331. Available from https://doi.org/10.1038/s41929-018-0057-x.

Wang, L., Chen, J., Cox, S. J., Liu, L., Sosso, G. C., Li, N., Gao, P., Michaelides, A., Wang, E., & Bai, X. (2021). Microscopic kinetics pathway of salt crystallization in graphene nanocapillaries. *Physical Review Letters*, *126*(13). Available from https://doi.org/10.1103/PhysRevLett.126.136001, https://journals.aps.org/prl/abstract/10.1103/PhysRevLett.126.136001.

Wang, Y., Qiu, L., Zhang, L., Tang, D.-M., Ma, R., Wang, Y., Zhang, B., Ding, F., Liu, C., & Cheng, H.-M. (2020). Precise identification of the active phase of cobalt catalyst for carbon nanotube growth by in situ transmission electron microscopy. *ACS Nano.*, *14*(12), 16823–16831. Available from https://doi.org/10.1021/acsnano.0c05542.

Wang, Z., Weng, H., Wu, Q., Dai, X., & Fang, Z. (2013). Three-dimensional Dirac semimetal and quantum transport in Cd_3 As_2. *Physical Review B.*, *88*(12). Available from https://doi.org/10.1103/physrevb.88.125427.

Wang, Z. J., Dong, J., Cui, Y., Eres, G., Timpe, O., Fu, Q., Ding, F., Schloegl, R., & Willinger, M. G. (2016). Stacking sequence and interlayer coupling in few-layer graphene revealed by in situ imaging. *Nature Communications.*, 7. Available from https://doi.org/10.1038/ncomms13256, http://www.nature.com/ncomms/index.html.

Ward, E. P. W., Yates, T. J. V., Fernández, J. J., Vaughan, D. E. W., & Midgley, P. A. (2007). Three-dimensional nanoparticle distribution and local curvature of heterogeneous catalysts revealed by electron tomography. *Journal of Physical Chemistry C, 111* (31), 11501–11505. Available from https://doi.org/10.1021/jp072441b.

Watanabe, K., Nagata, T., Oh, S., Wakayama, Y., Sekiguchi, T., Volk, J., & Nakamura, Y. (2016). Arbitrary cross-section SEM-cathodoluminescence imaging of growth sectors and local carrier concentrations within micro-sampled semiconductor nanorods. *Nature Communications.*, 7(1). Available from https://doi.org/10.1038/ncomms10609.

Watanabe, K., Nagata, T., Wakayama, Y., Sekiguchi, T., Erdélyi, R., & Volk, J. (2015). Band-gap deformation potential and elasticity limit of semiconductor free-standing nanorods characterized in situ by scanning electron microscope–cathodoluminescence nanospectroscopy. *ACS Nano.*, *9*(3), 2989–3001. Available from https://doi.org/10.1021/nn507159u.

Weiss, C., Wagner, C., Temirov, R., & Tautz, F. S. (2010). Direct imaging of intermolecular bonds in scanning tunneling microscopy. *Journal of the American Chemical Society, 132*(34), 11864–11865. Available from https://doi.org/10.1021/ja104332t.

Wen, C.-Y., Tersoff, J., Hillerich, K., Reuter, M. C., Park, J. H., Kodambaka, S., Stach, E. A., & Ross, F. M. (2011). Periodically changing morphology of the growth interface in Si, Ge, and GaP nanowires. *Physical Review Letters, 107*(2). Available from https://doi.org/10.1103/physrevlett.107.025503.

Wernsdorfer, W. (2009). From micro-to nano-SQUIDs: Applications to nanomagnetism. *Superconductor Science and Technology.*, *22*(6). Available from https://doi.org/10.1088/0953-2048/22/6/064013.

Weyland, M., & Midgley, P. A. (2004). Electron tomography. *Materials Today.*, 7(12), 32–40. Available from https://doi.org/10.1016/S1369-7021(04)00569-3.

Wieghold, S., & Nienhaus, L. (2020). Probing semiconductor properties with optical scanning tunneling microscopy. *Joule.*, *4*(3), 524–538. Available from https://doi.org/10.1016/j.joule.2020.02.003.

Wiesendanger, R., Güntherodt, H.-J., Güntherodt, G., Gambino, R. J., & Ruf, R. (1990). Observation of vacuum tunneling of spin-polarized electrons with the scanning tunneling microscope. *Physical Review Letters, 65*(2), 247–250. Available from https://doi.org/10.1103/physrevlett.65.247.

Wu, B., Heidelberg, A., & Boland, J. J. (2005). Mechanical properties of ultrahigh-strength gold nanowires. *Nature Materials, 4*(7), 525–529. Available from https://doi.org/10.1038/nmat1403, http://www.nature.com/nmat/.

Wu, J., Gross, A., & Yang, H. (2011). Shape and composition-controlled platinum alloy nanocrystals using carbon monoxide as reducing agent. *Nano Letters, 11*(2), 798–802. Available from https://doi.org/10.1021/nl104094p.

Xu, J., Zhu, X., Tan, S., Zhang, Y., Li, B., Tian, Y., Shan, H., Cui, X., Zhao, A., Dong, Z., Yang, J., Luo, Y., Wang, B., & Hou, J. G. (2021). Determining structural and chemical heterogeneities of surface species at the single-bond limit. *Science (New York, N.Y.), 371*(6531), 818–822. Available from https://doi.org/10.1126/science.abd1827.

Xue, J., Sanchez-Yamagishi, J., Bulmash, D., Jacquod, P., Deshpande, A., Watanabe, K., Taniguchi, T., Jarillo-Herrero, P., & LeRoy, B. J. (2011). Scanning tunnelling microscopy and spectroscopy of ultra-flat graphene on hexagonal boron nitride. *Nature Materials, 10*(4), 282–285. Available from https://doi.org/10.1038/nmat2968.

Yang, B., Chen, G., Ghafoor, A., Zhang, Y., Zhang, Y., Zhang, Y., Luo, Y., Yang, J., Sandoghdar, V., Aizpurua, J., Dong, Z., & Hou, J. G. (2020). Sub-nanometre resolution in single-molecule photoluminescence imaging. *Nature Photonics.*, *14*(11), 693–699. Available from https://doi.org/10.1038/s41566-020-0677-y.

Yang, H., Liang, A., Chen, C., Zhang, C., Schroeter, N. B. M., & Chen, Y. (2018). Visualizing electronic structures of quantum materials by angle-resolved photoemission spectroscopy. *Nature Reviews Materials.*, *3*(9), 341–353. Available from https://doi.org/10.1038/s41578-018-0047-2.

Yang, W. C. D., Wang, C., Fredin, L. A., Lin, P. A., Shimomoto, L., Lezec, H. J., & Sharma, R. (2019). Site-selective CO disproportionation mediated by localized surface plasmon resonance excited by electron beam. *Nature Materials*, *18*(6), 614–619. Available from https://doi.org/10.1038/s41563-019-0342-3, http://www.nature.com/nmat/.

Yi, J., Zhuang, H., Zou, Q., Wu, Z., Cao, G., Tang, S., Calder, S. A., Kent, P. R. C., Mandrus, D., & Gai, Z. (2017). Competing antiferromagnetism in a quasi-2D itinerant ferromagnet: Fe_3GeTe_2. *2D Materials.*, *4*(1). Available from https://doi.org/10.1088/2053-1583/4/1/011005.

Yoon, K., Hwang, G., Chung, J., Kim, H. G., Kwon, O., Kihm, K. D., & Lee, J. S. (2014). Measuring the thermal conductivity of residue-free suspended graphene bridge using null point scanning thermal microscopy. *Carbon*, *76*, 77–83. Available from https://doi.org/10.1016/j.carbon.2014.04.051.

Yoshida, H., Takeda, S., Uchiyama, T., Kohno, H., & Homma, Y. (2008). Atomic-scale in-situ observation of carbon nanotube growth from solid state iron carbide nanoparticles. *Nano Letters*, *8*(7), 2082–2086. Available from https://doi.org/10.1021/nl080452q.

Yu, M. F., Lourie, O., Dyer, M. J., Moloni, K., Kelly, T. F., & Ruoff, R. S. (2000). Strength and breaking mechanism of multiwalled carbon nanotubes under tensile load. *Science (New York, N.Y.)*, *287*(5453), 637–640. Available from https://doi.org/10.1126/science.287.5453.637.

Yu, X. Z., Onose, Y., Kanazawa, N., Park, J. H., Han, J. H., Matsui, Y., Nagaosa, N., & Tokura, Y. (2010). Real-space observation of a two-dimensional skyrmion crystal. *Nature*, *465*(7300), 901–904. Available from https://doi.org/10.1038/nature09124.

Yuan, W., Zhu, B., Fang, K., Li, X.-Y., Hansen, T. W., Ou, Y., Yang, H., Wagner, J. B., Gao, Y., Wang, Y., & Zhang, Z. (2021). In situ manipulation of the active Au-TiO_2 interface with atomic precision during CO oxidation. *Science (New York, N.Y.)*, *371*(6528), 517–521. Available from https://doi.org/10.1126/science.abe3558.

Yuan, W., Zhu, B., Li, X. Y., Hansen, T. W., Ou, Y., Fang, K., Yang, H., Zhang, Z., Wagner, J. B., Gao, Y., & Wang, Y. (2020). Visualizing H_2O molecules reacting at TiO_2 active sites with transmission electron microscopy. *Science (New York, N.Y.)*, *367* (6476), 428–430. Available from https://doi.org/10.1126/science.aay2474, https://science.sciencemag.org/content/367/6476/428.

Zewail, A. H. (2010). Four-dimensional electron microscopy. *Science (New York, N.Y.)*, *328*(5975), 187–193. Available from https://doi.org/10.1126/science.1166135.

Zhang, D., Zhu, Y., Liu, L., Ying, X., Hsiung, C.-E., Sougrat, R., Li, K., & Han, Y. (2018). Atomic-resolution transmission electron microscopy of electron beam–sensitive crystalline materials. *Science (New York, N.Y.)*, *359*(6376), 675–679. Available from https://doi.org/10.1126/science.aao0865.

Zhang, J., Chen, P., Yuan, B., Ji, W., Cheng, Z., & Qiu, X. (2013). Real-space identification of intermolecular bonding with atomic force microscopy. *Science (New York, N.Y.)*, *342*(6158), 611–614. Available from https://doi.org/10.1126/science.1242603.

Zhang, J., Liu, X., Blume, R., Zhang, A., Schlögl, R., & Su, D. S. (2008). Surface-modified carbon nanotubes catalyze oxidative dehydrogenation of *n*-butane. *Science (New York, N.Y.)*, *322*(5898), 73–77. Available from https://doi.org/10.1126/science.1161916.

Zhang, L., Zhang, J., Kuang, Q., Xie, S., Jiang, Z., Xie, Z., & Zheng, L. (2011). Cu^{2+} -assisted synthesis of hexoctahedral Au–Pd alloy nanocrystals with high-index facets. *Journal of the American Chemical Society*, *133*(43), 17114–17117. Available from https://doi.org/10.1021/ja2063617.

Zhang, R., Zhang, Y., Dong, Z. C., Jiang, S., Zhang, C., Chen, L. G., Zhang, L., Liao, Y., Aizpurua, J., Luo, Y., Yang, J. L., & Hou, J. G. (2013). Chemical mapping of a single molecule by plasmon-enhanced Raman scattering. *Nature*, *498*(7452), 82–86. Available from https://doi.org/10.1038/nature12151.

Zhang, Y., Luo, Y., Zhang, Y., Yu, Y.-J., Kuang, Y.-M., Zhang, L., Meng, Q.-S., Luo, Y., Yang, J.-L., Dong, Z.-C., & Hou, J. G. (2016). Visualizing coherent intermolecular dipole–dipole coupling in real space. *Nature*, *531*(7596), 623–627. Available from https://doi.org/10.1038/nature17428.

Zhang, Y., Zhu, W., Hui, F., Lanza, M., Borca-Tasciuc, T., & Muñoz Rojo, M. (2020). A review on principles and applications of scanning thermal microscopy (SThM). *Advanced Functional Materials*, *30*(18), 1900892. Available from https://doi.org/10.1002/adfm.201900892.

Zhao, W., Liu, P., & Jiang, K. (2020). High-throughput methods for evaluating the homogeneity of carbon nanotubes and graphene. *Journal of Physics D: Applied Physics*, *53*(40), 403001. Available from https://doi.org/10.1088/1361-6463/ab95bb.

Zhu, Y., Inada, H., Nakamura, K., & Wall, J. (2009). Imaging single atoms using secondary electrons with an aberration-corrected electron microscope. *Nature Materials*, *8*(10), 808–812. Available from https://doi.org/10.1038/nmat2532.

CHAPTER FIVE

Novel properties of low-dimensional materials

Various properties of low-dimensional materials depend on their sizes, as schematically shown in Fig. 5.1. These effects will be discussed in detail below.

5.1 Mechanical properties of low-dimensional materials

The mechanical properties of low-dimensional materials are the basis for their application as structural materials and the foundation of reliable functional devices. In addition, strain engineering is also an important approach to enhance the electronic properties and performance of catalysts and electronic devices. Mechanical properties mainly include reversible elastic deformation, irreversible plastic deformation, fatigue, and fracture. As the size is reduced to nanometer scale, the elasticity, plasticity, and fracture have different scaling behaviors, leading to the length scale—dependent mechanical properties, even brittle-to-ductile transitions.

5.1.1 Elastic deformation

Elastic deformation is the reversible strain of materials under mechanical stress, within the elastic limit. Since elastic stress directly changes the lattice constant, it is possible to modify the band structure and thus modulate the electrical properties, known as strain engineering. This technique has been applied as a major innovation to transistors in computer central processing units to improve carrier mobility, since the 90-nm node.

The elastic modulus is determined by the strength of the chemical bonds. Since carbon nanomaterials, such as carbon nanotubes and graphene, are bonded by strong sp^2 covalent bonds, they have a high in-plane elasticity modulus, reaching 1 TPa (Lee et al., 2008). As will be introduced in the following section, compared with the bulk materials,

Low-Dimensional Materials.
DOI: https://doi.org/10.1016/B978-0-443-29046-6.00011-0

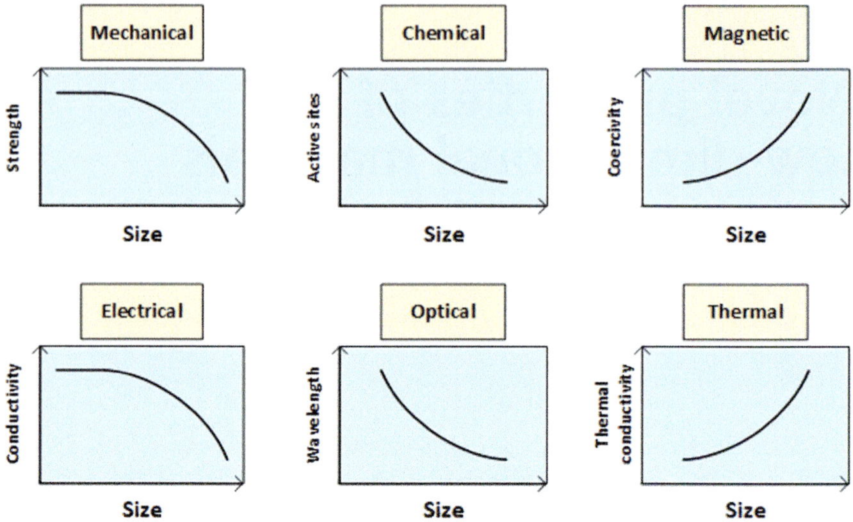

Figure 5.1 ***Size effects of low-dimensional materials' properties:*** as size decreases, fracture strength approaches theoretical strength; electrical conductivity approaches the quantized conductance of ballistic transport; the number of catalytically active sites increases exponentially; semiconductor bandgap and luminescence frequency blueshifts; magnetic coercivity decreases; and thermal conductivity is lowered by surface scattering.

low-dimensional materials typically have higher strength and therefore higher elastic deformation limit, due to the small size and defect-free structure (Zhu & Li, 2010). Lu et al. at City University of Hong Kong reported the high uniform elastic tensile strains of ~7% for single-crystal diamond bridge-shaped structures with length and width of 1 μm and 100 nm, respectively (Dang et al., 2021). The large elastic limit of the low-dimensional materials will enable them to engineer the properties and functions in a wider range compared with the bulk counterparts.

Because of the anisotropy of the crystalline structures of one- and two-dimensional (1D and 2D) materials, their elastic behaviors are different along different directions. Assuming that the length, width, and thickness of a piece of graphene are L, W, and t, respectively, its bending elasticity stiffness K as a double-clamped beam can be expressed as $K = 32YWt^3/L^3$, where Y is Young's modulus (Wolf, 2014). Since the bending stiffness is inversely proportional to the cube of the length/thickness ratio, apparently, it is much easier to bend than to elongate. The high tensile strength coupled with the low bending rigidity makes the 1D and 2D materials excellent candidates for flexible and wearable electronics.

It has been a challenge to precisely measure the elastic properties of the individual low-dimensional structures. By using the in situ transmission electron microscope (TEM) method, it is possible to establish the relation between the properties with the structures directly. We have applied an in situ TEM method to measure the high-order resonance to accurately measure Young's modulus of individual nanowires and nanotubes (Hsia et al., 2019; Zhou et al., 2019). Up to 10 resonance modes were observed on a silicon nanowire cantilever. The errors for the first- and fifth-order resonance frequencies are about 3.14% and 0.25%, respectively, from the predictions of the Euler–Bernoulli beam theory, showing the higher precision of the high-order resonance method. Using a coaxial nanowire model with a silicon core–silicon oxide surface layer, the elastic modulus of the silicon nanowire in the Choi et al. (2020) direction was calculated to be 169 GPa, which is comparable to that of the bulk silicon (Hsia et al., 2019).

In addition to the intrinsic elastic properties, by using the electron beam irradiation, it is possible to introduce defects and investigate the defect influence. As shown in Fig. 5.2, Young's modulus of boron nitride nanotubes was gradually reduced from 906 GPa of the intrinsic structure to 663 GPa due to the irradiation-induced voids (Zhou et al., 2019). Interestingly, with further increase of the defects, the modulus was kept at a high level, instead of continuously decreasing, indicating their potential applications as structure materials in harsh space environment.

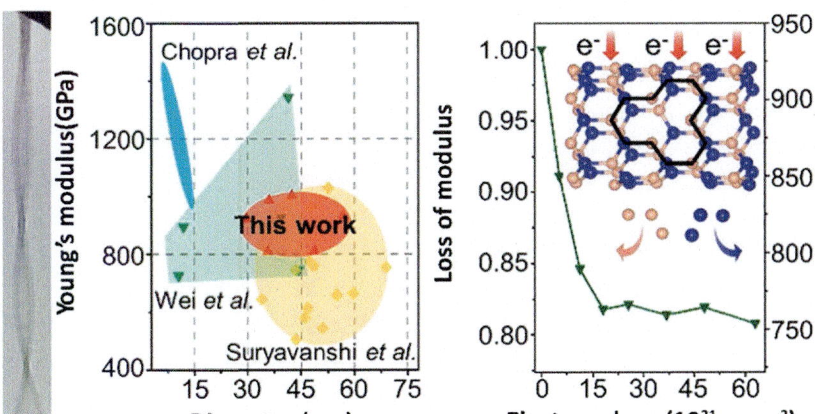

Figure 5.2 Intrinsic elastic modulus of boron nitride nanotubes and effect of defects measured by in situ transmission electron microscopy (Zhou et al., 2019).

5.1.2 Plastic deformation

When stress exceeding elastic limit is applied, irreversible plastic deformation or fracture will occur. In metals, the main mechanisms of plastic deformation are dislocations, twins, and grain boundary activities. Depending on the different stresses and temperatures, different mechanisms dominate. When the grain size reaches nanometer level, grain boundary diffusion may dominate plastic deformation. Lu et al. (2000) at the Institute of Metal Research, Chinese Academy of Sciences, found that high-purity nanocrystalline copper has an ultrahigh ductility of 5000%, attributed to the grain boundary activities.

On the other hand, for inorganic materials with covalent bonds, such as carbon materials and silicon, it is difficult to activate the dislocations at room temperature. Therefore the bulk materials of graphite, diamond, and silicon are brittle, similar to the glass. However, as the size of the materials is reduced to nanometer scale, because of the higher resistance for the propagation of cracks, under certain stress states, it becomes possible to initiate dislocations, resulting plastic deformation. By using in situ TEM manipulation, we observed that silicon nanowires are ductile under bending, while they are brittle under tension. For the carbon nanomaterials, Huang et al. (2006) found that superplastic deformation could happen for the Joule heated carbon nanotubes. As the diameter was shrunk 15 times, the length was increased by 280%. The plastic deformation mechanism was attributed to the climbing and sliding of dislocations in the 2D graphene lattice (Ding et al., 2007). Since different chiralities of the carbon nanotubes are related to each other by the dislocations, and chirality controls the electronic structures, in principle, by plastic deformation, it is possible to modify their electrical properties. Recently, by using in situ TEM manipulation and electrical measurements, we realized the feedback controlled in situ plastic deformation and successfully the controlled metal-to-semiconductor transition (Tang et al., 2018).

5.1.3 Fracture and toughness

As the stress continues to increase, materials will eventually fracture, corresponding to the breaking of chemical bonds with crack initiation and propagation. Under ideal conditions, defect-free materials undergo simultaneous chemical bond breakage under tensile stresses at the elastic limit. The theoretical fracture strength derived from the Orowan model is $\sigma_{\max} = \sqrt{\frac{E\gamma_s}{a_0}}$, where E, γ_s, and a_0 are Young's modulus, surface energy, and atomic spacing, respectively. Because of the unavoidable defects, in the Griffith model, the critical conditions for crack propagation were analyzed, and the associated fracture strength is $\sigma_c = \sqrt{\frac{2E\gamma_s}{\pi a}}$, which is inversely proportional to $\sqrt{a}$, where a is the featured size of the crack (Griffith, 1921). Because of the small size of the low-dimensional structures and the associated defects, it is expected that the fracture strength is much higher than the bulk counterparts, approaching the theoretical strength (Gao et al., 2003).

The ultrahigh fracture strength of the individual low-dimensional materials has been verified by direct mechanical tests. Hone et al. performed nanoindentation tests on monolayer graphene and found an intrinsic fracture strength of 42 N/m, close to the theoretical fracture strength of 130 GPa (Lee et al., 2008). Wang et al. (2010) measured the tensile strength of individual single-walled carbon nanotubes (SWCNTs) by in situ TEM tensile tests and found that the fracture strength was closely related with the defects and distributed between 25 and 100 GPa. Because of the general "smaller, stronger" trends of the low-dimensional materials, Zhu and Li (2010) defined the materials with the fracture strengths near their theoretical strengths ($\sim$1/10 of their elastic moduli) as "ultra-strength materials."

The mechanical properties of individual silicon nanowires were investigated by in situ TEM tensile test, and it was found that the fracture strength was close to the theoretical limit due to the size effects (Fig. 5.3) (Tang et al., 2012). In addition, the mechanical behaviors are related with the stress states. Under uniaxial tension state, the silicon nanowires are brittle, with a high tensile strength of up to 11.3 GPa. Under bending and compression stress states, plastic deformation up to 20% has been observed, due to the dislocation activities and amorphization. Such stress state-related brittle-to-ductile transition could be explained by the ratio of maximum shear and normal stresses, which is the so-called coefficient of softness (Tang et al., 2012).

In practice, the mechanical reliability depends on the fracture toughness, that is, the ability to resist crack propagation or the energy consumed during the process. Lou et al. measured the tensile properties of chemical vapor deposition grown graphene by in situ scanning electron microscope (SEM) method and found that the fracture stress was dramatically reduced due to the preexisting cracks, with a toughness of 4.0 MPa $m^{1/2}$ (Zhang et al., 2014). By using the same method, the fracture toughness of h-BN sheets was measured to be 8.7 MPa $m^{1/2}$. The higher toughness of h-BN compared with that of graphene was attributed to the deflection of the cracks during the propagation, because of the ionic-like B—N bonds (Yang et al., 2021).

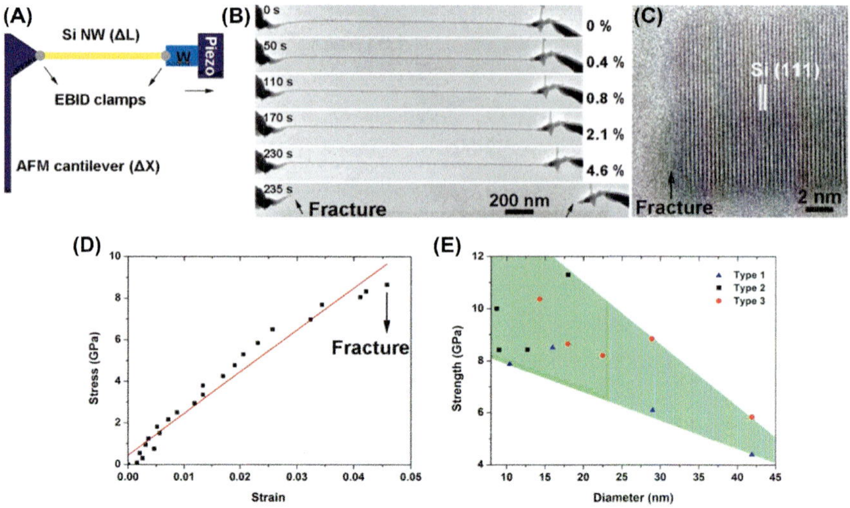

Figure 5.3 ***Tensile strength of silicon nanowires:*** (A,B) schematic and TEM images of in situ tensile test of an individual silicon nanowire; (C,D) TEM images of the tensile fracture and stress-strain curves showing brittle fracture; and (E) statistics of tensile strength versus diameter for different types of fracture, showing a "smaller, stronger" trend (Tang et al., 2012).

Fatigue is a critical mechanical property under periodic stress, for the applications such as nanoelectromechanical system (NEMS) devices. On the one hand, it is challenging to conduct the periodic loading for millions of times on the nanometer-sized low-dimensional structures. On the other hand, the high resonance frequency associated with the small size of the NEMS devices provides a way to investigate the fatigue properties. Zhang and Wei at Tsinghua University developed a noncontact acoustic resonance testing system to investigate the fatigue properties of individual carbon nanotubes with the length up to centimeter level (Bai et al., 2020). The fatigue strain limit and normalized fatigue strength corresponding to 10^7 cyclic loadings reached 0.90 and 0.95, respectively.

5.2 Electrical properties

The electrical properties of low-dimensional systems are determined by their electronic structures. As seen from Chapter 1, the electronic structure, such as the density of states, is closely related to the dimensionality and the length scale. Along with the advance of nanotechnology, emerging research directions, including molecular devices, 1D transistors, valleytronics, and twistronics based on 2D materials, have been developed. This section will introduce the electrical properties of low-dimensional materials and devices, as well as briefly discuss ion transport properties.

5.2.1 Electrical interface

As Nobel laureate Herbert Kroemer said, "The interface is the device." In electronic devices, interfaces can be classified into metal–semiconductor contacts, semiconductor heterojunctions, semiconductor–electrolyte interfaces, and metal electrode–electrolyte interfaces, and so on. These different types of electrostatic interfaces can be understood within a unified theoretical framework regarding the equilibrium of chemical potential. At equilibrium, a dynamic balance is established between diffusion current driven by chemical potential gradient and drift current determined by built-in electric field.

In three-dimensional (3D) systems, the electrical interfaces under the electrostatic equilibrium are equivalent to capacitors with 2D parallel plates. It is determined by the balance between diffusion and drift currents of charge carriers, and the potential difference is equal to the Fermi energy difference between contacted materials. When 2D materials form in-plane contacts, the interfaces become 1D. The electric field flux generated by the charge at the interface passes through the vacuum space and cannot be effectively screened (Fig. 5.4). Accordingly, the charge density decreases as $1/x$ with distance x away from the interface (Yu et al., 2016). For the electronic devices based on 1D carbon nanotubes, the interface becomes quasizero-dimensional. Calculations by Léonard and Tersoff showed that in this case, the charge density decays as $1/\ln x$. When the doping concentration is low, the thickness of depletion layer can reach micrometer scale (Leonard & Tersoff, 1999). Such dimensional effects on the electrical interfaces must be considered when designing low-dimensional transistors.

Metal–semiconductor contacts can lead to a Schottky barrier for carrier transport. In 3D systems, due to pinning effects from the surface states, the height of the Schottky barrier hardly changes with the types and Fermi levels of metals. 2D materials have no dangling bonds on their surfaces and therefore can avoid interface states and the resulting pinning effects. Liu et al. developed a direct transfer method to fabricate contacts

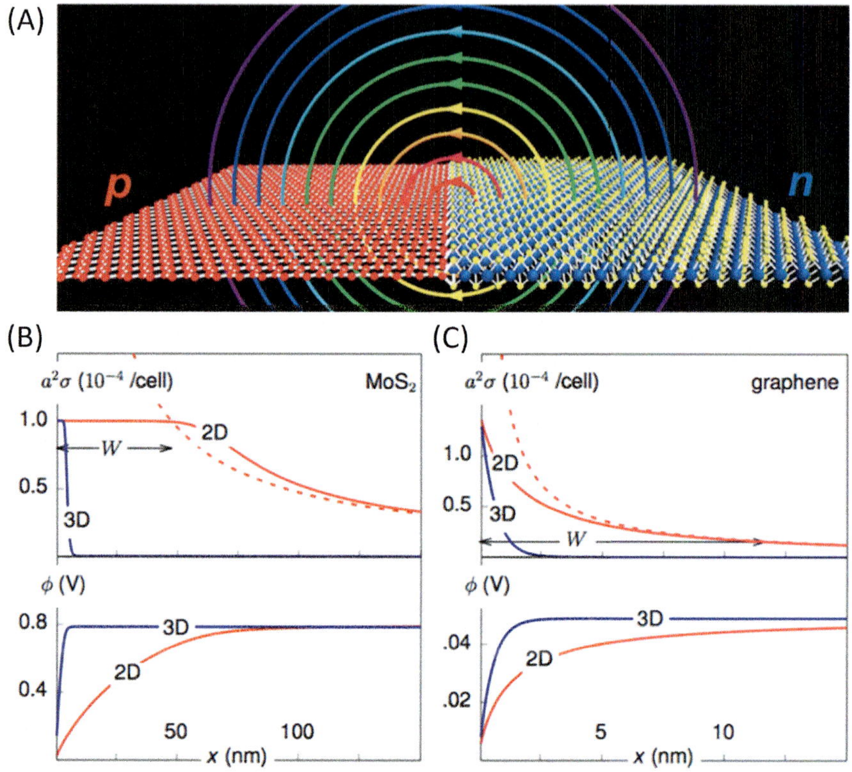

Figure 5.4 (A) Schematic diagram of one-dimensional heterojunction formed by two-dimensional materials and the generated electric field fluxes; (B,C) the variation of surface charge (top) and electrostatic potential (bottom) with distance for symmetrically doped p–n junctions of (B) MoS_2-MoS_2 and (C) graphene–graphene in two- and three-dimensional models (Yu et al., 2016).

between metals and molybdenum disulfide. Compared with the conventional electron beam evaporation, the transfer method greatly reduced the damage to the contact and channels. As a result, the Schottky barrier exhibits a linear relationship with the work function of the contact metals, approaching the Schottky–Mott limit (Liu et al., 2018).

2D materials can form heterojunctions along either in-plane or out-of-plane direction. In-plane heterojunctions include edge contacts between metal and semiconductor structures discussed above, heterophase homojunctions formed by metallic 1T (1T′) phase and semiconducting 1H phases of transition metal dichalcogenides (Cho et al., 2015), and p–n junctions formed by gate-controlled p- and n-type doping 2D semiconductors

(Pospischil et al., 2014). Vertical heterojunctions include p–n junctions consisting of different stacked 2D materials and tunnel field-effect transistors with 2D materials and highly doped bulk materials as channels and sources, respectively (Sarkar et al., 2015). Vertical stacking can introduce modulations to the periodic potential fields of the crystals, providing a new way for controlling the electronic states and transport behaviors of carriers. Stacked graphene with twisting angles is an intriguing system. At certain twisting angles, the kinetic energy of electrons can be strongly suppressed, resulting in flat bands in the band structure with significantly enhanced correlation effects. Novel phenomena emerge such as correlated insulating states and unconventional superconductivity (Cao et al., 2018a, 2018b).

5.2.2 Electron transport

The function of electronic devices is realized through controlling electron transport. Electron transport properties are closely related to dimensionality and length scale. Examples include Coulomb blockade in zero-dimensional quantum dots, ballistic and coherent transport in 1D and 2D systems, and unidirectional conduction related to the band topology of 2D systems.

In zero-dimensional systems such as clusters, quantum dots, and nanocrystals, the quantum confinement effects result in discrete energy levels and Coulomb blockade, which can be utilized to design single-electron field-effect transistors. The gate voltage controls the energy levels of quantum dots. The source–drain voltage adjusts chemical potentials of the contacts. When one unoccupied energy level of a quantum dot is within the chemical potential window of the source and drain, electrons can tunnel into the quantum dot and occupy the energy level. Due to the quantum capacitance effect, once the energy level is occupied, electron tunneling is blocked. Accordingly, electrons tunnel through the quantum dot one by one, and a single-electron transistor can be realized. In an extreme example, Fuechsle et al. demonstrated a single-electron transistor using a single phosphorus atom on a silicon surface as a quantum dot. Transport characteristics corresponding to the occupation of 0, 1, and 2 electrons were observed (Fuechsle et al., 2012).

In nanowires, the scattering of electron transport is mainly due to surface roughness. Gunawan et al. (2008) measured the electron and hole mobilities of silicon nanowires as $\sim$300 and 100 $cm^2/V\ s$, respectively. When the length of 1D conductive channels, such as nanotubes and nanowires, is shorter than the electron mean free path, ballistic transport occurs. If the radius of a silicon nanowire is smaller than 5 nm, quantization will result in discrete subbands. Yi et al. (2011) observed quantum oscillations of the transport behaviors in such ultrathin silicon nanowires at room temperature. In comparison to group IV semiconductors, group III–V semiconductors typically have lower effective mass, higher mobility, and larger Bohr radius, allowing for the observation of quantization effects at larger sizes. Chuang et al. (2013) observed ballistic transport in InAs nanowires at room temperature (Fig. 5.5), with an electron mean free path of approximately 150 nm.

The electronic structure of carbon nanotubes is closely related to graphene, where the axial periodicity and circumferential chirality jointly determine the band structure. The electronic states at the $\boldsymbol{K}$ and $\boldsymbol{K}'$ points in the Brillouin zone exhibit chiral characteristics, leading to suppressed backscattering. Additionally, the absence of dangling bonds on the surface reduces surface scattering, resulting in a larger electron mean free path. Liang et al. (2001) observed Fabry–Perot interference in carbon nanotubes acting as coherent electron waveguides, and Javey et al. (2003) reported ballistic and coherent transport in transistors with semiconductor carbon nanotubes as channels.

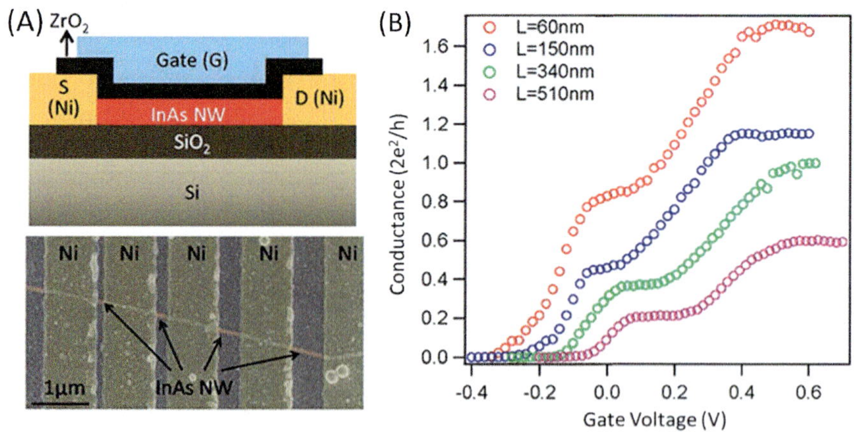

Figure 5.5 Ballistic transport properties of InAs nanowires at room temperature: (A) schematic diagram and scanning electron microscopy image of InAs nanowire field-effect transistor; (B) conductance–gate voltage curves for InAs nanowire transistors with different channel lengths, showing quantized conductance (Chuang et al., 2013).

Specific band structures of some 2D systems exhibit unique transport properties. For example, the electrons in graphene show behaviors similar to relativistic massless fermions, giving rise to the half-integer quantum Hall effect (Novoselov et al., 2005; Zhang et al., 2005). In this case, Landau levels are determined by $E_N = \sqrt{2e\hbar c_*^2 B(N + 1/2 \pm 1/2)}$, where N represents the Landau level, " $\pm$ " corresponds to the pseudo-spin of graphene sublattice, and c^* is the Fermi velocity of electrons in graphene. Carrier transport in graphene is influenced by scattering from phonons and interfaces. Chen et al. (2008) studied the transport properties of graphene on SiO_2 substrate. Under a carrier concentration of 1×10^{12} cm^2, the carrier mobility and mean free path limited by acoustic phonon scattering are 2×10^5 $cm^2/(V\ s)$ and 2 μm, at low temperature. When the temperature is above 200K, the carrier mobility is reduced to 4×10^4 $cm^2/(V\ s)$ by surface phonon scattering. Du et al. (2008) found that surface scattering in suspended graphene is effectively reduced, and a mobility of 2×10^6 $cm^2/(V\ s)$ at a doping concentration of 5×10^9 cm^{-2} is reported. Wang et al. (2013) fabricated 1D contact between metal electrodes and graphene, which greatly reduces the interface scattering. And a carrier mobility of 1.4×10^6 $cm^2/(V\ s)$ at room temperature and a ballistic transport mean free path of 15 μm at low temperature were achieved.

Monolayer MoS_2 exhibits broken inversion symmetry and strong spin–orbit coupling (Xiao et al., 2012). The broken inversion symmetry leads to the valley Hall effect, where carriers from different valleys flow toward opposite transverse boundaries when an in-plane electric field is applied. After considering the topological effects, an anomalous term proportional to the Berry curvature emerges in the equation of motion of electrons. Then, the electron velocity is given by

$$\vec{v}_n(k) = \frac{\nabla\vec{k}\varepsilon_n\left(\vec{k}\right)}{h} - \frac{e}{h}\vec{E} \times \vec{\Omega}_n\left(\vec{k}\right)$$

$$\vec{\Omega}_n(k) = i\nabla\vec{k}u_n\left(\vec{k}\right)| \times |\nabla\vec{k}u_n\left(\vec{k}\right)$$

where ε_n, u_n, and $\vec{E}$ represent the energy of the nth band, the periodic part of the Bloch function, and the electric field. The Berry curvature $\vec{\Omega}_n$ has the opposite signs for carriers at $\boldsymbol{K}$ and $\boldsymbol{K}'$ points, leading to the valley Hall effect. Spin–orbit coupling gives rise to significant spin splitting at the band edges, with a splitting up to 150 meV at the top of valence band. Meanwhile, the energy bands at the inequivalent $\boldsymbol{K}$ and $\boldsymbol{K}'$ points show opposite spin directions, due to time-reversal symmetry. The coupling between spin and valley degrees of freedom in the band edges enables potential applications in spintronics and valleytronics.

5.2.3 Ion transport

While electrons flow "within" materials, ion transport occurs in the "void" channels of the materials, including 1D channels of carbon nanotubes, 2D channels of 2D materials, and 3D channel networks in porous materials. Ion transport is closely related to ion size, charge, electrolyte, channel size, and surface charge, etc.

Lee et al. measured the ion current inside single-walled carbon nanotubes (SWCNTs) and found oscillatory phenomena related to the driving electric field, which is attributed to the coupling between channel blockage and ion diffusion limitation at the channel opening. The measured proton migration rates in carbon nanotubes are two orders of magnitude higher than those in aqueous solutions (5×10^2 S/cm), attributed to the atomic-level smoothness in the carbon nanotubes (Lee et al., 2010).

In 2D materials, ion channels include nanoscale pores that vertically penetrate the 2D plane and intervals parallel to the layers. Hu et al. found that monolayer graphene and h-BN have very high proton permeability in the vertical direction. For example, monolayer h-BN shows a proton resistivity of 10 Ω cm^2, corresponding to a 0.3 eV activation barrier (Hu et al., 2014). In addition, monolayer graphene and h-BN can separate hydrogen ions and isotopes with a separation factor of 10, surpassing that of current Girdler sulfide process (Lozada-Hidalgo et al., 2016).

Composite membranes composed of graphene and 2D materials can be obtained through bottom-up assembly. By placing 2D materials with different thicknesses between the top and bottom graphene layers, a channel space with controllable thickness ranging from a monolayer to tens of layers can be constructed, enabling the control of water transport in nanochannels. Due to high capillary pressure, large slip lengths, and improved order among water molecules, the flow rate of water can reach as high as 1 m/s (Radha et al., 2016). Qian et al. reported that thin films assembled from $CdPS_3$ nanosheets (Fig. 5.6) show a proton conductivity of 0.95 S/cm under conditions of 90°C and humidity of 98%, significantly surpassing that of the commercial proton membranes (~0.2 S/cm). Such high proton conductivity is attributed to the rich proton donor centers, facile proton adsorption, and good hydration provided by cadmium vacancies, showing

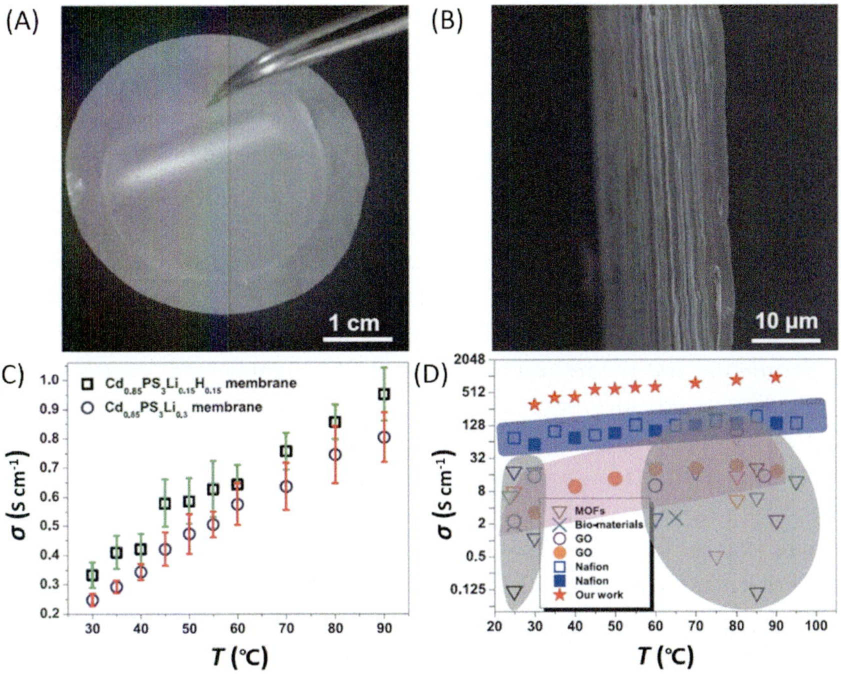

Figure 5.6 Ion transport properties of $CdPS_3$ nanosheet-assembled thin films: (A,B) optical and scanning electron microscopy images; (C) relationship between proton conductivity and temperature at 98% humidity; (D) comparison of the performance with typical proton transport materials at 90% humidity (Qian et al., 2020).

potential for applications as proton exchange membranes in fuel cells (Qian et al., 2020).

5.3 Magnetic properties

According to the Ising model, 2D and lower dimensional systems with continuous symmetry should not have ferromagnetic phase transition. Thermodynamic theory predicts that neither ferromagnetic phase transitions nor other long-range ordered structures can occur in 2D systems. The discovery of 2D magnetic materials with rich physical phenomena has challenged such prediction. Compared to bulk materials, the magnetism of 2D materials is more easily controlled by external fields, electronic doping, strain, and twisting (Huang et al., 2018; Lin et al., 2018; Webster & Yan, 2018).

Gong et al. (2017) and Huang et al. (2017a) independently discovered long-range intrinsic ferromagnetism in bilayer $Cr_2Ge_2Te_6$ and monolayer CrI_3, respectively. Although the Curie temperature (T_c) is relatively low in these systems (~28K and 45K, respectively), the appearance of intrinsic magnetism in 2D systems has greatly inspired further research interest. Deng et al. (2018) discovered the 2D ferromagnetic conductor Fe_3GeTe_2 and found the Curie temperature can be increased to near room temperature in trilayer samples through electric control. With the joint efforts of experimental and theoretical work, more 2D ferromagnetic materials have been discovered, including VSe_2, $MnSe_2$, and $MnBi_2Te_4$ (Bonilla et al., 2018; Li et al., 2019a; O'Hara et al., 2018).

Based on the lattice symmetry formed by transition metal ions, 2D magnetic materials can be roughly divided into hexagonal, triangular, and square lattices. Representative systems of hexagonal lattices include transition metal trihalides MX_3 (X is halogen element), ternary compound $Cr_2B_2Y_6$ (B = Ge or Si; Y is chalcogen element), and ternary transition metal phosphates $M_2P_2Y_6$ (M = V, Cr, Mn, Fe, Co, Ni, Cu, Zn, etc.; Y is chalcogen element). The representative systems of triangular lattices include $M^IM^{III}P_2Y_6$ ($M^I = Ag^+$ or Cu^+; M^{III} is V^{3+}, In^{3+}, Cr^{3+}, Sc^{3+}, Bi^{3+}; Y is a chalcogen element), Fe_3GeTe_2, transition metal disulfides or dibromides MX_2, MXenes (M_2X, M_2XT_2), $MnBi_2Te_4$, and MGa_2Y_4 (M is an early transition metal; Y is a chalcogen element). The representative systems of square lattices include FeY (Y is chalcogen element), MXY (M is transition metal; X is halogen element; Y is chalcogen element), and VOX_2 (X = Cl, Br, and I).

5.3.1 Magnetic coupling

5.3.1.1 Magnetoelectric coupling

The magnetoelectric coupling of magnetization and polarization can result in the novel multiferroic properties. Normally, ferroelectricity and ferromagnetism are incompatible. Magnetic properties typically originate from the ordered spin states of transition metal atoms with partially filled d or f orbitals, while ferroelectric properties usually arise from the deviation of transition metal atoms with empty d orbitals from the symmetry center (Khomskii, 2006). In recent years, scientists have explored various methods to construct 2D multiferroic materials. The first type of multiferroic compounds is called type I multiferroic compounds, where ferroelectricity and magnetism come from different origins (Khomskii, 2009). The second type generates ferroelectric polarization through noncentrosymmetric magnetic ordering. The third type combines the characteristics of types I and II, known as single-cation type I multiferroic crystals, where the orbitals contributing to magnetism and the distortion mode have different symmetries and can come from the same ions (Tan et al., 2019).

Theoretical predictions suggest that 2D transition metal phosphorus–sulfur compounds $CuM^{III}P_2X_6$ (M^{III} = Cr, V; X = S, Se) exhibit type I multiferroicity (Qi et al., 2018). The magnetic properties are introduced by partially occupied d orbitals of trivalent Cr/V atoms, while the copper atoms are not located in the central layer, resulting in an electric dipole moment perpendicular to the 2D plane. The coupling between magnetism and ferroelectricity in $CuCrP_2X_6$ has been experimentally verified (Lai et al., 2019). Fig. 5.7A shows the structures of the intralayer antiferroelectric and ferroelectric states of $CuCrP_2S_6$, distinguished by the position and symmetry of copper atoms. Fig. 5.7B depicts the lowest energy path for the ferroelectric–antiferroelectric phase transition obtained from

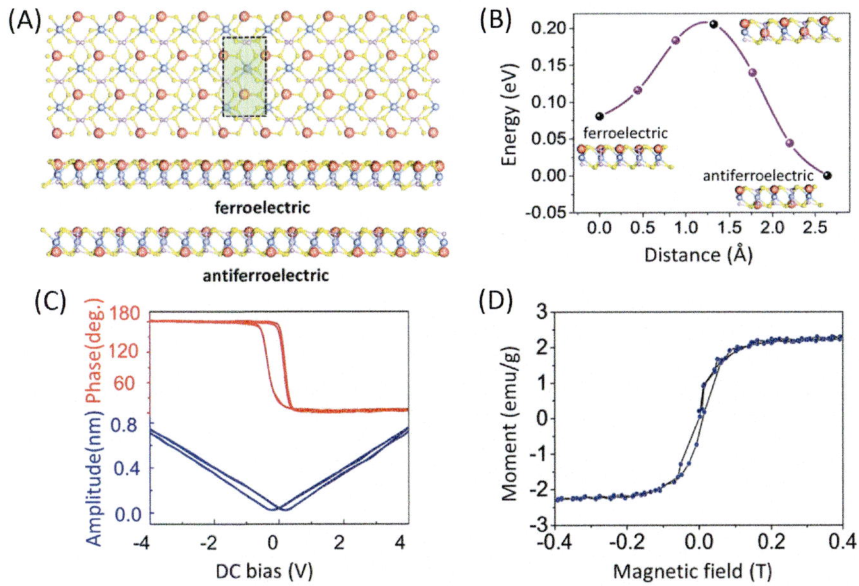

Figure 5.7 (A,B) The ferroelectric and antiferroelectric states of two-dimensional $CuCrP_2S_6$ and the lowest energy path for the transition; (C) the hysteresis loop of phase (top), the butterfly curve of amplitude (bottom); and (D) the magnetization of $CuCrP_2S_6$ nanosheets (Lai et al., 2019).

theoretical calculations, with an energy barrier of about 0.11 eV/f.u., sufficient to prevent spontaneous phase transition at room temperature. The energy barrier from antiferroelectric to ferroelectric state is about 0.21 eV/f.u., which can be achieved by an external electric field. Fig. 5.7C presents the hysteresis loop of phase and "butterfly" curve of polarized amplitude obtained from experimental measurements on a 13.3 nm thick $CuCrP_2S_6$ nanosheet, providing strong evidence for the existence of ferroelectricity in 2D $CuCrP_2S_6$. Furthermore, for both ferroelectric and antiferroelectric states, their calculations show that the ground state of monolayer $CuCrP_2S_6$ is ferromagnetic. Fig. 5.7D shows the magnetic hysteresis loop of $CuCrP_2S_6$ nanosheets at 10K, providing evidence for the presence of ferromagnetism.

5.3.1.2 Magneto-topological coupling

The coupling between spin and topological properties gives rise to peculiar phenomena, including quantum anomalous Hall effect. Theoretical calculations by Li et al. (2019a) suggest that 2D $MnBi_2Te_4$ has inherent magnetic topological insulator properties. As shown in Fig. 5.8A, first-principles calculations indicate that the ground state of $MnBi_2Te_4$ is intralayer ferromagnetic and interlayer antiferromagnetic. For the antiferromagnetic phase, the

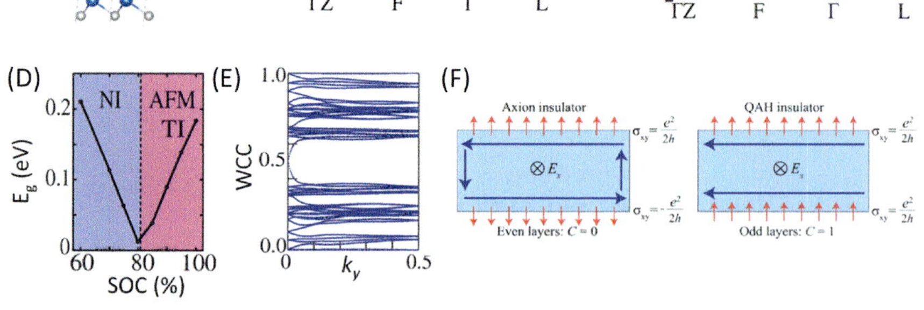

Figure 5.8 (A) $MnBi_2Te_4$ structure with antiferromagnetic interlayer coupling; (B,C) band structures (B) without and (C) with considering spin–orbit coupling. " + " and " − " represent the parity of the band edges at the Γ point; (D) variation of the bandgap at the Γ point with the strength of SOC; (E) WCC calculated in the $k_z = 0$ plane; (F) topological properties of even (left) and odd (right) layered $MnBi_2Te_4$ thin films (Li et al., 2019a). *SOC*, Spin–orbit coupling; *WCC*, Wannier charge center.

parities of the valence band maximum and conduction band minimum at the Γ point are opposite, and the spin–orbit coupling effect can reverse their parity (Fig. 5.8B and C). This band inversion can be understood by the variation of the bandgap with the strength of the spin–orbit coupling (Fig. 5.8D). The accompanying topological phase transition can be verified by the Wannier charge center at $k_z = 0$ plane, with a Z_2 index of 1 (Fig. 5.8E). Fig. 5.8F shows the thickness-dependent topological properties of $MnBi_2Te_4$ films. For $MnBi_2Te_4$ thin films with even layer, the contributions from the top and bottom surfaces of the half-integer Hall conductivity cancel each other out, resulting in an axion-insulating state with zero Chern number. For odd-layered $MnBi_2Te_4$, the contributions from the top and bottom surfaces add up, corresponding to a quantum anomalous Hall state with a Chern number of 1. Otrokov et al. (2019) clearly observed the gapped Dirac surface states using low-temperature photoemission spectroscopy. Liu et al. (2020) obtained direct evidence of the axion-insulating state in a six-layer $MnBi_2Te_4$ device, showing a large longitudinal resistance and zero Hall plateau. And zero-field quantum anomalous Hall effect was observed in a five-layer $MnBi_2Te_4$ sample (Deng et al., 2020).

5.3.1.3 Spin–phonon coupling

The A_{1g} phonon mode in CrI_3, corresponding to out-of-phase vibrations between two adjacent iodine layers in the vertical direction, strongly couples with interlayer magnetism. In bilayer CrI_3, the interlayer van der Waals interaction splits the A_{1g} phonon mode into two modes, which exhibit different Raman signals when coupled with interlayer magnetism as shown in Fig. 5.9 (Huang et al., 2020). Without external magnetic field, the antiferromagnetic ground state of bilayer CrI_3 is characterized by two Raman peaks at 126.7 and 128.8 cm^{-1} in the XY and XX Raman spectra, respectively. When a sufficiently strong external field induces ferromagnetic interlayer coupling, only a single peak is detected at 128.8 cm^{-1} in both the XX and XY Raman spectra. Due to the centrosymmetric structure of bilayer CrI_3, these two modes can be classified as an odd-parity mode (u, with frequency of 126.7 cm^{-1}) and an even-parity mode (g, with frequency of 128.8 cm^{-1}). For the case of interlayer ferromagnetic coupling, the centrosymmetry is preserved, and the Raman spectra resemble those of a ferromagnetic monolayer, allowing for the observation of the g mode in both XX and XY Raman spectra, while the u mode is inactive. However, when interlayer antiferromagnetic coupling

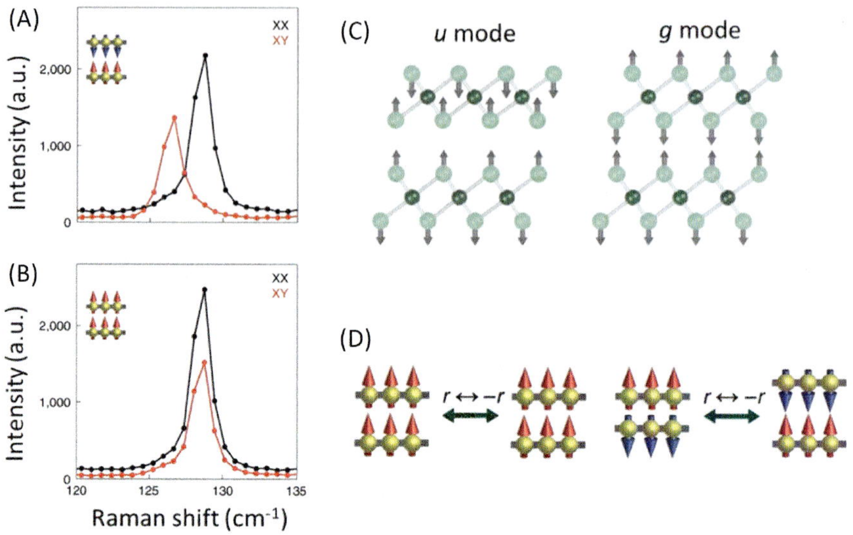

Figure 5.9 (A) Raman spectra of bilayer CrI_3 with interlayer (A) antiferromagnetic and (B) ferromagnetic coupling; (C) schematic representation of two Davydov-split A_{1g} modes, with the *u* mode at $\sim 126.7\ cm^{-1}$ and the *g* mode at $\sim 128.8\ cm^{-1}$; (D) illustration of the inversion operation for ferromagnetic and antiferromagnetic states (Huang et al., 2020).

is present, the antiparallel spin configuration breaks the inversion symmetry, making the *u* mode active in the XY Raman spectrum.

5.3.2 External control of magnetism

5.3.2.1 Electrical control

The control of magnetism in 2D magnets by gate voltage can be easily applied to field-effect spintronic devices. By measuring the magnetization hysteresis loop of $Cr_2Ge_2Te_6$ at different gate voltages, Wang et al. (2018) discovered the high sensitivity of the magnetization of $Cr_2Ge_2Te_6$ to gate voltage. The saturation magnetization M_s (represented by saturation Kerr rotation) of $Cr_2Ge_2Te_6$ is enhanced under electron or hole doping, which can be explained by the enhancement of the magnetic moments of Cr ion upon doping. First-principles calculations show that the normalized spin magnetization $(M_s - M_{s0})/M_{s0}$ as a function of carrier doping concentration matches well with the experimental measurements (Fig. 5.10A).

In addition, analysis for the spin-polarized electron density reveals that the variation in magnetization is caused by the changes in the Fermi energy level and electron occupation. $Cr_2Ge_2Te_6$ is a magnetic bipolar

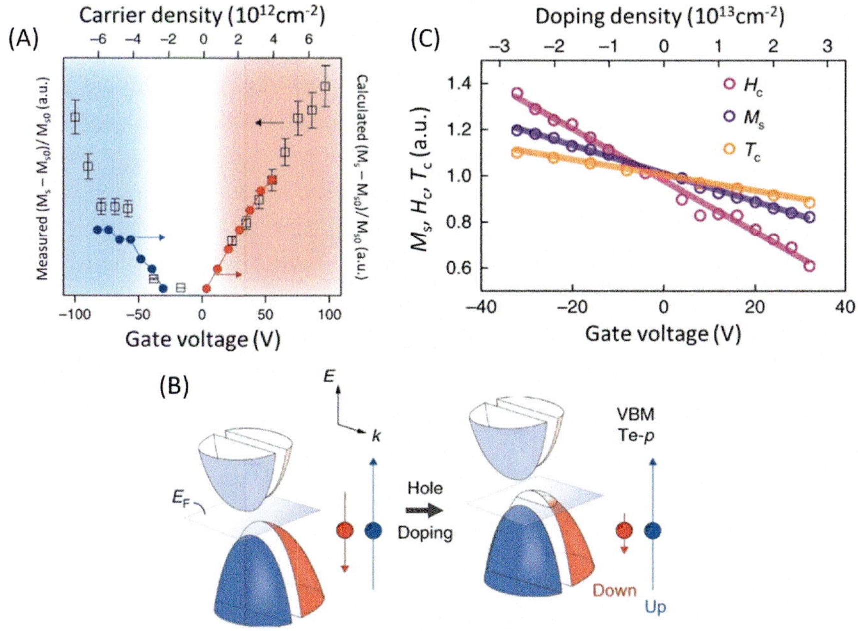

Figure 5.10 (A) The normalized saturation magnetization (M_s) of $Cr_2Ge_2Te_6$ obtained from experimental measurements (black square) and theoretical calculations (solid circle) as a function of gate voltage or equivalent carrier concentration. (B) Schematic diagrams showing the spin-polarized energy bands and corresponding occupation before (left) and after (right) hole doping. (C) Variation of the saturation magnetization (M_s), coercivity (H_c), and Curie temperature (T_c) of CrI_3 with gate voltage or doping density (Jiang et al., 2018a; Wang et al., 2018).

semiconductor, with its valence band top and conduction band bottom in different spin channels, and at neutral state, the spin-up states are occupied by more electrons. Hole doping lowers the energy of the Fermi level, making it cross the spin-down valence bands, but still lies higher than the spin-up valence bands top. Therefore the decrease in the occupation of spin-down states leads to an increase in unpaired spins and thus magnetic moments. Similarly, for electron doping, the Fermi level intersects with the spin-up conduction bands, increasing the occupation of spin-up states and magnetic moments (Fig. 5.10B).

Monolayer CrI_3 exhibits similar electrically tunable magnetic properties, but the response to hole and electron doping is opposite (Fig. 5.10C) (Jiang et al., 2018a) because the valence band top and conduction band bottom have the same spin direction. Accordingly, inducing electron (hole) doping into CrI_3 through positive (negative) gate voltage can weaken (enhance) the saturation magnetization, coercivity, and Curie temperature. For few-layer CrI_3 with a monoclinic structure, gating or carrier doping significantly affects the interlayer coupling, inducing a transition from antiferromagnetic to ferromagnetic phases (Huang et al., 2018; Jiang et al., 2018b; Xu & Zou, 2020).

The Curie temperature of the ferromagnetic metal Fe_3GeTe_2 can be enhanced by electric control. Deng et al. (2018) applied carrier doping through ion gating, resulting in the increase of the Curie temperature of a three-layer Fe_3GeTe_2 sample from around 100–300K. When its inversion symmetry is broken, it is possible to control the magnetic structures and properties with electric current through the combination of spin–orbit torque and out-of-plane magnetic anisotropy, as observed in Fe_3GeTe_2/heavy metal heterostructures (Johansen et al., 2019). Due to strong spin–orbit coupling, spin-polarized currents can be generated in the heavy metal layer. When the current-induced spin torque is large enough, it can switch the spin orientation in Fe_3GeTe_2 between two opposite out-of-plane directions (Wang et al., 2019).

5.3.2.2 Mechanical control

Stress can change the overall structural symmetries of 2D crystals or distort their local structures, thereby controlling their physical properties. Both theoretical and experimental studies have shown that monolayer CrI_3 is an Ising ferromagnet, while monoclinic bilayer CrI_3 is an A-type antiferromagnet. The interlayer exchange interaction is much weaker than intralayer exchange interaction; thus the interlayer coupling can be effectively modulated by pressure (Li et al., 2019b). Fig. 5.11A illustrates the hysteresis behaviors of the tunneling conductance as a function of pressure in a bilayer CrI_3 tunnel junction at a temperature of 1.7K. Without pressure, the tunneling conductance abruptly increases at the spin-flip critical field B_{sf} (approximately 0.75 T). However, when a pressure of 1 GPa is applied, both the critical field B_{sf} and the saturated tunneling conductance value increase. When the pressure is further increased to 1.8 GPa, the hysteresis in the tunneling conductance is almost completely suppressed. After releasing the pressure, the properties of the tunnel junction do not fully recover. Polarized Raman spectroscopy indicates that the significant suppression of the hysteresis in the tunneling conductance under high pressure (1.8 GPa) is attributed to an irreversible structural phase transition of bilayer CrI_3 from monoclinic to rhombohedral stackings (Fig. 5.11B).

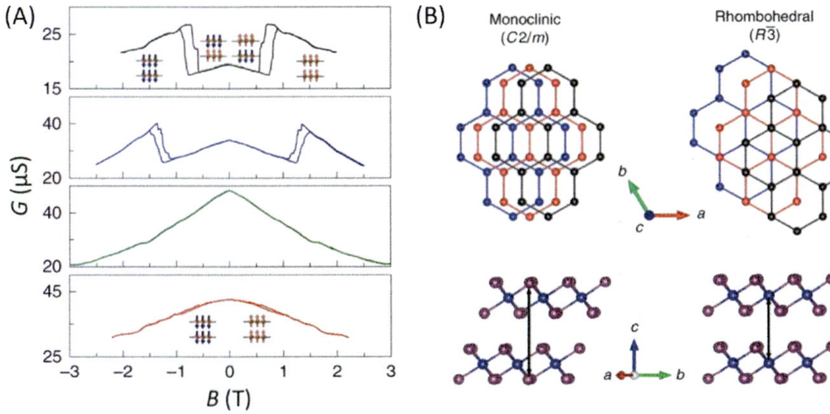

Figure 5.11 (A) The tunneling conductance for a bilayer CrI_3 tunnel junction as a function of the applied magnetic field at 1.7K under different pressures (from top to bottom: without pressure, 1, 1.8 GPa, and after pressure release). (B) Schematic diagrams of the structures of bilayer CrI_3 with (left) monoclinic and (right) rhombohedral stacking (Li et al., 2019b).

5.4 Thermal properties

Thermal properties of materials and thermal management of devices are critical for various applications in energy and electronics. Rocket fairings, turbine engines, thermoelectric conversion, refrigerators, and smart buildings require heat-insulating materials with low thermal conductivity, while computers, lasers, energy storage devices, and temperature management of human body need heat-conducting materials with high thermal conductivity. Particularly, with the development of modern electronics industry, the dissipation of heat in the nanoscale becomes a bottleneck of the device performance.

Heat dissipates mainly through three ways, that is, thermal radiation as electromagnetic waves in vacuum, thermal convection in fluids, and thermal conduction through electrons and phonons in solids. In low-dimensional materials, when the characteristic scale of systems becomes comparable to the wavelengths of electromagnetic or lattice waves, it is necessary to consider phase-related phenomena as shown below. Moreover, Fong et al. (2019) reported that when the distance between solid surfaces reaches nanometer scale, phonons could transport through Casimir force coupling, due to quantum fluctuation effects in vacuum.

5.4.1 Heat radiation in vacuum

Heat transports in vacuum mainly through the radiation of electromagnetic waves. It can be described by the black body heat radiation model, based on the thermal equilibrium between black body and photons. According to Stephan–Boltzmann law, energy radiated by black body per unit area and unit time is proportional to T^4 (T, temperature). Exchanged energy between two objects through heat radiation is expressed by $q \propto \sigma(T_1^4 - T_2^4)$, with σ Stephan–Boltzmann constant. Yu et al. proposed that the upper limit of electromagnetic waves is determined by the size of black body and introduced a size-related correction function. Based on this model, the heat radiation efficiency of low-dimensional materials decreases with size (Yu et al., 2010).

Black body is an ideal model, which can absorb all radiation from electromagnetic waves of any wavelength, and has the strongest electromagnetic radiation capability. Practically, there is no absolute black body, and any materials would reflect some electromagnetic waves. Mizuno et al. (2009) found that the absorption of vertically grown SWCNT array in the wavelength range between 200 nm and 200 μm can reach 98%, closely resembling the behaviors of black body. The unique absorption properties of SWCNT array are related to both the intrinsic electronic and optical properties, as well as the optical traps formed by the sparse and vertical arrays, in which light can be reflected and absorbed multiple times.

Classical heat radiation theory does not consider about the structures of materials. However, radiation of electromagnetic waves usually results from the electron transition between energy levels. The dimension and size of materials have great influences on the electronic structures, which significantly alters the heat radiation behaviors. Due to the 1D confinement, the density of states of carbon nanotubes (CNTs) shows sharp peaks corresponding to the Van Hove singularities. Nishihara et al. (2018) compared heat radiation spectra from single metallic and semiconducting CNTs at 1000–2000K and found ultranarrow energy distribution of radiated photons from semiconducting CNTs due to thermal generated excitons.

Heat can also be transported between solid surfaces through electromagnetic wave radiation. It is noted that classical theory can only be applied to the case when the distance between surfaces is larger than the wavelength of electromagnetic waves, which is called the far-field transmission. When the distance decreases to nanoscale, the evanescent waves propagating near the surfaces become important, which can increase the heat transfer efficiency by several orders of magnitude (Polder & Vanhove, 1971). The enhanced heat radiation in the near field has been verified by a recent experiment (Narayanaswamy et al., 2008).

5.4.2 Heat transfer in fluids

Heat transfer in fluids includes thermal convection and thermal conduction. While thermal conduction originates from the random collision of carriers, thermal convection results from the energy transfer during the movement of fluids.

Thermal convection between solid surfaces and fluids can be described by Newton cooling law as $Q = hA(T_s - T_f)$, where Q, h, A, T_s, and T_f are thermal convection efficiency, thermal convection coefficient, surface area, temperature of solid surface, and temperature of fluid, respectively. Thermal convection coefficient is closely related with the type of fluids, flow condition, and geometry. For a long cylinder, Peirs et al. (1998) found the relation between thermal convection coefficient h and diameter d as $h = \left(2.68 + \frac{0.11}{\sqrt{d}}\right)^2$. As the size decreases, specific surface area increases, so does the thermal convection efficiency. Huang et al. (2017b) measured thermal transport properties of confined gases with micrometer length and nanometer width surrounded by solids and found thermal convection coefficient could be as high as 1242 W/(m^2 K), 50 times higher than that at macroscale.

Thermal conduction coefficient of gases is correlated with Knudson number (Kn), which is the ratio between mean free path of gas molecules λ and characteristic length of system L. Kn is closely related to temperature (T), radius of gas molecules (d), and pressure (p) and can be expressed as $Kn = \frac{\lambda}{L} = \frac{k_B T}{\sqrt{2}\pi d^2 pL}$. When Kn is very small, corresponding to the case of large gas pressure and system size, the thermal conduction coefficient of gases is constant. When the system size is close to or even smaller than mean free path of gases, the transport of gases becomes ballistic. In this case, Kn is very large, and thermal conduction coefficient decreases exponentially as the size decreases (Denpoh, 1998).

5.4.3 Heat conduction in solids

In solids, heat conduction is mainly carried by electrons and phonons. Classical thermal conduction theory is based on the random scattering of particles, of which the thermal energy is determined by the temperature. This theory leads to the famous Fourier's law, which states that thermal flux q is proportional to temperature gradient ∇T, and the coefficient is the thermal conductivity k, that is, $q = k\nabla T$.

According to the kinetic theory of gases, thermal conductivity is expressed as $k = \frac{1}{3}\sum C_n v_n l_n$, where C_n, v_n, and l_n are specific heat capacity, phonon velocity (defined as $\frac{\partial \omega_n}{\partial q}$, that is, the derivative of frequency with respect to wave vector q), and mean free path of the nth phonon mode, respectively. When the system size is smaller than the mean free path of phonon scattering, thermal conduction becomes ballistic.

The thermal conduction can be analyzed by different scattering mechanisms with different mean free paths, characteristic relaxation time τ, and relaxation rate $1/\tau$. Phonon scattering mainly includes phonon–phonon Umklapp scattering (*U*), impurity scattering (*im*), boundary scattering (*B*), and phonon–electron scattering (*ph–e*). According to Matthiessen's rule, total relaxation time τ_C is described as

$$\frac{1}{\tau_C} = \frac{1}{\tau_U} + \frac{1}{\tau_{im}} + \frac{1}{\tau_B} + \frac{1}{\tau_{ph-e}}$$

where Umklapp scattering represents the process where the wavevector of the scattered phonon is outside of the first Brillouin zone, which is equivalent to phonon transport with the reversed direction. For three-phonon process, $\vec{k}_1 + \vec{k}_2 = \vec{k}_3 + \vec{G}$, the scattering rate is proportional to the square of phonon frequency ω^2. For impurity scattering, its scattering rate is proportional to ω^4, $\frac{1}{\tau_{im}} = \frac{V_0\omega^4}{4\pi v_g^3}$, and accordingly, the short-wave phonons are significantly influenced. Boundary scattering rate is inversely proportional to the sample size W, that is, $\frac{1}{\tau_B} = \frac{v_g}{W}$, and it is the dominant scattering mechanism in low-dimensional materials.

In the following, the influence of dimensionality on thermal properties will be discussed. First, dimensionality could influence the specific heat capacity. Because quantum confinement has significant influences on phonon density of states, the specific heat capacity of materials with different dimensionalities shows different scaling laws versus temperature. Following the method introduced in Chapter 1, specific heat capacities in 3D, 2D, and 1D systems are proportional to T^3, T^2, and T, respectively.

Second, dimensionality can affect phonon group velocity v_g. When solving the elasticity equation, different from the periodic boundary condition in bulk materials, the force component along the direction perpendicular to the surfaces of low-dimensional materials should be zero, and different boundary conditions lead to different phonon dispersion. Balandin and Wang (1998) calculated the phonon dispersion for 2D systems and found their phonon group velocity is about 1/3 of those of bulk materials.

Third, dimensionality could impact phonon mean free path. The calculations from Balandin and Wang (1998) showed that in superlattice structures, the thermal conductivity of 10-nm thick quantum well is lower by one order of magnitude than that of bulk systems because of the change of phonon band structure and the increase of scattering. Li et al. (2003) found that when the diameter of silicon nanowires decreases from 115 to 22 nm, the thermal conductivity decreases from ~40 to 9 W/(m K). Lim et al. investigated the influences of surface scattering of silicon nanowires on their thermal properties and found that thermal conductivity is mainly determined by surface roughness rather than radius. As surface roughness increases, its room-temperature thermal conductivity decreases from 25 to 5 W/(m K), about one order of magnitude smaller than that of single-crystalline bulk silicon (~125 W/(m K)) (Lim et al., 2012).

Thermal conductivity of carbon materials is strongly dependent on the microstructures. The room-temperature thermal conductivity varies from 10^{-2} W/(m K) for amorphous carbon to 10^3 W/(m K) for carbon nanotubes (Balandin, 2011). Because of their dangling bond-free structures, the intrinsic thermal conductivity of carbon nanotubes and graphene can reach 3000 W/(m K) (Balandin et al., 2008; Kim et al., 2001). Chen et al. measured the thermal conductivity of graphene containing different concentrations of ^{12}C and ^{13}C isotopes and found that the thermal conductivity of graphene (99.9% isotope purity) could be as high as 4000 W/(m K). And when the concentrations of ^{12}C and ^{13}C isotopes are close to equal, the thermal conductivity decreases by half (Chen et al., 2012). Regarding graphene nanoribbons, Bae et al. (2013) showed that the thermal conductivity (k) is affected by both the width (W) and edge disorder, described as $k \propto W^{1.8 \pm 0.3}$.

For polycrystalline graphene, Ma et al. discovered that when the grain size decreases from 1 μm to 200 nm, thermal conductivity decreases from 5230 to 610 W/(m K). By separating the boundary scattering from phonon–phonon scattering, the measured boundary thermal conductivity from grain boundaries is 3.8×10^9 W/(m^2 K), consistent with theoretical value obtained by nonequilibrium Green function calculations (Fig. 5.12) (Ma et al., 2017).

Many applications require macroscale assembly, and the thermal conductivity is related not only to the intrinsic thermal properties of component materials but also to the arrangement order and interface structures. Xin et al. (2015) assembled fibers using graphene flakes with different sizes followed by graphitization heat treatment, and the measured thermal conductivity of carbon fibers is as high as 1290 W/(m K). Peng et al. fabricated graphene membranes with microscale ripples by thermal treatment of aligned graphene oxide membranes. These graphene membranes show high thermal conductivity (1940 W/(m K)) and flexibility, demonstrating

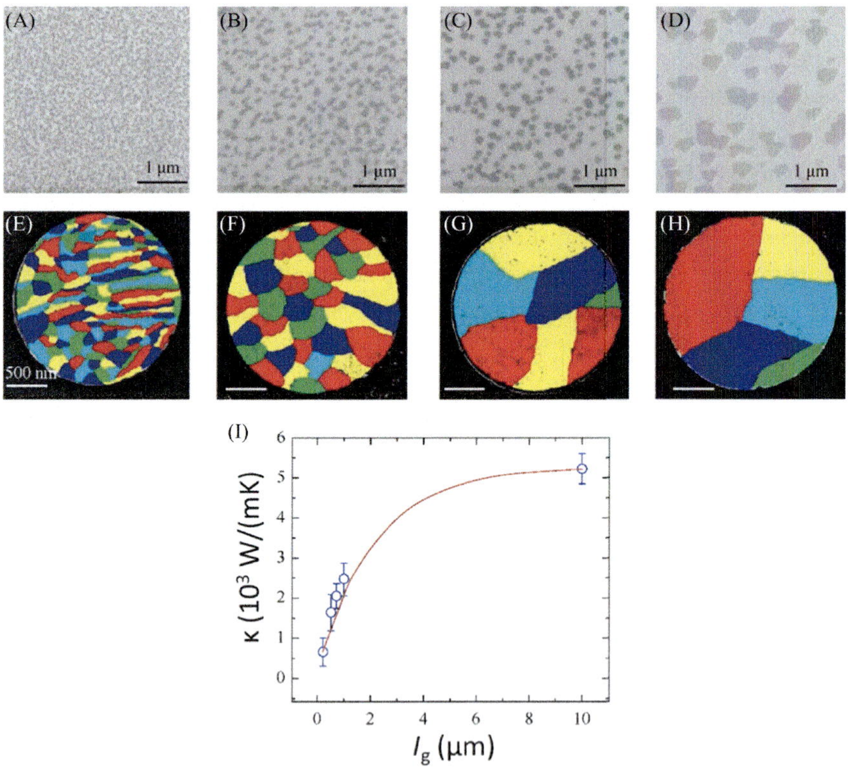

Figure 5.12 Thermal transport properties of polycrystalline graphene: (A to D) TEM images for graphene grains obtained at 900°C, 950°C, 1000°C, and 1040°C; (E to H) false-color images of polycrystalline graphene with different grain sizes; and (I) thermal conductivity as a function of grain size (Ma et al., 2017).

great potential for thermal management in electronics devices (Peng et al., 2017).

5.5 Chemical properties

Catalysts are crucial for the modern chemical industry, to accelerate reactions by reducing the activation energy. The difference of reaction activation energies for different reactions determines the product selectivity. The performance of catalysts depends on the intrinsic activity, that is, the electronic structure of active sites, and the density and number of active sites. Low-dimensional materials are attractive for catalysis applications for the large specific surface area and high ratio of surface atoms as active sites. Meanwhile, quantum confinement, alloy composition, crystal structures, surface structures, and strain are effective to modulate the electronic structures to improve the intrinsic activity.

The advance of synthesis of low-dimensional materials makes it possible to accurately control the size, facet, and morphology of nanoparticles and mesoporous materials. Meanwhile, by applying advanced characterization methods, including spectroscopy and electron microscopy, it is possible to do molecular-level characterization for deep understanding of catalytic processes. In particular, valuable insight could be obtained by in situ investigations of oxidation states, coordination numbers, crystal surface, the bonding, and orientation of adsorbates (Ye et al., 2016). In this section, after the introduction of Sabatier's principle and activity volcano plot, we will discuss the catalytic properties of metal nanoparticles, single-atom catalysts, and metal-free catalysts, emphasizing the dimensionality and size effects.

5.5.1 Volcano plot in heterogenous catalysis

Heterogenous catalytic reactions involve multiple steps from the adsorption, activation, diffusion, and reaction of reactants on the surfaces of catalysts, to the desorption of intermediates and final products. The key to design and optimize catalysts is to understand the formation and breaking of chemical bonds, exchange of matter, and change in energy of every reaction step. In 1920 French chemist Paul Sabatier proposed the famous Sabatier's principle for the design of heterogenous catalysts. It is stated that the reaction rate is closely related to the adsorption energies of intermediates on catalyst surfaces, and the adsorption energies should be neither too low nor too high. If the adsorption energy is too low, the coverage of reactants or intermediates is too low. If the adsorption energy is too high, the reactants cannot diffuse or desorb. Accordingly, the reaction rate as a function of the adsorption energy shows a volcano-like shape (Rothenberg, 2008).

Volcano plot provides a simple model to understand the catalytic reaction rate, but it faces various challenges in practical applications. First, as key descriptors, the adsorption energies of intermediates are difficult to measure experimentally. Second, volcano plot is suitable for optimizing activity, but it cannot predict selectivity and stability. Third, it is difficult to apply for the optimization of the reactions with multiple steps and a large parameter space. One reaction consists of several elementary reactions, accompanied by several intermediates. The adsorption energies of these intermediates usually have the so-called linear scaling relation (Greeley, 2016), which makes it challenging to optimize several reaction steps simultaneously. The key to break the linear scaling relation is to decouple the coupled processes. Low-dimensional materials provide opportunities to design high performance catalysts by using the low-dimensional effects, including size effect, crystal facets, dopants, alloying, heterogenous interface, strain engineering, and external fields, as will be discussed in the following sections.

5.5.2 Thermocatalytic reactions

5.5.2.1 Size effect

Low temperature CO oxidization is a key reaction for the treatment of vehicle exhaust, which relies heavily on noble metals such as Pt catalysts. It is an urgent research topic to improve the catalytic efficiency of Pt and develop non-noble-metal catalysts to replace Pt. Active catalytic sites are usually unsaturated atoms with low coordination number on catalyst surfaces. When the size of catalyst particles decreases, the number of active sites increases. The limit of catalyst size is the isolated single atoms, which in principle leads to the highest efficiency. However, as the catalyst size decreases, the specific surface area increases, and the thermal stability is also reduced, resulting in agglomeration driven by surface energy. In 2011 Qiao et al. (2011) found new method to stabilize the single-atom catalyst by loading isolated Pt atoms on iron oxides, showing high activity and stability for CO oxidization. Single-atom catalysts have become an active research frontier widely explored for various catalytic reactions.

The discovery of Au nanocatalysts suggests that size effect not only increases the number of active sites but also changes the intrinsic chemical properties. In 1987 Haruta et al. (1987, 1989) found that Au nanoparticles loaded on TiO_2 are highly active for low-temperature CO oxidization, while bulk Au and Au loaded on inert oxides show no activity. Valden (1998) investigated the change of structures and electronic structures of Au nanoparticles on single-crystalline TiO_2 by scanning tunneling microscope, providing solid evidence for the activation of Au nanoparticles for CO oxidization, with the highest activity observed for the nanoparticles with the size around 3.5 nm.

5.5.2.2 Confinement effect

When the catalyst particles are confined in a nanometer-scale space, the size of the particle is limited by the size of the pore, and the electronic structure could be modified by the interaction at the interface. Carbon nanotubes can not only confine catalysts to subnanometer scale, but also their π electrons can generate electrostatic field associated with the curvature, leading to the distinct interaction and charge transfer with the metal particles inside and outside the nanotube channels (Pan & Bao, 2011; Xiao et al., 2015). Pan et al. utilized carbon nanotubes to confine Rh particles, significantly enhancing their activities for the synthesis of ethanol from CO and H_2. In particular, the confined catalysts show a reaction efficiency one order of magnitude higher than those on the nanotube surfaces, which is attributed to the modulation of adsorption energies of reaction intermediates by the interaction between metal particles and nanotube walls (Pan et al., 2007).

5.5.2.3 Strain effect

Strain can change lattice constants of catalysts, influence the overlap among atomic orbitals, alter electronic structures of catalysts, and thus modulate their catalytic properties. Under tensile strain, the distance between atoms increases, and the overlap between d orbitals of neighboring atoms decreases, resulting in the narrowing of the bandwidth composed of d orbitals and the increase of the energy of d band center. Following the d band theory, this can lead to the enhancement of the interaction between reaction intermediates and catalyst surfaces.

Because of the anisotropy of crystal surfaces, different adsorption sites have different coordinations and symmetries and respond to strain differently. For a CH_2 group adsorbed on Cu (1 1 0), when it locates at the bridge site of two atoms with C2 symmetry, it introduces positive stress for neighboring atoms. On the other hand, when it sits at the hollow site of four atoms with C4 symmetry, it generates negative stress for neighboring atoms. Therefore, by adsorbing intermediates on different sites, it is possible to generate asymmetric strain effects, which can be used to break the linear scaling relation, and to improve reaction activity and selectivity (Khorshidi et al., 2018).

5.5.3 Electrochemical reactions

Electrochemical reactions are the chemical reactions involving the conversion of chemical energy and electron current on the surface of electrodes. These kinds of reactions are the foundation for many energy storage and conversion devices, including fuel cells, metal ion batteries, and supercapacitors. Therefore electrochemical reactions are crucial for clean energy, carbon neutrality, and generally sustainable development. In this section, the principles and catalysts for some of the most important electrochemical reactions will be introduced, including hydrogen evolution reaction (HER), oxygen reduction reaction (ORR), CO_2 reduction reaction (CO_2RR), and nitrogen reduction reaction (NRR).

5.5.3.1 Hydrogen evolution reaction

Hydrogen is an attractive secondary energy, since only water is produced by burning hydrogen. Electrochemical HER is one key step to produce hydrogen. HER is a typical two-electron process, through either Volmer–Heyrovsky or Volmer–Tafel mechanism. Reaction dynamics is closely related to the adsorption energy of intermediate *H ($\Delta G_{*\mathrm{H}}$). If the bonding between hydrogen and catalysts is strong, Heyrovsky or Tafel process is the limiting step. If the bonding between hydrogen and catalysts is weak, Volmer process is the rate-determining step. Reaction rate versus ΔG_{*H} shows the classical volcano shape (Nørskov et al., 2005), and noble Pt, the most efficient HER catalyst, is indeed close to the volcano vertex.

The unique structures of low-dimensional materials provide opportunities for modifying the electronic structure to discover alternative catalyst to replace the expensive Pt catalyst. In 2005 Hinnemann et al. discovered that MoS_2 nanoflakes can effectively catalyze HER with overpotential in the range of 0.1–0.2 V. Theoretical calculations showed that $\Delta G_{*\mathrm{H}}$ on

S-oriented edge with 50% S coverage is only 0.08 eV (Hinnemann et al., 2005), while that on bulk surface is 1.92 eV (Tsai et al., 2014). Accordingly, active sites mainly locate at the edges. Jaramillo et al. (2007) utilized a scanning tunneling microscope to characterize MoS_2 at atomic scale and found the linear scaling between HER exchange current density and the length of edges rather than the area of the surface, confirming that the active sites of MoS_2 nanoflakes are the edge atoms. In addition, Voiry et al. (2013) compared HER performance of MoS_2 nanoflakes in different phases and found that metallic 1T-MoS_2 shows the highest activity with a Tafel slope of only 40 mV/dec (Fig. 5.13).

5.5.3.2 Oxygen reduction reaction

ORR is a key reaction in fuel cells and metal–air batteries, and one descriptor of the volcano plot is the adsorption energy of oxygen atom (Nørskov et al., 2004). Currently, the commercial ORR catalysts are Pt nanoparticles loaded on porous carbon with a high specific surface area. However, the overpotential is as high as 0.3–0.4 V. To optimize the ORR performance, a lot of efforts have been made to develop alloy nanocatalysts. Alloying can change the electronic structures by composition and strain to increase the activity and stability. Specifically, alloying can control the occupancy and position of d band to optimize the adsorption properties of intermediates (Greeley et al., 2009).

Stamenkovic et al. (2007) synthesized a series of Pt_3M alloy nanoparticles (M = Ni, Co, Fe, Ti, and V) and found that Pt_3Co has the highest activity, as its d band center position is at the peak of the volcano plot. Another interesting alloy is the so-called Pt-skin catalysts, with Pt at the

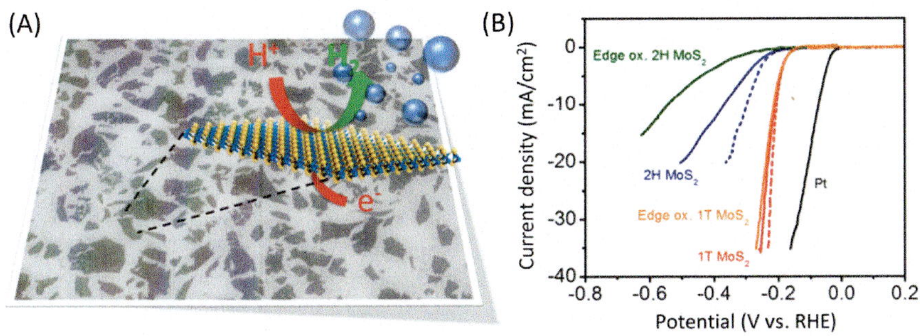

Figure 5.13 (A) SEM image and schematic for HER on MoS_2 nanoflakes. (B) Reaction polarization curves for MoS_2 catalysts in different phases compared with Pt (Voiry et al., 2013). *HER*, Hydrogen evolution reaction.

surface showing high activity and transition metals such as Ni at the core maintaining high structural stability. Pt−Ni core−shell nanoparticles with superior catalytic performance have been fabricated by acid treatment and annealing of Pt−Ni solid solution (Wang et al., 2011).

5.5.3.3 CO_2 reduction reaction

Since the industrial revolution, we human beings have been burning a lot of carbon, and emitting large amounts of CO_2 into the atmosphere, resulting in global warming and climate changes. CO_2RR is an important reaction of the reverse process, contributing to carbon neutrality.

Compared with two-electron HER, CO_2RR have multiple reaction paths, involving 2, 4, 6, 8, and 12 electrons, and producing up to 16 products, including organic molecules with 1, 2, and 3 carbon atoms (C_1, C_2, and C_3). Cu is the mostly used catalyst for CO_2RR, producing multiple-carbon compounds through C−C coupling reactions. A possible explanation for such activity is that Cu has negative *CO adsorption energy and positive *H adsorption energy (Bagger et al., 2017). Notably, the reaction path of CO_2RR is dependent on the surfaces. Cu (1 1 1) surface mainly produces methane (CH_4), and Cu (1 0 0) mainly generates ethylene (C_2H_4). By theoretical calculations, Luo et al. (2016) revealed that on Cu (1 1 1), *CO is preferably reduced to *COH, which is transformed into *CH_2 through deoxidization, and finally methane and ethylene through further hydrogenation. On Cu (1 0 0), *CO is mainly reduced to *CHO, and the coupling of two *CHO forms C_2 intermediates. Further hydrogenation of C_2 intermediates on Cu (1 0 0) requires high overpotential, but the (1 1 1) step could help reduce the overpotential, improving ethylene productivity. Choi et al. (2020) found Cu nanowires with rich step structures have a high Faraday efficiency of 70% to C_2H_4 and high stability over 200 hours. Asadi et al. (2014) discovered that vertically aligned MoS_2 in ionic liquids could convert CO_2 to CO with an overpotential of only 54 mV, and the current density reaches 65 mA/cm^2 at −0.764 V relative to reversible hydrogen electrode, which is attributed to high density of active sites at the edges.

Another important type of CO_2RR catalyst is carbon-based materials, which can be classified into metal-containing and metal-free ones. Metal-containing carbon-based catalysts are mainly TM-N_4 (TM, transition metal) structures, and metal-free carbon-based catalysts are mainly nitrogen-doped graphene quantum dots (N-GQDs). N-GQDs show high Faraday efficiency of 90% in reducing CO_2, with C_1 product mainly

CH_4, and the selectivity to ethylene and ethane can also be as high as 45% (Wu et al., 2016). For metal-containing carbon-based materials, Ju et al. (2017) calculated the adsorption energies of typical intermediates on metal and nitrogen-doped porous carbon materials and combined experiments to show that $Co-N_x$ is favorable for HER, while $Fe-N_x$ and Ni–N tend to produce CO.

5.5.3.4 Nitrogen reduction reaction

Ammonia is an important industrial raw material, widely used in the production of fertilizers, chemical products, and energy storage (Foster et al., 2018). In industry, ammonia is mainly produced by Haber process under high temperature (500°C) and high pressure (200 atm) conditions. Because of the harsh reaction conditions, as much as 1% of total energy consumption of the whole world is consumed in this process, and large amounts of greenhouse gases are released at the same time. Searching for a substitute method is important to reduce the energy consumption and the carbon emission. Electrochemical catalysis is one promising route, and high-efficiency catalyst is the key.

NRR mainly includes three elementary reactions, activation of N≡ N ($N_2 + * + H^+ + e^- \rightarrow {}^*N_2H$), reduction of $^*NH$ to $^*NH_2$ ($^*NH + H^+ + e^- \rightarrow {}^*NH_2$), and reduction of $^*NH_2$ to NH_3 ($^*NH_2 + H^+ + e^- \rightarrow NH_3(g) + {}^*$) (Montoya et al., 2015). Researchers use adsorption energy of nitrogen atom as the descriptor to plot volcano pattern, and to screen for effective catalysts (Medford et al., 2015; Skúlason et al., 2012). According to prediction, Mo, Fe, Rh, and Ru are the most efficient catalysts for NRR (Fig. 5.14). However, HER is a competing reaction in aqueous environment, and it is still a great challenge to improve the selectivity of NRR.

5.5.4 Photoelectrochemical reaction

Photoelectrochemical reaction is the fundamental process of artificial photosynthesis, including two steps, photon–electron excitation and electrochemical reaction. Different from the metal electrodes in electrochemical reaction involving electrons with continuous energy distribution, photoelectrochemical reaction uses semiconductor electrode with a bandgap, and light excitation is needed to generate nonequilibrium electrons and holes.

Semiconductors used for photoelectrochemical reactions should have bandgaps matching the solar spectra and band edge positions matching the

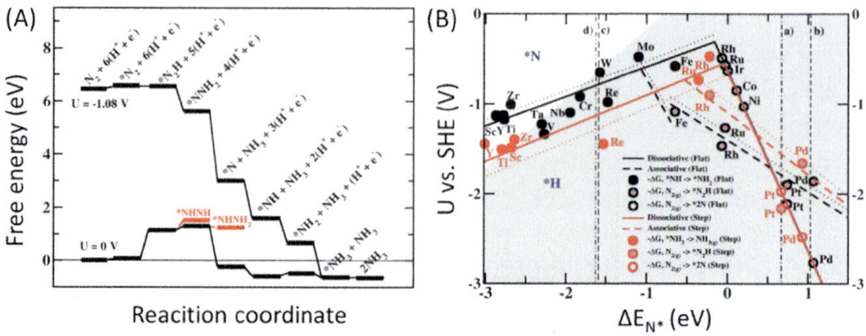

Figure 5.14 (A) Free energy diagrams for NRR on Ru(0 0 0 1) calculated from density functional theory. (B) Volcano plot for NRR on various metals (Skúlason et al., 2012). *NRR*, Nitrogen reduction reaction.

redox potentials. Meanwhile, they should be stable in electrolytes. Single material usually cannot fulfill all these requirements. In particular, stability and reactivity are usually contradictory to each other. Liu et al. designed a hybrid system of silicon and TiO_2 (Fig. 5.15), which absorbs visible and ultraviolet light to make full use of the solar spectra. Light generates electron–hole pairs in the two semiconductors and participate the HER and oxygen evolution reaction on the silicon and TiO_2 sides, respectively. At the interface, holes in silicon recombine with electrons in TiO_2. As a result, the conversion efficiency is 0.12% under simulated solar light, close to that of photosynthesis under natural light (Liu et al., 2013a).

5.6 Optical properties

Optical properties of low-dimensional materials determine their performance in lighting, display, communication, and quantum devices. For metals, interaction between near-free electrons and electromagnetic waves forms plasmons. For semiconductors, their optical behaviors are determined by electronic band structure, including type, width of bandgaps, and positions of band edges. For insulators, the atomic defects have electronic structures similar as individual atoms. And their interaction with light dominates the optical properties of insulators. Dimensionality and size effects of low-dimensional materials can effectively modulate electronic structures, and thus the optical properties.

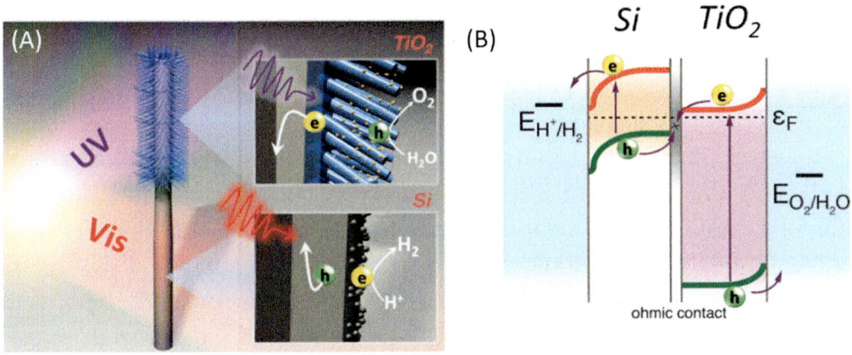

Figure 5.15 (A) Schematic for the hybrid system of Si nanocrystals and TiO_2 nanowire absorbing ultraviolet and visible light for OER and HER, respectively. (B) Energy diagram showing the transport and reaction processes of photo-generated electrons and holes (Liu et al., 2013a). *OER*, Oxygen evolution reaction.

Optical properties of materials can be discussed in either classical or quantum ways. The former considers the interaction between a continuous medium with electromagnetic waves, described by Maxwell's equations. The latter quantizes both electrons and photons, as described by quantum electrodynamics. This section will introduce plasmon resonance in metal, luminescence from band edges in semiconductors, and single-photon emission in quantum optics.

5.6.1 Surface plasmon resonance

Electromagnetic waves in dielectric materials are described as $E \propto e^{i\omega t}e^{i\boldsymbol{K}\cdot\boldsymbol{r}}$, which can be substituted into wave equation to obtain the dispersion relation $\epsilon(\omega, \boldsymbol{K})\omega^2 = c^2K^2$. When relative dielectric constant $\epsilon > 0$, waves can propagate in dielectrics. If $\epsilon < 0$, waves will be absorbed and damped.

In free electron gas model, dielectric function is expressed as $\epsilon(\omega) = 1 - \frac{\omega_p^2}{\omega^2}$, with plasmon frequency $\omega_p = \left(\frac{ne^2}{\epsilon_0 m}\right)^{1/2}$. When the frequency of incident light is equal to ω_p, $\epsilon = 0$, collective fluctuation of free electron gas at the interface between metal surfaces and dielectrics is excited, forming surface plasmon resonance (SPR).

In low-dimensional materials, electrons are confined in small space, and thus plasmon modes are also confined, becoming local surface plasmon (LSP). The resonance condition of LSP is dependent on the size and shape of low-dimensional structures, as well as environments. For spherical particles, the scattering strength is calculated as $C_{sca}(\omega) = 4\pi R^2[k(\epsilon_H)^{1/2}R]^4\left|\frac{\epsilon_p(\omega)-\epsilon_H}{\epsilon_p(\omega)+2\epsilon_H}\right|^2$.

When $\epsilon_p(\omega) + 2\epsilon_H = 0$, resonance scattering emerges (Agrawal et al., 2018). As temperature and surface adsorbates change, the resonance frequency shifts. This is the fundamental principle for LSP sensing (Fig. 5.16) (Culver et al., 2018; Willets & Van Duyne, 2007).

In quasi-1D CNTs, sp^2-hybridized electrons are confined in an atomic shell, and the Coulomb interaction among electrons becomes stronger, showing significant influences on absorption peaks (Ando, 1997). Sfeir et al. (2004) developed Rayleigh scattering technique to measure the excited states of a single-carbon nanotube. Rayleigh scattering belongs to resonance scattering (Yu & Brus, 2001), and peaks in Rayleigh spectra correspond to those peaks in absorption spectra. Liu et al. (2012) combined Rayleigh scattering and electron diffraction to characterize 206 CNTs, analyzed 522 excitation peaks, and built a database of absorption energies and chirality, which becomes a guide for the characterization of CNT chirality using optical spectra. They further used polarization-based homodyne technique to enhance the signals in absorption (Liu et al., 2014), optical imaging, and reflection spectra (Liu et al., 2013b) of CNTs. The developed high-throughput characterization method was used to analyze the chirality distribution of CNTs in devices (Yao et al., 2016).

The enhanced Coulomb interaction in CNTs renders the excitonic properties determining their optical behaviors. Using two-photon excitation spectroscopy, Wang et al. (2005) measured exciton binding energy in CNTs with 8-nm diameters to be as high as 400 meV, significantly higher

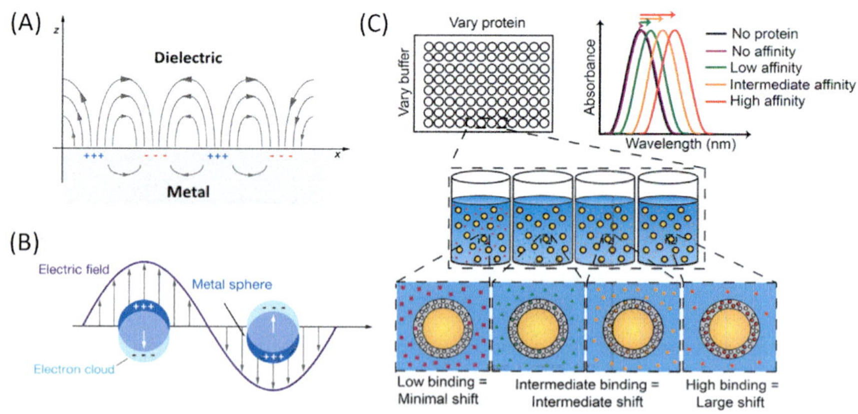

Figure 5.16 (A,B) Schematic of the principles of surface plasmons and localized surface plasmon resonance (Willets & Van Duyne, 2007). (C) Protein sensing based on local surface plasmon on Au nanoparticles (Culver et al., 2018).

than those in 3D Si and GaAs. Moreover, the strong correlation in quasi-1D CNTs is expected to give rise to Luttinger liquid behaviors (Brus, 2014). Wang's group in University of California at Berkeley adopted infrared scanning near-field optical microscopy to show that while the plasmons in metallic CNTs act as linear Luttinger liquid characteristics (Shi et al., 2015), those in semiconducting CNTs exhibit nonlinear Luttinger liquid properties (Wang et al., 2020).

5.6.2 Band edge luminescence in semiconductors

The luminescence processes in low-dimensional materials include the excitation of electrons by light absorption, transport, and recombination of electrons and holes with photon emission. According to types of excitations, they can be classified as photoluminescence, electroluminescence, and stimulated luminescence. Dimensionality and size effects can be used to modulate electronic structures, through the quantum confinement effect, blueshift of energy bands, and light polarization effects.

The wavelength of fluorescence in semiconductor quantum dots can be controlled by changing their size and chemical composition (Smith & Nie, 2010). The development of the growth of nanocrystals makes it possible to synthesize II–VI semiconductors with uniform sizes, such as CdSe and PbSe nanocrystals (Du et al., 2002; Peng & Peng, 2001). By controlling the precursor ratio, Weidman et al. obtained uniform PbS nanocrystals with size distribution range within 5%. The diameter of quantum dots is tuned from 4.3 to 8.4 nm, and the corresponding wavelength of fluorescence at 1000–1800 nm (1.25 to 0.70 eV), with the full width at half maximum (FWHM) of only 20 meV (Weidman et al., 2014).

Kovalenko et al. reported the synthesis and optical properties of lead halide nanoparticles. By changing the composition and size of nanoparticles, the wavelength of fluorescence could cover from ultraviolet to near-infrared range. The FWHMs for green and red lights are 20 and 30 nm, respectively. The color gamut ranges for red, green, and blue reach 140% of the National Television System Committee specification (Kovalenko et al., 2017).

Low-dimensional structures have a high ratio of surface atoms, and surface defects can act as recombination center for electrons and holes, decreasing luminescence efficiency. An effective strategy is to grow heterogeneous core–shell structures, where electrons are confined in semiconductors with smaller bandgap to avoid the influences of surface states (Efros & Nesbitt, 2016). Zhou et al. delicately controlled the growth of CdSe–CdS core–shell structures, which enabled the quantitative study of the effects of core size, shell thickness, and surface ligands. It is found that nanoparticles with 3–8 monolayers of CdS shell show ideal luminescence properties, with nonblinking, antibleaching, monoexponential PL decay, etc. (Zhou et al., 2017).

The electronic structures and optical properties of nanowires can be modulated by chemical composition. For example, the luminescence wavelength of III–V $GaAs_xP_{1-x}$ nanowire array can be tuned in the range of 550–900 nm (Mårtensson et al., 2004). Zhang et al. (2015) used solution method to grow $CsPbX_3$ (X = Cl, Br, and I) nanowires, and the emission wavelength covers from blue to near-infrared range. Yang et al. (2014) introduced gradient change in composition of a single CdS–CdSe nanowire, and the photoluminescence wavelength can be continuously tuned from 507 to 633 nm (Fig. 5.17). Based on this, ultraminiaturized single-nanowire spectrometer can be fabricated to perform spectra analysis for both visible-range monochromatic and broadband light reconstruction (Yang et al., 2019).

The size effects of low-dimensional materials can not only induce blueshift of bandgap but also alter the relative position of energy bands due to the crystal anisotropy, leading to the change of types of band structures. Bulk MoS_2 is an indirect semiconductor with a bandgap of 1.3 eV. Surprisingly, monolayer MoS_2 is a direct semiconductor with a bandgap of 1.9 eV. Accordingly, monolayer MoS_2 shows significantly enhanced photoluminescence, with an intensity 10^4 times as that of bulk one (Mak et al., 2010; Splendiani et al., 2010).

5.6.3 Single-photon source

Single-photon sources are the key components in quantum communication. For a simplified two-level system, when an electron occupies the excited state, another electron cannot be excited into the same excited level, because of the Pauli exclusion principle for Fermions. Therefore it can only emit one photon during the lifetime of spontaneous emission, corresponding to a single-photon source. Until now, it is still challenging to realize such ideal single-photon source, and intensive efforts have been devoted to search for promising candidates in solid-state materials (Aharonovich et al., 2016).

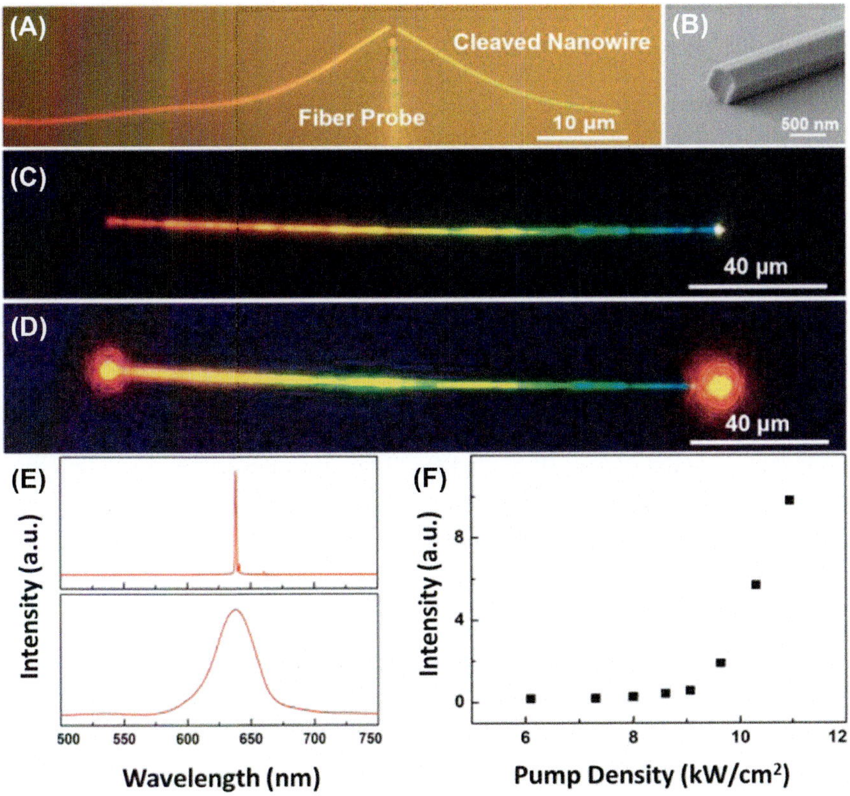

Figure 5.17 CdS–CdSe nanowire with gradient change in composition and bandgap. (A) Continuous change in color of CdS–CdSe nanowire with gradient change in composition. (B) Cleavage of the nanowire. (C,D) Optical microscope images of CdS–CdSe nanowire under low and high pump power density, showing fluorescence and lasing, respectively. (E) Spectra for the nanowire emitting fluorescence and laser. (F) Laser strength as a function of pump power density (Yang et al., 2014).

To realize the single-photon emission, it requires there is only one luminescence center under excitation. Therefore it is necessary to control the luminescence centers precisely. There are two sources for single-photon emission, that is, quantum dots of semiconductors and color centers of insulators. The formers are semiconductor nanostructures, and the transition between band edges of intrinsic band structures contributes to the emission. The latter are atomic defects in wide-gap semiconductors, and the transition between deep defect levels gives rise to the emission. Single-photon emission has been realized in quantum dots (Utzat et al., 2019), carbon nanotubes (Ishii et al., 2018; Ma et al., 2015), and 2D materials (He et al., 2015; Tran et al., 2016a).

The single-photon emission can be analyzed by second-order correlation function, defined as $g^2(\tau) = \frac{\langle \hat{a}^+(t)\hat{a}^+(t+\tau)\hat{a}(t)\hat{a}(t+\tau)\rangle}{\langle \hat{a}^+(t)\hat{a}(t)\rangle}$, which represents when the first photon is detected at time t, the conditional probability of detecting the second photon at $t+\tau$. Usually, $g^2(\tau) < 1/2$ corresponds to the single-photon state.

2D h-BN has a large bandgap of $\sim$6 eV and high chemical stability, which provides a stable inert environment for defect luminescence centers. Bright and stable single-photon emissions with narrow bandwidth have been observed in h-BN, covering visible to near-infrared range (Tran et al., 2016b). There are many discussions on the sources of single-photon emitting centers in h-BN. In one case, Tran et al. measured $g^2(\tau)$ and found it is smaller than 0.5 at $\tau = 0$, confirming its single-photon emission characteristics (Fig. 5.18). The luminescence wavelength is 1.99 eV, and by combining first-principles calculations, the luminescence center is proposed to be N_BV_N defect (Tran et al., 2016a, 2016b).

5.6.4 Nonlinear optics

The propagation of light in dielectrics can induce polarization, including both linear and nonlinear polarization, that is, $\vec{P}(t) = \chi^{(1)}\vec{E}(t) + \chi^{(2)}\vec{E}(t)^2 + \chi^{(3)}\vec{E}(t)^3 + \cdots$, where $\vec{E}(t)$ is the electric field from light, $\vec{P}(t)$ is the polarization, and $\chi^{(n)}$ is the n-order susceptibility. The first term is linear polarization, describing optical

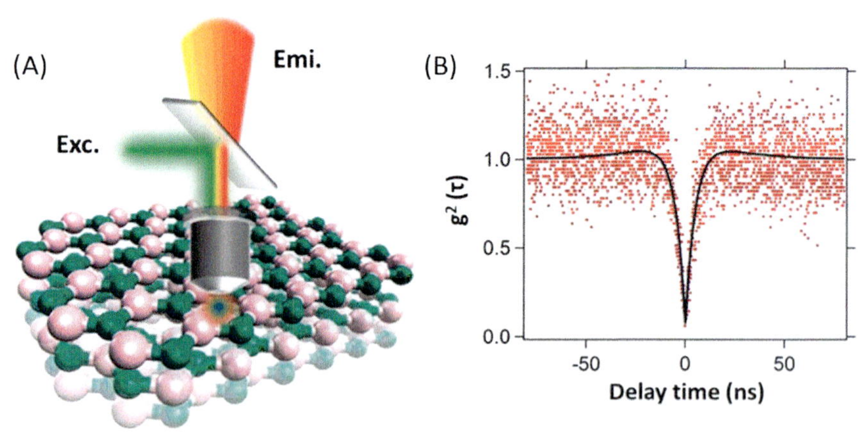

Figure 5.18 (A) Schematic of luminescence center in h-BN and experimental setup. (B) Second-order correlation function for single-photon emission in h-BN (Tran et al., 2016b).

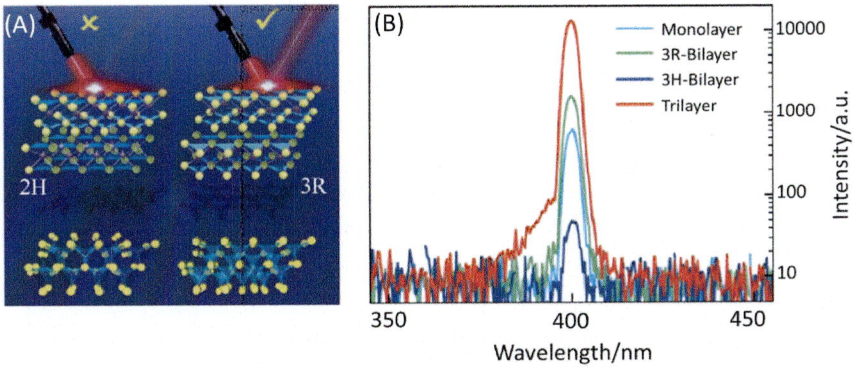

Figure 5.19 (A) The relation between SHG and stacking symmetry. (B) SHG spectra of MoS_2 with different numbers of layer and stacking manners (Li et al., 2018). *SHG*, Second harmonic generation.

phenomena such as reflection, interference, and diffraction, while the other terms are nonlinear polarization, representing processes involving multiple photons. Typical second-order nonlinear optics include second harmonic generation (SHG), sum frequency generation (SFG), and difference frequency generation (DFG) (Autere et al., 2018). Nonlinear optics is sensitive to crystal structures and symmetries. When crystals have inversion center, the even-order nonlinear susceptibility is zero. Therefore nonlinear optics can be used to detect the crystal structures, stacking, and magnetic structures of low-dimensional materials.

Transition metal dichalcogenides like MoS_2 have two different interlayer stackings, that is, 2H and 3R. Since 2H phase has inversion symmetry, the second-order susceptibility is zero (Li et al., 2013). Li et al. (2018) grow MoS_2 nanoribbons on monolayer MoS_2 using vapor–liquid–solid method and found the SHG signal intensity of MoS_2 nanoribbons is significantly smaller than that of monolayer MoS_2, indicating the stacking between MoS_2 nanoribbons and monolayer is 2H type (Fig. 5.19).

5.7 Summary

This chapter discussed how geometrical and intrinsic effects influence the physical and chemical properties of low-dimensional materials, giving rise to many intriguing properties. Classical examples include mechanical strength close to theoretical limit, ballistic transport in electrical and thermal behaviors, the coupling between ferroelectricity and ferromagnetism, catalytically active edges of MoS_2 nanoflakes, shape-sensitive SPR in metal nanoparticles, and single-photon emission of defects.

Superior properties of low-dimensional materials could find promising applications in mechanically enhanced composites, nanoelectronics and information devices, energy storage, and conversion. Furthermore, the coupling between different properties can lead to novel principles, functionalities, and applications, including the control of electron mobility of low-dimensional semiconductors and catalytic activities by strain, the modulation of magnetic properties by electric fields, and the coupling between surface plasmon and catalytic reactions.

References

Agrawal, A., Cho, S. H., Zandi, O., Ghosh, S., Johns, R. W., & Milliron, D. J. (2018). Localized surface plasmon resonance in semiconductor nanocrystals. *Chemical Reviews*, *118*(6), 3121–3207.

Aharonovich, I., Englund, D., & Toth, M. (2016). Solid-state single-photon emitters. *Nature Photonics*, *10*(10), 631–641.

Ando, T. (1997). Excitons in carbon nanotubes. *Journal of the Physical Society of Japan*, *66* (4), 1066–1073.

Asadi, M., Kumar, B., Behranginia, A., Rosen, B. A., Baskin, A., Repnin, N., Pisasale, D., Phillips, P., Zhu, W., Haasch, R., Klie, R. F., Král, P., Abiade, J., & Salehi-Khojin, A. (2014). Robust carbon dioxide reduction on molybdenum disulphide edges. *Nature Communications*, *5*(1), 4470.

Autere, A., Jussila, H., Dai, Y. Y., Wang, Y. D., Lipsanen, H., & Sun, Z. P. (2018). Nonlinear optics with 2D layered materials. *Advanced Materials*, *30*(24), 1705963.

Bae, M. H., Li, Z. Y., Aksamija, Z., Martin, P. N., Xiong, F., Ong, Z. Y., Knezevic, I., & Pop, E. (2013). Ballistic to diffusive crossover of heat flow in graphene ribbons. *Nature Communications*, *4*, 1734.

Bagger, A., Ju, W., Varela, A. S., Strasser, P., & Rossmeisl, J. (2017). Electrochemical CO_2 reduction: A classification problem. *Chemphyschem: A European Journal of Chemical Physics and Physical Chemistry*, *18*(22), 3266–3273.

Bai, Y. X., Yue, H. J., Wang, J., Shen, B. Y., Sun, S. L., Wang, S. J., Wang, H. D., Li, X. D., Xu, Z. P., Zhang, R. F., & Wei, F. (2020). Super-durable ultralong carbon nanotubes. *Science (New York, N.Y.)*, *369*(6507), 1104–1106.

Balandin, A. A. (2011). Thermal properties of graphene and nanostructured carbon materials. *Nature Materials*, *10*(8), 569–581.

Balandin, A. A., Ghosh, S., Bao, W. Z., Calizo, I., Teweldebrhan, D., Miao, F., & Lau, C. N. (2008). Superior thermal conductivity of single-layer graphene. *Nano Letters*, *8* (3), 902–907.

Balandin, A., & Wang, K. L. (1998). Significant decrease of the lattice thermal conductivity due to phonon confinement in a free-standing semiconductor quantum well. *Physical Review B*, *58*(3), 1544–1549.

Bonilla, M., Kolekar, S., Ma, Y. J., Diaz, H. C., Kalappattil, V., Das, R., Eggers, T., Gutierrez, H. R., Phan, M. H., & Batzill, M. (2018). Strong room-temperature ferromagnetism in VSe_2 monolayers on van der Waals substrates. *Nature Nanotechnology, 13*(4), 289–293.

Brus, L. (2014). Size, dimensionality, and strong electron correlation in nanoscience. *Accounts of Chemical Research, 47*(10), 2951–2959.

Cao, Y., Fatemi, V., Demir, A., Fang, S., Tomarken, S. L., Luo, J. Y., Sanchez-Yamagishi, J. D., Watanabe, K., Taniguchi, T., Kaxiras, E., Ashoori, R. C., & Jarillo-Herrero, P. (2018a). Correlated insulator behaviour at half-filling in magic-angle graphene superlattices. *Nature, 556*(7699), 80–84.

Cao, Y., Fatemi, V., Fang, S., Watanabe, K., Taniguchi, T., Kaxiras, E., & Jarillo-Herrero, P. (2018b). Unconventional superconductivity in magic-angle graphene superlattices. *Nature, 556*(7699), 43–50.

Chen, J. H., Jang, C., Xiao, S. D., Ishigami, M., & Fuhrer, M. S. (2008). Intrinsic and extrinsic performance limits of graphene devices on SiO_2. *Nature Nanotechnology, 3*(4), 206–209.

Chen, S. S., Wu, Q. Z., Mishra, C., Kang, J. Y., Zhang, H. J., Cho, K. J., Cai, W. W., Balandin, A. A., & Ruoff, R. S. (2012). Thermal conductivity of isotopically modified graphene. *Nature Materials, 11*(3), 203–207.

Cho, S., Kim, S., Kim, J. H., Zhao, J., Seok, J., Keum, D. H., Baik, J., Choe, D. H., Chang, K. J., Suenaga, K., Kim, S. W., Lee, Y. H., & Yang, H. (2015). Phase patterning for ohmic homojunction contact in $MoTe_2$. *Science (New York, N.Y.), 349*(6248), 625–628.

Choi, C., Kwon, S., Cheng, T., Xu, M., Tieu, P., Lee, C., Cai, J., Lee, H. M., Pan, X., Duan, X., Goddard, W. A., & Huang, Y. (2020). Highly active and stable stepped Cu surface for enhanced electrochemical CO_2 reduction to C_2H_4. *Nature Catalysis, 3*(10), 804–812.

Chuang, S., Gao, Q., Kapadia, R., Ford, A. C., Guo, J., & Javey, A. (2013). Ballistic InAs nanowire transistors. *Nano Letters, 13*(2), 555–558.

Culver, H. R., Wechsler, M. E., & Peppas, N. A. (2018). Label-free detection of tear biomarkers using hydrogel-coated gold nanoshells in a localized surface plasmon resonance-based biosensor. *ACS Nano, 12*(9), 9342–9354.

Dang, C. Q., Chou, J. P., Dai, B., Chou, C. T., Yang, Y., Fan, R., Lin, W. T., Meng, F. L., Hu, A., Zhu, J. Q., Han, J. C., Minor, A. M., Li, J., & Lu, Y. (2021). Achieving large uniform tensile elasticity in microfabricated diamond. *Science (New York, N.Y.), 371*(6524), 76–78.

Deng, Y. J., Yu, Y. J., Song, Y. C., Zhang, J. Z., Wang, N. Z., Sun, Z. Y., Yi, Y. F., Wu, Y. Z., Wu, S. W., Zhu, J. Y., Wang, J., Chen, X. H., & Zhang, Y. B. (2018). Gate-tunable room-temperature ferromagnetism in two-dimensional Fe_3GeTe_2. *Nature, 563*(7729), 94–99.

Deng, Y. J., Yu, Y. J., Shi, M. Z., Guo, Z. X., Xu, Z. H., Wang, J., Chen, X. H., & Zhang, Y. B. (2020). Quantum anomalous Hall effect in intrinsic magnetic topological insulator $MnBi_2Te_4$. *Science (New York, N.Y.), 367*(6480), 895–900.

Denpoh, K. (1998). Modeling of rarefied gas heat conduction between wafer and susceptor. *IEEE Transactions on Semiconductor Manufacturing, 11*(1), 25–29.

Ding, F., Jiao, K., Wu, M. Q., & Yakobson, B. I. (2007). Pseudoclimb and dislocation dynamics in superplastic nanotubes. *Physical Review Letters, 98*(7), 075503.

Du, H., Chen, C. L., Krishnan, R., Krauss, T. D., Harbold, J. M., Wise, F. W., Thomas, M. G., & Silcox, J. (2002). Optical properties of colloidal PbSe nanocrystals. *Nano Letters, 2*(11), 1321–1324.

Du, X., Skachko, I., Barker, A., & Andrei, E. Y. (2008). Approaching ballistic transport in suspended graphene. *Nature Nanotechnology, 3*(8), 491–495.

Efros, A. L., & Nesbitt, D. J. (2016). Origin and control of blinking in quantum dots. *Nature Nanotechnology, 11*(8), 661–671.

Fong, K. Y., Li, H. K., Zhao, R. K., Yang, S., Wang, Y., & Zhang, X. (2019). Phonon heat transfer across a vacuum through quantum fluctuations. *Nature, 576*(7786), 243–247.

Foster, S. L., Bakovic, S. I. P., Duda, R. D., Maheshwari, S., Milton, R. D., Minteer, S. D., Janik, M. J., Renner, J. N., & Greenlee, L. F. (2018). Catalysts for nitrogen reduction to ammonia. *Nature Catalysis, 1*(7), 490–500.

Fuechsle, M., Miwa, J. A., Mahapatra, S., Ryu, H., Lee, S., Warschkow, O., Hollenberg, L. C. L., Klimeck, G., & Simmons, M. Y. (2012). A single-atom transistor. *Nature Nanotechnology*, 7(4), 242–246.

Gao, H., Ji, B., Jäger, I. L., Arzt, E., & Fratzl, P. (2003). Materials become insensitive to flaws at nanoscale: Lessons from nature. *Proceedings of the National Academy of Sciences, 100*(10), 5597–5600.

Gong, C., Li, L., Li, Z. L., Ji, H. W., Stern, A., Xia, Y., Cao, T., Bao, W., Wang, C. Z., Wang, Y. A., Qiu, Z. Q., Cava, R. J., Louie, S. G., Xia, J., & Zhang, X. (2017). Discovery of intrinsic ferromagnetism in two-dimensional van der Waals crystals. *Nature, 546*(7657), 265–269.

Greeley, J. (2016). Theoretical heterogeneous catalysis: Scaling relationships and computational catalyst design. *Annual Review of Chemical and Biomolecular Engineering*, 7, 605–635.

Greeley, J., Stephens, I. E. L., Bondarenko, A. S., Johansson, T. P., Hansen, H. A., Jaramillo, T. F., Rossmeisl, J., Chorkendorff, I., & Nørskov, J. K. (2009). Alloys of platinum and early transition metals as oxygen reduction electrocatalysts. *Nature Chemistry, 1*(7), 552–556.

Griffith, A. A. (1921). VI. The phenomena of rupture and flow in solids. *Philosophical Transactions of the Royal Society of London. Series A, Containing Papers of a Mathematical or Physical Character, 221*, 163–198.

Gunawan, O., Sekaric, L., Majumdar, A., Rooks, M., Appenzeller, J., Sleight, J. W., Guha, S., & Haensch, W. (2008). Measurement of carrier mobility in silicon nanowires. *Nano Letters, 8*(6), 1566–1571.

Haruta, M., Yamada, N., Kobayashi, T., & Iijima, S. (1989). Gold catalysts prepared by coprecipitation for low-temperature oxidation of hydrogen and of carbon monoxide. *Journal of Catalysis, 115*(2), 301–309.

Haruta, M., Kobayashi, T., Sano, H., & Yamada, N. (1987). Novel gold catalysts for the oxidation of carbon monoxide at a temperature far below 0°C. *Chemistry Letters, 16* (2), 405–408.

He, Y. M., Clark, G., Schaibley, J. R., He, Y., Chen, M. C., Wei, Y. J., Ding, X., Zhang, Q., Yao, W., Xu, X. D., Lu, C. Y., & Pan, J. W. (2015). Single quantum emitters in monolayer semiconductors. *Nature Nanotechnology, 10*(6), 497–502.

Hinnemann, B., Moses, P. G., Bonde, J., Jørgensen, K. P., Nielsen, J. H., Horch, S., Chorkendorff, I., & Nørskov, J. K. (2005). Biomimetic hydrogen evolution: MoS_2 nanoparticles as catalyst for hydrogen evolution. *Journal of the American Chemical Society, 127*(15), 5308–5309.

Hsia, F.-C., Tang, D.-M., Jevasuwan, W., Fukata, N., Zhou, X., Mitome, M., Bando, Y., Nordling, T. E. M., & Golberg, D. (2019). Realization and direct observation of five normal and parametric modes in silicon nanowire resonators by *in situ* transmission electron microscopy. *Nanoscale Advances, 1*(5), 1784–1790.

Hu, S., Lozada-Hidalgo, M., Wang, F. C., Mishchenko, A., Schedin, F., Nair, R. R., Hill, E. W., Boukhvalov, D. W., Katsnelson, M. I., Dryfe, R. A. W., Grigorieva, I. V., Wu, H. A., & Geim, A. K. (2014). Proton transport through one-atom-thick crystals. *Nature, 516*(7530), 227–230.

Huang, B., Clark, G., Klein, D. R., MacNeill, D., Navarro-Moratalla, E., Seyler, K. L., Wilson, N., McGuire, M. A., Cobden, D. H., Xiao, D., Yao, W., Jarillo-Herrero, P.,

& Xu, X. D. (2018). Electrical control of 2D magnetism in bilayer CrI_3. *Nature Nanotechnology*, *13*(7), 544–548.

Huang, B., Clark, G., Navarro-Moratalla, E., Klein, D. R., Cheng, R., Seyler, K. L., Zhong, D., Schmidgall, E., McGuire, M. A., Cobden, D. H., Yao, W., Xiao, D., Jarillo-Herrero, P., & Xu, X. D. (2017a). Layer-dependent ferromagnetism in a van der Waals crystal down to the monolayer limit. *Nature*, *546*(7657), 270–273.

Huang, Z. X., Wang, J. Q., Bai, S. Y., Guan, J. W., Zhang, F. T., & Tang, Z. N. (2017b). Size effect of heat transport in microscale gas gap. *IEEE Transactions on Industrial Electronics*, *64*(9), 7387–7391.

Huang, B., Cenker, J., Zhang, X. O., Ray, E. L., Song, T. C., Taniguchi, T., Watanabe, K., McGuire, M. A., Xiao, D., & Xu, X. D. (2020). Tuning inelastic light scattering *via* symmetry control in the two-dimensional magnet CrI_3. *Nature Nanotechnology*, *15* (3), 212–216.

Huang, J. Y., Chen, S., Wang, Z. Q., Kempa, K., Wang, Y. M., Jo, S. H., Chen, G., Dresselhaus, M. S., & Ren, Z. F. (2006). Superplastic carbon nanotubes. *Nature*, *439* (7074), 281.

Ishii, A., He, X. W., Hartmann, N. F., Machiya, H., Htoon, H., Doorn, S. K., & Kato, Y. K. (2018). Enhanced single-photon emission from carbon-nanotube dopant states coupled to silicon microcavities. *Nano Letters*, *18*(6), 3873–3878.

Jaramillo, T. F., Jørgensen, K. P., Bonde, J., Nielsen, J. H., Horch, S., & Chorkendorff, I. (2007). Identification of active Edge sites for electrochemical H_2 evolution from MoS_2 nanocatalysts. *Science (New York, N.Y.)*, *317*(5834), 100–102.

Javey, A., Guo, J., Wang, Q., Lundstrom, M., & Dai, H. J. (2003). Ballistic carbon nanotube field-effect transistors. *Nature*, *424*(6949), 654–657.

Jiang, S. W., Li, L. Z., Wang, Z. F., Mak, K. F., & Shan, J. (2018a). Controlling magnetism in 2D CrI_3 by electrostatic doping. *Nature Nanotechnology*, *13*(7), 549–553.

Jiang, S. W., Shan, J., & Mak, K. F. (2018b). Electric-field switching of two-dimensional van der Waals magnets. *Nature Materials*, *17*(5), 406–410.

Johansen, O., Risinggard, V., Sudbo, A., Linder, J., & Brataas, A. (2019). Current control of magnetism in two-dimensional Fe_3GeTe_2. *Physical Review Letters*, *122*(21), 217203.

Ju, W., Bagger, A., Hao, G.-P., Varela, A. S., Sinev, I., Bon, V., Roldan Cuenya, B., Kaskel, S., Rossmeisl, J., & Strasser, P. (2017). Understanding activity and selectivity of metal-nitrogen-doped carbon catalysts for electrochemical reduction of CO_2. *Nature Communications*, *8*(1), 944.

Khomskii, D. I. (2006). Multiferroics: Different ways to combine magnetism and ferroelectricity. *Journal of Magnetism and Magnetic Materials*, *306*(1), 1–8.

Khomskii, D. (2009). Trend: Classifying multiferroics: Mechanisms and effects. *Physics*, *2*, 20.

Khorshidi, A., Violet, J., Hashemi, J., & Peterson, A. A. (2018). How strain can break the scaling relations of catalysis. *Nature Catalysis*, *1*(4), 263–268.

Kim, P., Shi, L., Majumdar, A., & McEuen, P. L. (2001). Thermal transport measurements of individual multiwalled nanotubes. *Physical Review Letters*, *87*(21), 215502.

Kovalenko, M. V., Protesescu, L., & Bodnarchuk, M. I. (2017). Properties and potential optoelectronic applications of lead halide perovskite nanocrystals. *Science (New York, N.Y.)*, *358*(6364), 745–750.

Lai, Y. F., Song, Z. G., Wan, Y., Xue, M. Z., Yang, W. Y., Du, H. L., & Yang, J. B. (2019). Two-dimensional ferromagnetism and driven ferroelectricity in van der Waals $CuCrP_2S_6$. *Nanoscale*, *11*(12), 5163–5170.

Lee, C., Wei, X. D., Kysar, J. W., & Hone, J. (2008). Measurement of the elastic properties and intrinsic strength of monolayer graphene. *Science (New York, N.Y.)*, *321*(5887), 385–388.

Lee, C. Y., Choi, W., Han, J. H., & Strano, M. S. (2010). Coherence resonance in a single-Walled carbon nanotube ion channel. *Science (New York, N.Y.)*, *329*(5997), 1320–1324.

Leonard, F., & Tersoff, J. (1999). Novel length scales in nanotube devices. *Physical Review Letters*, *83*(24), 5174–5177.

Li, D. Y., Wu, Y. Y., Kim, P., Shi, L., Yang, P. D., & Majumdar, A. (2003). Thermal conductivity of individual silicon nanowires. *Applied Physics Letters*, *83*(14), 2934–2936.

Li, J. H., Li, Y., Du, S. Q., Wang, Z., Gu, B. L., Zhang, S. C., He, K., Duan, W. H., & Xu, Y. (2019a). Intrinsic magnetic topological insulators in van der Waals layered $MnBi_2Te_4$-family materials. *Science Advances*, *5*(6), eaaw5685.

Li., Lin, S. S., Zhao, Y. C., Wu, W., Wang, J., Hu, Z., Shen, Z. H., Tang, Y. D., Wang, D. M., Zhang, J. Y., Zhu, Q., Chu, H., Zhao, L. Q., Liu, W. J., Sun, C., Taniguchi, Z. P., Osada, T., Chen, M., Xu, W., Wee, Q. H., et al. (2018). Vapour–liquid–solid growth of monolayer MoS_2 nanoribbons. *Nature Materials*, *17* (6), 535–542.

Li, T. X., Jiang, S. W., Sivadas, N., Wang, Z. F., Xu, Y., Weber, D., Goldberger, J. E., Watanabe, K., Taniguchi, T., Fennie, C. J., Mak, K. F., & Shan, J. (2019b). Pressure-controlled interlayer magnetism in atomically thin CrI_3. *Nature Materials*, *18*(12), 1303–1308.

Li, Y. L., Rao, Y., Mak, K. F., You, Y. M., Wang, S. Y., Dean, C. R., & Heinz, T. F. (2013). Probing symmetry properties of few-layer MoS_2 and h-BN by optical second-harmonic generation. *Nano Letters*, *13*(7), 3329–3333.

Liang, W. J., Bockrath, M., Bozovic, D., Hafner, J. H., Tinkham, M., & Park, H. (2001). Fabry-Perot interference in a nanotube electron waveguide. *Nature*, *411*(6838), 665–669.

Lim, J. W., Hippalgaonkar, K., Andrews, S. C., Majumdar, A., & Yang, P. D. (2012). Quantifying surface roughness effects on phonon transport in silicon nanowires. *Nano Letters*, *12*(5), 2475–2482.

Lin, Z. S., Lohmann, M., Ali, Z. A., Tang, C., Li, J. X., Xing, W. Y., Zhong, J. N., Jia, S., Han, W., Coh, S., Beyermann, W., & Shi, J. (2018). Pressure-induced spin reorientation transition in layered ferromagnetic insulator $Cr_2Ge_2Te_6$. *Physical Review Materials*, *2*(5), 051004.

Liu, C., Tang, J. Y., Chen, H. M., Liu, B., & Yang, P. D. (2013a). A fully integrated nanosystem of semiconductor nanowires for direct solar water splitting. *Nano Letters*, *13*(6), 2989–2992.

Liu, K. H., Hong, X. P., Zhou, Q., Jin, C. H., Li, J. H., Zhou, W. W., Liu, J., Wang, E. G., Zettl, A., & Wang, F. (2013b). High-throughput optical imaging and spectroscopy of individual carbon nanotubes in devices. *Nature Nanotechnology*, *8*(12), 917–922.

Liu, C., Wang, Y. C., Li, H., Wu, Y., Li, Y. X., Li, J. H., He, K., Xu, Y., Zhang, J. S., & Wang, Y. Y. (2020). Robust axion insulator and Chern insulator phases in a two-dimensional antiferromagnetic topological insulator. *Nature Materials*, *19*(5), 522–527.

Liu, K. H., Deslippe, J., Xiao, F. J., Capaz, R. B., Hong, X. P., Aloni, S., Zettl, A., Wang, W. L., Bai, X. D., Louie, S. G., Wang, E. G., & Wang, F. (2012). An atlas of carbon nanotube optical transitions. *Nature Nanotechnology*, 7(5), 325–329.

Liu, K. H., Hong, X. P., Choi, S., Jin, C. H., Capaz, R. B., Kim, J., Wang, W. L., Bai, X. D., Louie, S. G., Wang, E. G., & Wang, F. (2014). Systematic determination of absolute absorption cross-section of individual carbon nanotubes. *Proceedings of the National Academy of Sciences USA*, *111*(21), 7564–7569.

Liu, Y., Guo, J., Zhu, E. B., Liao, L., Lee, S. J., Ding, M. N., Shakir, I., Gambin, V., Huang, Y., & Duan, X. F. (2018). Approaching the Schottky–Mott limit in van der Waals metal-semiconductor junctions. *Nature*, *557*(7707), 696–700.

Lozada-Hidalgo, M., Hu, S., Marshall, O., Mishchenko, A., Grigorenko, A. N., Dryfe, R. A. W., Radha, B., Grigorieva, I. V., & Geim, A. K. (2016). Sieving hydrogen isotopes through two-dimensional crystals. *Science (New York, N.Y.)*, *351*(6268), 68–70.

Lu, L., Sui, M. L., & Lu, K. (2000). Superplastic extensibility of nanocrystalline copper at room temperature. *Science (New York, N.Y.)*, *287*(5457), 1463–1466.

Luo, W., Nie, X., Janik, M. J., & Asthagiri, A. (2016). Facet dependence of CO_2 reduction paths on Cu electrodes. *ACS Catalysis*, *6*(1), 219–229.

Ma, T., Liu, Z. B., Wen, J. X., Gao, Y., Ren, X. B. A., Chen, H. J., Jin, C. H., Ma, X. L., Xu, N. S., Cheng, H. M., & Ren, W. C. (2017). Tailoring the thermal and electrical transport properties of graphene films by grain size engineering. *Nature Communications*, *8*, 14486.

Ma, X. D., Hartmann, N. F., Baldwin, J. K. S., Doorn, S. K., & Htoon, H. (2015). Room-temperature single-photon generation from solitary dopants of carbon nanotubes. *Nature Nanotechnology*, *10*(8), 671–675.

Mak, K. F., Lee, C., Hone, J., Shan, J., & Heinz, T. F. (2010). Atomically thin MoS_2: A new direct-gap semiconductor. *Physical Review Letters*, *105*(13)136805.

Mårtensson, T., Svensson, C. P. T., Wacaser, B. A., Larsson, M. W., Seifert, W., Deppert, K., Gustafsson, A., Wallenberg, L. R., & Samuelson, L. (2004). Epitaxial III–V nanowires on silicon. *Nano Letters*, *4*(10), 1987–1990.

Medford, A. J., Vojvodic, A., Hummelshøj, J. S., Voss, J., Abild-Pedersen, F., Studt, F., Bligaard, T., Nilsson, A., & Nørskov, J. K. (2015). From the Sabatier principle to a predictive theory of transition-metal heterogeneous catalysis. *Journal of Catalysis*, *328*, 36–42.

Mizuno, K., Ishii, J., Kishida, H., Hayamizu, Y., Yasuda, S., Futaba, D. N., Yumura, M., & Hata, K. (2009). A black body absorber from vertically aligned single-walled carbon nanotubes. *Proceedings of the National Academy of Sciences USA*, *106*(15), 6044–6047.

Montoya, J. H., Tsai, C., Vojvodic, A., & Nørskov, J. K. (2015). The challenge of electrochemical ammonia synthesis: A new perspective on the role of nitrogen scaling relations. *ChemSusChem*, *8*(13), 2180–2186.

Narayanaswamy, A., Shen, S., & Chen, G. (2008). Near-field radiative heat transfer between a sphere and a substrate. *Physical Review B*, *78*(11), 115303.

Nishihara, T., Takakura, A., Miyauchi, Y., & Itami, K. (2018). Ultra-narrow-band near-infrared thermal exciton radiation in intrinsic one-dimensional semiconductors. *Nature Communications*, *9*(1), 3144.

Nørskov, J. K., Rossmeisl, J., Logadottir, A., Lindqvist, L., Kitchin, J. R., Bligaard, T., & Jónsson, H. (2004). Origin of the overpotential for oxygen reduction at a fuel-cell cathode. *The Journal of Physical Chemistry. B*, *108*(46), 17886–17892.

Nørskov, J. K., Bligaard, T., Logadottir, A., Kitchin, J. R., Chen, J. G., Pandelov, S., & Stimming, U. (2005). Trends in the exchange current for hydrogen evolution. *Journal of the Electrochemical Society*, *152*(3), J23.

Novoselov, K. S., Geim, A. K., Morozov, S. V., Jiang, D., Katsnelson, M. I., Grigorieva, I. V., Dubonos, S. V., & Firsov, A. A. (2005). Two-dimensional gas of massless Dirac fermions in graphene. *Nature*, *438*(7065), 197–200.

O'Hara, D. J., Zhu, T. C., Trout, A. H., Ahmed, A. S., Luo, Y. K., Lee, C. H., Brenner, M. R., Rajan, S., Gupta, J. A., McComb, D. W., & Kawakami, R. K. (2018). Room temperature intrinsic ferromagnetism in epitaxial manganese selenide films in the monolayer limit. *Nano Letters*, *18*(5), 3125–3131.

Otrokov, M. M., Klimovskikh, I. I., Bentmann, H., Estyunin, D., Zeugner, A., Aliev, Z. S., Gass, S., Wolter, A. U. B., Koroleva, A. V., Shikin, A. M., Blanco-Rey, M., Hoffmann, M., Rusinov, I. P., Vyazovskaya, A. Y., Eremeev, S. V., Koroteev, Y. M., Kuznetsov, V. M., Freyse, F., Sanchez-Barriga, J., et al. (2019). Prediction and observation of an antiferromagnetic topological insulator. *Nature*, *576*(7787), 416–422.

Pan, X., Fan, Z., Chen, W., Ding, Y., Luo, H., & Bao, X. (2007). Enhanced ethanol production inside carbon-nanotube reactors containing catalytic particles. *Nature Materials*, *6*(7), 507–511.

Pan, X., & Bao, X. (2011). The effects of confinement inside carbon nanotubes on catalysis. *Accounts of Chemical Research*, *44*(8), 553–562.

Peirs, J., Reynaerts, D., Van Brussel, H. (1998). *Proceedings of the 1998 IEEE international conference on robotics and automation (Cat. No.98CH36146)*. 2, 1516–1521.

Peng, L., Xu, Z., Liu, Z., Guo, Y., Li, P., & Gao, C. (2017). Ultrahigh thermal conductive yet superflexible graphene films. *Advanced Materials, 29*(27), 1700589.

Peng, Z. A., & Peng, X. G. (2001). Formation of high-quality CdTe, CdSe, and CdS nanocrystals using CdO as precursor. *Journal of the American Chemical Society, 123*(1), 183–184.

Polder, D., & Vanhove, M. (1971). Theory of radiative heat transfer between closely spaced bodies. *Physical Review B, 4*(10), 3303–3314.

Pospischil, A., Furchi, M. M., & Mueller, T. (2014). Solar-energy conversion and light emission in an atomic monolayer p–n diode. *Nature Nanotechnology, 9*(4), 257–261.

Qi, J. S., Wang, H., Chen, X. F., & Qian, X. F. (2018). Two-dimensional multiferroic semiconductors with coexisting ferroelectricity and ferromagnetism. *Applied Physics Letters, 113*(4), 043102.

Qian, X. T., Chen, L., Yin, L. C., Liu, Z. B., Pei, S. F., Li, F., Hou, G. J., Chen, S. M., Song, L., Thebo, K. H., Cheng, H. M., & Ren, W. C. (2020). $CdPS_3$ nanosheets-based membrane with high proton conductivity enabled by Cd vacancies. *Science (New York, N.Y.), 370*(6516), 596–600.

Qiao, B., Wang, A., Yang, X., Allard, L. F., Jiang, Z., Cui, Y., Liu, J., Li, J., & Zhang, T. (2011). Single-atom catalysis of CO oxidation using Pt_1/FeO_x. *Nature Chemistry, 3*(8), 634–641.

Radha, B., Esfandiar, A., Wang, F. C., Rooney, A. P., Gopinadhan, K., Keerthi, A., Mishchenko, A., Janardanan, A., Blake, P., Fumagalli, L., Lozada-Hidalgo, M., Garaj, S., Haigh, S. J., Grigorieva, I. V., Wu, H. A., & Geim, A. K. (2016). Molecular transport through capillaries made with atomic-scale precision. *Nature, 538*(7624), 222–225.

Rothenberg, B. H. G. (2008). *Catalysis-concepts and green applications*. Wiley-VCH.

Sarkar, D., Xie, X. J., Liu, W., Cao, W., Kang, J. H., Gong, Y. J., Kraemer, S., Ajayan, P. M., & Banerjee, K. (2015). A subthermionic tunnel field-effect transistor with an atomically thin channel. *Nature, 526*(7571), 91–95.

Sfeir, M. Y., Wang, F., Huang, L. M., Chuang, C. C., Hone, J., O'Brien, S. P., Heinz, T. F., & Brus, L. E. (2004). Probing electronic transitions in individual carbon nanotubes by Rayleigh scattering. *Science (New York, N.Y.), 306*(5701), 1540–1543.

Shi, Z. W., Hong, X. P., Bechtel, H. A., Zeng, B., Martin, M. C., Watanabe, K., Taniguchi, T., Shen, Y. R., & Wang, F. (2015). Observation of a Luttinger-liquid plasmon in metallic single-walled carbon nanotubes. *Nature Photonics, 9*(8), 515–519.

Skúlason, E., Bligaard, T., Gudmundsdóttir, S., Studt, F., Rossmeisl, J., Abild-Pedersen, F., Vegge, T., Jónsson, H., & Norskov, J. K. (2012). A theoretical evaluation of possible transition metal electro-catalysts for N_2 reduction. *Physical Chemistry Chemical Physics: PCCP, 14*(3), 1235–1245.

Smith, A. M., & Nie, S. M. (2010). Semiconductor nanocrystals: Structure, properties, and band gap engineering. *Accounts of Chemical Research, 43*(2), 190–200.

Splendiani, A., Sun, L., Zhang, Y. B., Li, T. S., Kim, J., Chim, C. Y., Galli, G., & Wang, F. (2010). Emerging photoluminescence in monolayer MoS_2. *Nano Letters, 10*(4), 1271–1275.

Stamenkovic, V. R., Mun, B. S., Arenz, M., Mayrhofer, K. J. J., Lucas, C. A., Wang, G., Ross, P. N., & Markovic, N. M. (2007). Trends in electrocatalysis on extended and nanoscale Pt-bimetallic alloy surfaces. *Nature Materials, 6*(3), 241–247.

Tan, H. X., Li, M. L., Liu, H. T., Liu, Z. R., Li, Y. C., & Duan, W. H. (2019). Two-dimensional ferromagnetic-ferroelectric multiferroics in violation of the d^0 rule. *Physical Review B, 99*(19), 195434.

Tang, D.-M., Kvashnin, D. G., Cretu, O., Nemoto, Y., Uesugi, F., Takeguchi, M., Zhou, X., Hsia, F.-C., Liu, C., Sorokin, P. B., Kawamoto, N., Mitome, M., Cheng,

H.-M., Golberg, D., & Bando, Y. (2018). Chirality transitions and transport properties of individual few-walled carbon nanotubes as revealed by *in situ* TEM probing. *Ultramicroscopy*, *194*, 108–116.

Tang, D.-M., Ren, C.-L., Wang, M.-S., Wei, X., Kawamoto, N., Liu, C., Bando, Y., Mitome, M., Fukata, N., & Golberg, D. (2012). Mechanical properties of Si nanowires as revealed by *in situ* transmission electron microscopy and molecular dynamics simulations. *Nano Letters*, *12*(4), 1898–1904.

Tran, T. T., Bray, K., Ford, M. J., Toth, M., & Aharonovich, I. (2016a). Quantum emission from hexagonal boron nitride monolayers. *Nature Nanotechnology*, *11*(1), 37–41.

Tran, T. T., Elbadawi, C., Totonjian, D., Lobo, C. J., Grosso, G., Moon, H., Englund, D. R., Ford, M. J., Aharonovich, I., & Toth, M. (2016b). Robust multicolor single photon emission from point defects in hexagonal boron nitride. *ACS Nano*, *10*(8), 7331–7338.

Tsai, C., Chan, K. R., Abild-Pedersen, F., & Norskov, J. K. (2014). Active edge sites in $MoSe_2$ and WSe_2 catalysts for the hydrogen evolution reaction: A density functional study. *Physical Chemistry Chemical Physics: PCCP*, *16*(26), 13156–13164.

Utzat, H., Sun, W. W., Kaplan, A. E. K., Krieg, F., Ginterseder, M., Spokoyny, B., Klein, N. D., Shulenberger, K. E., Perkinson, C. F., Kovalenko, M. V., & Bawendi, M. G. (2019). Coherent single-photon emission from colloidal lead halide perovskite quantum dots. *Science (New York, N.Y.)*, *363*(6431), 1068–1072.

Valden, M. (1998). Onset of catalytic activity of gold clusters on titania with the appearance of nonmetallic properties. *Science (New York, N.Y.)*, *281*(5383), 1647–1650.

Voiry, D., Salehi, M., Silva, R., Fujita, T., Chen, M. W., Asefa, T., Shenoy, V. B., Eda, G., & Chhowalla, M. (2013). Conducting MoS_2 nanosheets as catalysts for hydrogen evolution reaction. *Nano Letters*, *13*(12), 6222–6227.

Wang, C., Chi, M. F., Li, D. G., Strmcnik, D., van der Vliett, D., Wang, G. F., Komanicky, V., Chang, K. C., Paulikas, A. P., Tripkovic, D., Pearson, J., More, K. L., Markovic, N. M., & Stamenkovic, V. R. (2011). Design and synthesis of bimetallic electrocatalyst with multilayered Pt-skin surfaces. *Journal of the American Chemical Society*, *133*(36), 14396–14403.

Wang, F., Dukovic, G., Brus, L. E., & Heinz, T. F. (2005). The optical resonances in carbon nanotubes arise from excitons. *Science (New York, N.Y.)*, *308*(5723), 838–841.

Wang, L., Meric, I., Huang, P. Y., Gao, Q., Gao, Y., Tran, H., Taniguchi, T., Watanabe, K., Campos, L. M., Muller, D. A., Guo, J., Kim, P., Hone, J., Shepard, K. L., & Dean, C. R. (2013). One-dimensional electrical contact to a two-dimensional material. *Science (New York, N.Y.)*, *342*(6158), 614–617.

Wang, M.-S., Golberg, D., & Bando, Y. (2010). Tensile tests on individual single-walled carbon nanotubes: linking nanotube strength with its defects. *Advanced Materials*, *22* (36), 4071–4075.

Wang, S., Zhao, S. H., Shi, Z. W., Wu, F. Q., Zhao, Z. Y., Jiang, L. L., Watanabe, K., Taniguchi, T., Zettl, A., Zhou, C. W., & Wang, F. (2020). Nonlinear Luttinger liquid plasmons in semiconducting single-walled carbon nanotubes. *Nature Materials*, *19*(9), 986–991.

Wang, X., Tang, J., Xia, X. X., He, C. L., Zhang, J. W., Liu, Y. Z., Wan, C. H., Fang, C., Guo, C. Y., Yang, W. L., Guang, Y., Zhang, X. M., Xu, H. J., Wei, J. W., Liao, M. Z., Lu, X. B., Feng, J. F., Li, X. X., Peng, Y., et al. (2019). Current-driven magnetization switching in a van der Waals ferromagnet Fe_3GeTe_2. *Science Advances*, *5*(8), eaaw8904.

Wang, Z., Zhang, T. Y., Ding, M., Dong, B. J., Li, Y. X., Chen, M. L., Li, X. X., Huang, J. Q., Wang, H. W., Zhao, X. T., Li, Y., Li, D., Jia, C. K., Sun, L. D., Guo, H. H., Ye, Y., Sun, D. M., Chen, Y. S., Yang, T., et al. (2018). Electric-field control of magnetism in a few-layered van der Waals ferromagnetic semiconductor. *Nature Nanotechnology, 13*(7), 554−559.

Webster, L., & Yan, J. A. (2018). Strain-tunable magnetic anisotropy in monolayer $CrCl_3$, $CrBr_3$, and CrI_3. *Physical Review B, 98*(14), 144411.

Weidman, M. C., Beck, M. E., Hoffman, R. S., Prins, F., & Tisdale, W. A. (2014). Monodisperse, air-stable PbS nanocrystals *via* precursor stoichiometry control. *ACS Nano, 8*(6), 6363−6371.

Willets, K. A., & Van Duyne, R. P. (2007). Localized surface plasmon resonance spectroscopy and sensing. *Annual Review of Physical Chemistry, 58*, 267−297.

Wolf, E. L. (2014). *Graphene: A new paradigm in condensed matter and device physics*. Oxford University Press.

Wu, J., Ma, S., Sun, J., Gold, J. I., Tiwary, C., Kim, B., Zhu, L., Chopra, N., Odeh, I. N., Vajtai, R., Yu, A. Z., Luo, R., Lou, J., Ding, G., Kenis, P. J. A., & Ajayan, P. M. (2016). A metal-free electrocatalyst for carbon dioxide reduction to multi-carbon hydrocarbons and oxygenates. *Nature Communications*, 7(1), 13869.

Xiao, D., Liu, G. B., Feng, W. X., Xu, X. D., & Yao, W. (2012). Coupled spin and valley physics in monolayers of MoS_2 and other group-VI dichalcogenides. *Physical Review Letters, 108*(19), 196802.

Xiao, J. P., Pan, X. L., Guo, S. J., Ren, P. J., & Bao, X. H. (2015). Toward fundamentals of confined catalysis in carbon nanotubes. *Journal of the American Chemical Society, 137* (1), 477−482.

Xin, G. Q., Yao, T. K., Sun, H. T., Scott, S. M., Shao, D. L., Wang, G. K., & Lian, J. (2015). Highly thermally conductive and mechanically strong graphene fibers. *Science (New York, N.Y.), 349*(6252), 1083−1087.

Xu, R. Z., & Zou, X. L. (2020). Electric field-modulated magnetic phase transition in van der Waals CrI_3 bilayers. *Journal of Physical Chemistry Letters, 11*(8), 3152−3158.

Yang, Y. C., Song, Z. G., Lu, G. Y., Zhang, Q. H., Zhang, B. Y., Ni, B., Wang, C., Li, X. Y., Gu, L., Xie, X. M., Gao, H. J., & Lou, J. (2021). Intrinsic toughening and stable crack propagation in hexagonal boron nitride. *Nature, 594*(7861), 57−61.

Yang, Z. Y., Wang, D. L., Meng, C., Wu, Z. M., Wang, Y., Ma, Y. G., Dai, L., Liu, X. W., Hasan, T., Liu, X., & Yang, Q. (2014). Broadly defining lasing wavelengths in single bandgap-graded semiconductor nanowires. *Nano Letters, 14*(6), 3153−3159.

Yang, Z. Y., Albrow-Owen, T., Cui, H. X., Alexander-Webber, J., Gu, F. X., Wang, X. M., Wu, T. C., Zhuge, M. H., Williams, C., Wang, P., Zayats, A. V., Cai, W. W., Dais, L., Hofmann, S., Overend, M., Tong, L. M., Yang, Q., Sun, Z. P., & Hasan, T. (2019). Single-nanowire spectrometers. *Science (New York, N.Y.), 365* (6457), 1017−1020.

Yao, F. R., Tang, J. Y., Wang, F., & Liu, K. H. (2016). Structure−property relations in individual carbon nanotubes [Invited]. *Journal of the Optical Society of America. B, Optical Physics, 33*(7), 102−107.

Ye, R., Hurlburt, T. J., Sabyrov, K., Alayoglu, S., & Somorjai, G. A. (2016). Molecular catalysis science: Perspective on unifying the fields of catalysis. *Proceedings of the National Academy of Sciences, 113*(19), 5159−5166.

Yi, K. S., Trivedi, K., Floresca, H. C., Yuk, H., Hu, W., & Kim, M. J. (2011). Room-temperature quantum confinement effects in transport properties of ultrathin Si nanowire field-effect transistors. *Nano Letters, 11*(12), 5465−5470.

Yu, H., Kutana, A., & Yakobson, B. I. (2016). Carrier delocalization in two-dimensional coplanar p-n junctions of graphene and metal dichalcogenides. *Nano Letters, 16*(8), 5032−5036.

Yu, S. J., Youn, S. J., & Kim, H. (2010). Size effect of thermal radiation. *Physica B, 405* (2), 638−641.

Yu, Z. H., & Brus, L. (2001). Rayleigh and Raman scattering from individual carbon nanotube bundles. *The Journal of Physical Chemistry B*, *105*(6), 1123–1134.

Zhang, D. D., Eaton, S. W., Yu, Y., Dou, L. T., & Yang, P. D. (2015). Solution-phase synthesis of cesium lead halide perovskite nanowires. *Journal of the American Chemical Society*, *137*(29), 9230–9233.

Zhang, P., Ma, L. L., Fan, F. F., Zeng, Z., Peng, C., Loya, P. E., Liu, Z., Gong, Y. J., Zhang, J. N., Zhang, X. X., Ajayan, P. M., Zhu, T., & Lou, J. (2014). Fracture toughness of graphene. *Nature Communications*, *5*, 3782.

Zhang, Y. B., Tan, Y. W., Stormer, H. L., & Kim, P. (2005). Experimental observation of the quantum Hall effect and Berry's phase in graphene. *Nature*, *438*(7065), 201–204.

Zhou, J. H., Zhu, M. Y., Meng, R. Y., Qin, H. Y., & Peng, X. G. (2017). Ideal CdSe/CdS core/shell nanocrystals enabled by entropic ligands and their core size-, shell thickness-, and ligand-dependent photoluminescence properties. *Journal of the American Chemical Society*, *139*(46), 16556–16567.

Zhou, X., Tang, D.-M., Mitome, M., Bando, Y., Sasaki, T., & Golberg, D. (2019). Intrinsic and defect-related elastic moduli of boron nitride nanotubes as revealed by *in situ* transmission electron microscopy. *Nano Letters*, *19*(8), 4974–4980.

Zhu, T., & Li, J. (2010). Ultra-strength materials. *Progress in Materials Science*, *55*(7), 710–757.

CHAPTER SIX

Applications of low-dimensional materials

Currently, fossil fuel is dispensable as the main source of energy, critical raw materials for chemical industry, and in the meantime the main source of greenhouse gases and air pollution. In recent years, more and more frequently extreme weathers have been reported in the news, warning us about the limitation of the tolerance of environment on the planet earth. In addition, in information technology, feature lengths of transistors have been scaled down to $\sim$10 nm, approaching the physical limit, and challenging the economic growth model over the last century. What's more, in the 3 years from 2020 to 2023, the whole world has been profoundly impacted by COVID-19 pandemic, which has cost millions of people's lives, revealing that the civilization is still fragile. To tackle the global challenges in energy, environment, information, and health, history has taught us that the key is destructive technologies that are based on breakthroughs in fundamental science. As summarized in Fig. 6.1, due to the unique structures and extraordinary properties, low-dimensional materials are the foundations and will be the key solutions.

6.1 Low-dimensional information materials

After tribal, agrarian, and industrial societies, mankind entered information society at the end of the 20th century and is currently in transition to the "smart society". In abstraction, information, that is, knowledge, can be defined as a quantity that reduces the uncertainty of a system. Its basic unit is a "bit," which can be represented as 0 or 1. Information and intelligent technologies developed on the basis of semiconductors are at the heart of information revolution and have become the most important driving forces of the world economy, with sales reaching US$440 billion in 2020.

Low-Dimensional Materials.
DOI: https://doi.org/10.1016/B978-0-443-29046-6.00012-2

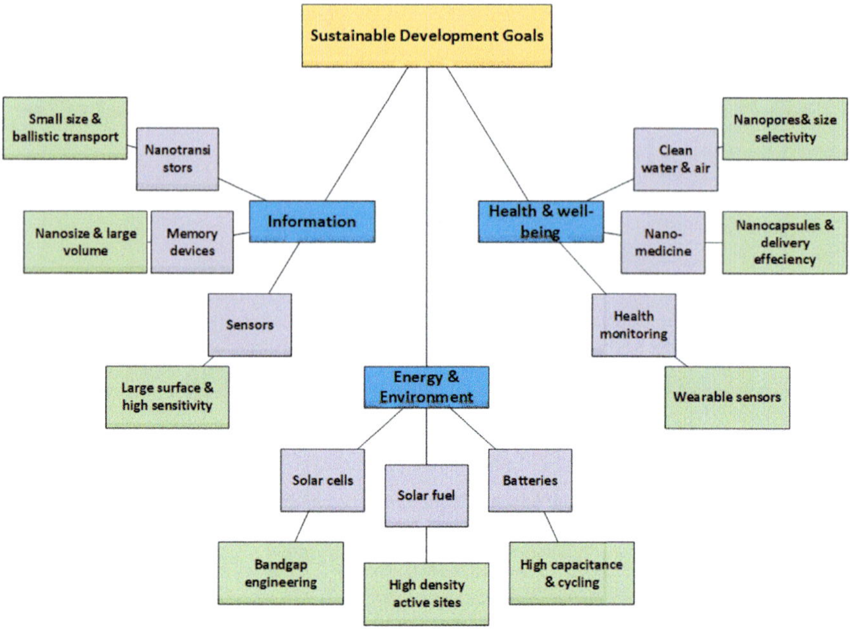

Figure 6.1 ***Technologies enabled by low-dimensional materials.*** Applications in energy, information, environment, and health, as crucial solutions to tackle the SDGs. *SDGs*, Sustainable development goals.

Information technology includes information acquisition by sensors, processing by microprocessors, storage by memory devices, transmission by optoelectronic devices, and execution by actuators. In a typical control system, physical signals are converted into digital information by sensors. The signal is then passed to the processors and analyzed by control algorithms to calculate the feedback control signals that are executed by the actuators, forming a closed loop. In recent years, such process is used to construct and train a model of the relations between input and output, leading to machine learning and artificial intelligence (AI).

For different processes and purposes, there are different requirements, and therefore different information carriers are being used. Information transmission is carried out by laser in the optical fibers, where photons belonging to bosons could be in the same coherent state and transported at the speed of light. For processing information, it is required to manipulate the state efficiently with a high speed and low power consumption. Currently, charges and currents of electrons are being controlled to fulfill the logical computations, because their states could be easily controlled by electrical field in capacitors. For information storage, which requires a high density and a long retention time, electron spin and its condensed ferromagnetic phase is a suitable choice.

Since the 1960s, the size, density, integration scale, and performance of semiconducting devices, such as transistors, have evolved following the so-called Moore's law. As a result, electronic devices have been more powerful, more compact, and in the meantime cheaper, leading to larger scale of applications and revenues for further technical developments. Such a cycle strongly drives both the technological advancement and the economic growth. However, as the device size further decreases, silicon-based transistors have started to suffer from so-called short-channel effects, resulting in increased power consumption and difficulties in heat dissipation. It is important to explore the novel low-dimensional materials and electronic devices in post-Moore era.

6.1.1 Information-processing technology

Transistors are the essential logic devices for processing digital information. They are building blocks for integrated circuits that are widely used in information devices, such as the central processing unit (CPU), memory, amplifiers for wireless communications, and control switches of displays. Transistors are three-terminal semiconductor devices with the channel resistance between the source and drain terminals regulated by a third gate terminal to achieve switching and amplifying functions. For different purposes, there are different requirements for the transistors.

For processing information in microprocessors, transistors are logic switches, requiring high speed and low power consumption, known as high performance transistors. On the other hand, for information communications in radio frequency (RF) devices, transistors are used as power amplifiers, where the main merit is the operating frequency for power gain. For transistors used to control the backlight in liquid-crystal display (LED), uniformity in large area and high current is required to achieve a homogeneous brightness, and thin-film transistors are developed for such applications.

6.1.1.1 High performance transistors

The dependence of field-effect transistor (FET) performance and material properties can be analyzed in two ways. One is in the vertical direction, where the carrier concentration is modulated to change the conductivity by an electric field applied by the gate electrode. The modulation effect is related to the thickness, dielectric constant, and interface quality of the dielectric material. The other direction is along the horizontal direction, where the carrier transport under source–drain bias is related to the electron mobility of the channel materials and the contact quality at source and drain terminals.

Up to now, Moore's law has been extended to below 100 nm, and even close to 10 nm, thanks to continued industrial innovations in the above two directions. For example, along the vertical direction, introduction of high-κ dielectric layers in 2007 and fin-gate technology in 2011 has greatly improved gate control of the channel. And along the channel direction, strain engineering technology introduced for the 90-nm node in 2003 has effectively enhanced the carrier mobility.

Low-dimensional materials, such as one-dimensional carbon nanotubes and silicon nanowires, two-dimensional graphene, and MoS_2, have several advantages for applications as the channels of FETs. Due to their nanoscale or even atomic-scale thickness, low-dimensional materials have excellent electrostatic properties in the vertical direction. On the other hand, due to reduced surface scattering, low-dimensional materials exhibit a high mobility and even ballistic transport in the horizontal direction.

Fabrication processes of semiconductor nanowire-based FETs are compatible with existing semiconductor industry. Because of the small thickness and circular cross-section, gate-all-around (GAA) FETs could be fabricated, offering superior gate control. GAA FETs have been successfully realized for nanowire transistors with silicon, silicon–germanium, and gallium nitride nanowires as channels, where the subthreshold slopes have approached the theoretical limit (Li et al., 2018; Razavi & Fagas, 2013; Singh et al., 2006). In addition to extending Moore's law to smaller length scales, novel device configurations have been invented for nanowire transistors. In conventional semiconductor devices, junctions are basic components, including p–n junctions, metal–semiconductor junctions, and heterojunctions. However, as the feature sizes shrink to 10 nm or even smaller scale, it becomes a great challenge to control the doping distributions for forming such junctions. Junctionless transistors could avoid the complexity of fabricating junctions. Due to the small thickness, it is still effective to modulate the channel transport by gate electrostatic field. Colinge et al. used silicon-on-insulator technology to fabricate junctionless transistors with homogeneously doped silicon nanowires as the channels. Such junctionless transistors with a thickness of 10 nm and a width of 30 nm demonstrated an On/Off ratio of 1 million and a near-ideal subthreshold slope of 64 mV/dec (Colinge et al., 2010).

Carbon nanotubes have unique one-dimensional tubular structures that can be seen as seamlessly rolled-up graphene sheets. Symmetric complementary metal–oxide semiconductor (CMOS) could be fabricated because of the symmetric dispersion relations for electrons and holes. There are no dangling bonds on sp^2-bonded surface; therefore electron scattering is greatly reduced. Moreover, electrons can travel in carbon nanotube (CNT) channels with a high mobility and even ballistically. On the vertical direction, the thickness of CNTs is around 1 nm, facilitating effective gate modulation. In addition, CNTs are mechanically tough and flexible, chemically stable in air, and highly conductive for heat dissipation. These features have made CNTs one of the most promising channel materials in post-Moore era (Peng et al., 2014).

Since Delft University of Technology and IBM reported on first CNT transistors in 1998 (Martel et al., 1998; Tans et al., 1998), there have been tremendous progresses. By using palladium to realize Ohmic contact, Dai group in Stanford University realized ballistic transport in CNT transistors (Javey et al., 2003). Peng group in Peking University used scandium to make n-type Ohmic contact and fabricated CNT-based ballistic CMOS devices and integrated circuits (Zhang et al., 2007). In 2017 Peng group used graphene as a "Dirac" electrode to fabricate carbon nanotube transistors with the channel length as short as 5 nm, demonstrating superior performance than silicon transistors of the same channel length, featuring faster operation, lower operating voltage, and smaller subthreshold slope (Qiu et al., 2017). Around the same time, Cao et al. (2017) in Stanford University reported CNT transistors with a 40-nm footprint, providing a higher current under the same working voltage compared to silicon counterparts, showing the possibility of replacing silicon channels at the end of Moore's law.

Improvements in individual transistors have led to a success of CNT-based logic and integrated circuits. In 2017 Han et al. in IBM obtained semiconductor-enriched carbon nanotubes by solution separation methods with a purity higher than 99.9%, to fabricate p-type and n-type transistors and CMOS structures with palladium and scandium contacts, respectively. Oscillating frequency up to 2.82 GHz was demonstrated for the fabricated five-stage ring oscillator circuits (Han et al., 2017). In 2020 Peng group obtained semiconducting single-walled carbon nanotubes with uniform diameters (1.45 ± 0.23 nm), and a high semiconductor purity (6 N), by using a multistage polymer-assisted separation method. The nanotubes were assembled into highly oriented (9°), high density (100–200/μm), and large-area (4-in.) wafers to fabricate a five-stage ring oscillator with a maximum frequency as high as 8 GHz (Liu et al., 2020).

Microprocessors and computers based on CNT transistors have been developed. In 2013 Shulaker et al. reported the first CNT computer constructed from carbon nanotube transistors, containing 178 transistors to form 60 logic units that can perform logic operations and simple functions such as counting and sorting (Shulaker et al., 2013). In 2019 Shulaker et al. fabricated the first microprocessors of moder configurations composed of carbon nanotube CMOS using a standard industrial process. The microprocessor consisted of 3762 logic units from 14,702 transistors, could execute 16-bit standard codes to run a simple program and output "Hello, World" (Hills et al., 2019). In 2020 Shulaker's team achieved uniform carbon nanotube assembly and transistor fabrication on 200mm wafers using existing production lines in a commercial foundry, demonstrating the compatibility with the current semiconductor industry. The yield was nearly 100% for 14,400 transistors, with an average On/Off ratio of 4000 and a subthreshold slope of 109 mV/dec, showing the potential of carbon nanotube electronics for industrial scale applications (Bishop et al., 2020).

For transistors, there is an important feature to define the reachable miniaturization, which is the "naturel length" defined as $\lambda \propto \sqrt{\frac{t_s t_{ox} \varepsilon_s}{\varepsilon_{ox}}}$, where the thickness of the semiconducting channel (t_s) is an important factor (Chhowalla et al., 2016; Liu et al., 2021). Two-dimensional materials have thicknesses of a single atom or unit cell. In principle, by using two-dimensional semiconductors as the channels, it is possible to fabricate the shortest transistor that is an important task near the scaling limit of silicon transistors.

Transition metal dichalcogenides is a two-dimensional material system, with a common chemical formula of MX_2, where M is a transition group metal (Mo, W), and X is a chalcogen element (S, Se, and Te). In 2011 Kis group at Swiss Federal Institute of Technology in Lausanne (EPFL) reported MoS_2 transistors with an On/Off ratio of up to 10^8, a subthreshold slope of 74 mV/dec, and an electron mobility of 200 cm^2/V s. In 2019 Wang group of Nanjing University used a monolayer of molecular crystals as a seeding layer to grow high-quality high-κ dielectric layers of HfO_2 on 2D materials. MoS_2 transistors with a 20-nm channel length achieved a large On/Off ratio of 10^7 and a small subthreshold slope of 73 mV/dec (Li, Zhou, et al., 2019).

Monolayer black phosphorus (BP) is a two-dimensional material with an anisotropic structure. The electrical properties are also anisotropic, with an effective electron mass of only 0.17 m_0 in the "armchair" direction, leading to a high electron mobility (Lam et al., 2014). The bandgap of bulk BP is ~0.3 eV. As the thickness is reduced to a monolayer, the bandgap is blue-shifted to ~1 eV due to quantum confinement effects (Das et al., 2014). In 2014 Zhang et al. from Fudan University reported BP transistors with an On/Off ratio of 10^5 and an electron mobility as high as 1000 $cm^2/V\ s$ at room temperature (Li, Yu, et al., 2014).

In addition to transistors with one type of low-dimensional materials as channels, the combination of hybrid-dimensional materials may lead to device structures and functions that are different from those of conventional devices. In 2016 Duan group at UCLA used nanowires as templates to prepare MoS_2 transistors with a channel length shorter than 100 nm (Liu et al., 2016). Ali Javey et al. at UC Berkeley reported MoS_2 transistors using an individual single-walled carbon nanotube as the gate electrode with a width of only ~1 nm, limiting the effective channel length to ~3.9 nm. The ultrashort transistors exhibited a subthreshold slope of 65 mV/dec and a switching ratio of 10^6 (Desai et al., 2016).

6.1.1.2 Radio frequency transistors

In wireless communications, the relationship among achievable channel capacity (C), bandwidth (W), and signal-to-noise ratio (S/N) is described by Shannon's formula:$C = W\log_2(1 + S/N)$, where information capacity is proportional to the bandwidth (Shannon, 1949). Currently, the carrier frequency for the mainstream 4G communications is around 6 GHz. The frequency will be increased to ~100 GHz and ~1 THz for the 5G technology being applied and the 6G communication being developed, respectively, greatly enhancing the bandwidth and therefore the channel capacity (Dang et al., 2020).

With the dramatic increase in channel capacity, communication contents have been changed from voice in the 1980s, to text messaging in the 1990s, pictures and multimedia messaging service in the early 2000s, and video conversations supported by the current 4G technology. Further increase of channel capacity for the 5G and 6G communications will enable future technologies such as virtual reality and autonomous driving, becoming important infrastructures for the coming smart society.

RF transistors are crucial components in communication electronics for power amplification. The most important performance indicators are the cut-off frequency f_T and the maximum frequency f_{max}, corresponding to the limit frequencies of current amplification and power gain, respectively. f_T is mainly dependent on intrinsic properties of the semiconductor:$f_T = \frac{1}{2\pi}\left(\frac{1}{\tau}\right) = \frac{\nu_F}{2\pi L}$, where τ, ν_F, and L are the channel transit time, Fermi velocity, and gate length, respectively. On the other hand, f_{max} is closely related to device structure, such as electrode resistance (R_G, R_S, and R_D), channel resistance (R_i), and gate-drain capacitance ($C_{gd)}$. To enhance the working frequencies of RF transistors, it is necessary to increase the Fermi velocity and to optimize the device structure by shortening the gate length and reducing the resistance and capacitance.

Typical channel materials for RF transistor are high mobility III–V and SiGe semiconductors. The cut-off frequencies of InAs transistors (Chang et al., 2013) and silicon-based metal–oxide–semiconductor field-effect transistor (MOSFET) (Lee et al., 2007) have reached 710 and 485 GHz, respectively. SiGe heterojunction bipolar transistors with a 90-nm gate length have demonstrated a maximum frequency of 0.7 THz (Lachner, 2014; Pekarik et al., 2014). In addition, high mobility InP transistors with a gate length of 25 nm and a maximum frequency of 1.5 THz have been reported by Northrop Grumman (Mei et al., 2015). For conventional high mobility transistors, it is a challenge to further scale the channel length down to 20 nm while keeping high performance due to the so-called short-channel effects. On the other hand, two-dimensional materials with an atomic-level thickness and one-dimensional carbon nanotubes with a high Fermi velocity are attractive candidates for overcoming short-channel effects to further increase the operating frequency of RF transistors.

Among low-dimensional materials, semiconductor nanowires have the advantage of being compatible with the semiconductor industry. Lattice epitaxy techniques could be used to obtain highly crystalline junction structures. Li et al. reported dual-gate transistors with a horizontal GaAs nanowire array as the channel. With a density about 1.5/μm, and a length of about 150 nm, the cut-off frequency and maximum frequency reached 33 and 75 GHz, respectively (Miao et al., 2015). Yu et al. improved the quality of dielectric interface by epitaxially growing Ga_2O_3 on GaN nanowires and fabricated transistors with a channel length as short as 50 nm, where f_T and f_{max} reached 150 and 180 GHz, respectively (Yu et al., 2012). By developing a selective regrowth method, Zota et al. have fabricated high-density InGaAs nanowire arrays with a triangular cross-section and a pitch of ~100 nm. With a gate length of 32 nm, f_T and f_{max} reached sub-THz region, to be 280 and 312 GHz, respectively (Zota et al., 2014).

Carbon nanotubes have extraordinary electrical properties, such as ballistic transport, high electron mobility, high saturation velocity, and low intrinsic capacitance. Hasan et al. at Purdue University modeled the circuit of ideal, ballistic CNT RF transistors and found that the cut-off frequency f_T was related to the gate length L as follows: $f_T \approx \frac{v_F}{2\pi L} \approx 130$ GHz/L(μm), which predicted that the frequency could reach 1.3 THz with a channel length of 100 nm (Hasan et al., 2006). To achieve the predicted performance, high density, directionally alignment, and purely semiconducting carbon nanotubes are required (Rutherglen et al., 2009). The IBM Avouris group obtained aligned CNTs with the concentration of semiconducting nanotubes to be 99.6% by using the electrophoretic deposition. With a density of 50/μm, and a channel length of 100 nm, the cut-off frequency f_T was estimated to be 153 GHz (Steiner et al., 2012). Zhou group at the University of Southern California assembled CNTs with higher contents of semiconducting tubes (99.9%) with a density of ~40/μm, to fabricate transistors with a channel length of 100 nm, demonstrating a maximum frequency f_{max} of 70 GHz (Cao et al., 2016; Joo et al., 2014). In 2019 Peng group at Peking University obtained carbon nanotubes with a semiconductor purity of 99.99%, using a polymer separation method, and fabricated transistors with a gate length of 30 nm. The cut-off and maximum frequencies were 103 and 107 GHz, respectively (Zhong et al., 2019), as shown in Fig. 6.2. In 2023 by treating the aligned CNTs with ultraviolet ozone to remove residual organic molecules, both direct current (DC) and radio frequency (RF) performance of the CNT transistors have been effectively enhanced, with the cut-off and maximum frequencies reaching 376 and 318 GHz, respectively (Zhou et al., 2023).

As a semimetal, graphene lacks a suitable bandgap required for logic transistors; however, it is promising for RF transistors, because of the unique electronic structures and electrical properties, such as linear dispersion relationship, zero effective electrons mass, and extremely high carrier mobility. In 2009 Avouris group reported graphene RF transistors with a cut-off frequency at GHz level (Lin et al., 2009). In 2010 as the gate length was reduced from 360 nm to 240 nm, the cut-off frequency was increased to 100 GHz (Lin et al., 2010). And in 2012 transistors with the channel length of 40 nm were fabricated using graphene epitaxially grown on SiC wafers, where the cut-off frequency was increased to 300 GHz (Wu, Jenkins, et al., 2012). Using an individual nanowire as a gate template, Duan group at the University of California, Los Angeles (UCLA) fabricated graphene transistors

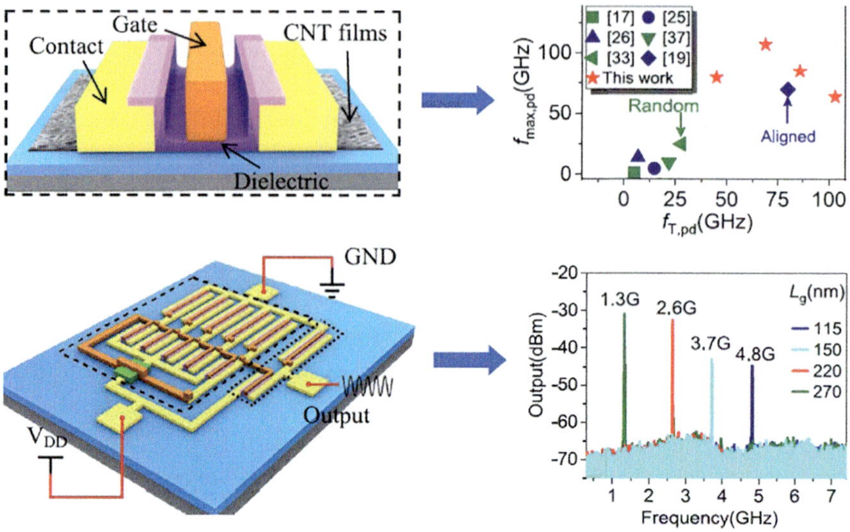

Figure 6.2 ***Carbon nanotube devices for RF electronics applications.*** Carbon nanotube RF transistor and five-stage ring oscillator with the maximum frequency and oscillation frequency to be 103 and 5 GHz, respectively (Zhong et al., 2019). *RF*, Radio frequency.

with a gate length of 140 nm, resulting in a cut-off frequency of 300 GHz (Liao et al., 2010). By using a self-aligned process, the channel length was further reduced to 67 nm, and accordingly, the cut-off frequency was increased to 427 GHz (Cheng et al., 2012).

In summary, the cut-off and maximum frequency of low-dimensional material RF transistors have reached 100 GHz level but are still an order of magnitude away from those of conventional high-mobility semiconductor transistors. Further improvement from 100 GHz to THz requires continued efforts in optimization the quality of the channel material and device structure, including semiconductor purity, crystallinity, alignment, density, contact resistance, and gate capacitance.

6.1.1.3 Thin-film transistors

In recent years, there have been many "smart" electronic devices such as smartphones and watches that combined a lot of functions, from telephone, camera, and navigation, to health check. It is amazing that such powerful multifunctional devices are so small and light, thanks to the great progresses in semiconducting transistors and processors, energy storage devices such as lithium-ion batteries. It could be expected that in the future personal electronic devices will be more powerful, where wearable electronic devices, such as flexible displays, are essential. For such devices, thin-film transistors are one of the key components. In addition to the electrical performance of conventional transistors, there are extra requirements for their optical, mechanical, and chemical properties. It is desired to develop new materials that combine mechanical flexibility, optical transparency, high electron mobility, and stable chemical properties. And printing techniques on large-area plastic substrates are preferred for the fabrication of transistors and circuits. Low-dimensional materials are intrinsically flexible, because of the small thickness and therefore are promising candidates for applications of flexible electronics.

Two-dimensional materials are extremely thin with only an atom or molecular unit in thickness. They are typically bonded by strong covalent bonds within plane and weak van der Waals forces out of the plane. There is a large 2D material family, with electrical properties ranging from metal, semiconductor, to insulators. Therefore two-dimensional materials and their heterojunction structures are ideal for fabricating large-area thin-film transistors. Hone et al. at Columbia University combined a variety of two-dimensional materials to fabricate a flexible transistor, with insulating h-BN as the dielectric layer, semimetal of graphene as the gate electrode, and MoS_2 as transistor channel. The electron mobility was maintained at 45 cm^2/V s at 1.5% strain, comparable to that of polycrystalline silicon thin-film transistors (Lee et al., 2013). Duan group at UCLA used a molecular intercalation and dispersion method to obtain colloidal dispersions of various two-dimensional materials to prepare thin-film transistors on a 10-cm wafer scale and logic circuits on plastic substrates with the I_{ON}/I_{OFF} ratios as high as 10^6 and an electron mobilities of 10 cm^2/V s.

Carbon nanotubes have been intensively investigated for building flexible electronic devices, because of the unique one-dimensional structure, extraordinary electrical, mechanical, chemical, and thermal properties. Depending on the diameter, semiconducting carbon nanotubes have a bandgap of ~1 eV. Young's modulus is as high as ~1 TPa due to the strong C–C bonds, and nanotubes are flexible due to the small diameter. In addition, they are chemically stable in air and under high temperature.

In 2008 Rogers group at the University of Illinois at Urbana-Champaign fabricated carbon nanotube thin-film transistors on a flexible substrate. The transistor with a gate length of 5 μm has a high I_{ON}/I_{OFF} ratio (10^5) and a low subthreshold slope (<140 mV/dec). Medium-scale integrated circuits with oscillation frequencies at kHz level were constructed with no significant deterioration in performance under bending at a radius of curvature of 5 mm, demonstrating their promising applications in flexible electronics (Cao et al., 2008).

There is an intrinsic challenge of carbon nanotube applications in transistors that is the coexistence of metallic (1/3) and semiconducting (2/3) nanotubes, depending on the chirality. Electrical properties of nanotube network further depend on the network parameters, such as the density and alignment. When the density of nanotube network is low, metallic carbon nanotubes do not form a continuous network and do not affect the overall electrical properties. In 2011 Ohno group in Nagoya University took advantage of such features to build flexible transistors and integrated circuits based on a low-density CNT network. The I_{ON}/I_{OFF} ratio was as high as 10^6, and the mobility was 35 cm^2/V s. A 21-stage ring oscillator with an oscillation frequency of 2.0 kHz was demonstrated (Sun et al., 2011). In 2018 Sun and Liu et al. in Institute of Metal Research, Chinese Academy of Sciences realized continuous growth, collection, and transfer of CNT networks on plastic substrates Fig. 6.3. A 110-stage ring oscillator was fabricated, demonstrating their potential applications in complex circuits in flexible and transparent electronics (Wang et al., 2018).

6.1.2 Information storage technology

Continued progress of human civilization is based on the knowledge and information, including creation, recording, transmission, and dissemination. The carrier of information has evolved along the history, such as knotted ropes of ancient times, clay-sintered cuneiform of ancient Babylon, and oracle-bone inscriptions of Yin and Shang dynasties in China. The invention of movable-type printing made it possible to reproduce information rapidly on a low-cost medium of papers. As one of China's four great inventions, the dissemination of this technology to Europe gave impetus to the Renaissance, speeding up the development of modern civilization. Now we are living in the society of digital information, and information is stored in states of electrons, including spins and charges.

The basic unit of digital information is a "bit," mathematically represented as 0 or 1. Physically to store a bit, it requires a two-level state that can be measured (read) and switched (write), such as "positive" or "negative" for charges, "up" or "down" for spins, "high" or "low" for resistance, and "high" or "low" of bias.

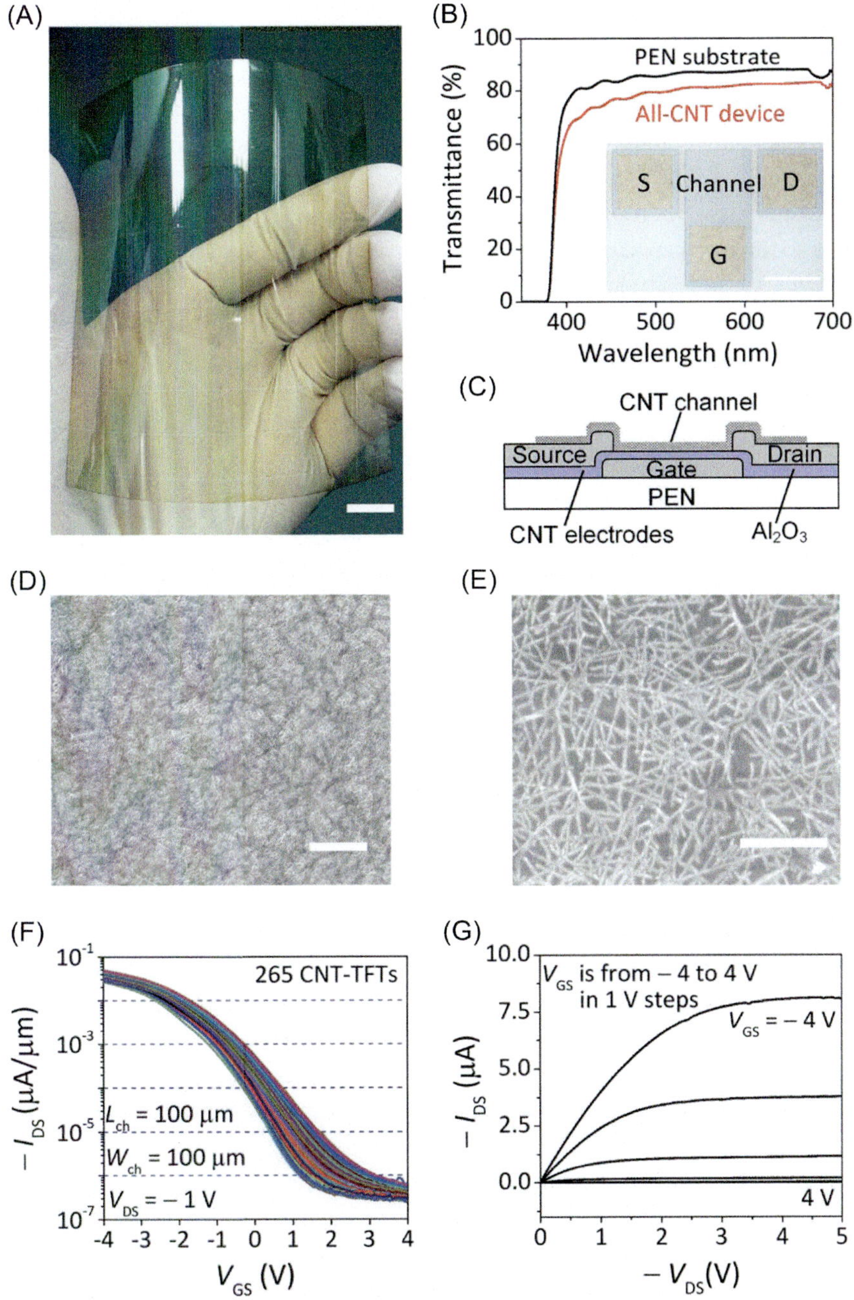

Figure 6.3 ***Carbon nanotube-based transparent flexible thin-film transistors.*** (A and B) Fabrication of carbon nanotube transistors on a PEN dicarboxylate substrate; (C and D) scanning electron microscope (SEM) image of homogeneous carbon nanotube network; and (F and G) transfer characteristics of 256 carbon nanotube transistors (Wang et al., 2018). *PEN*, Polyethylene naphthalene.

There are multiple storage systems in a modern electronic computer, including the cache, memory, external storage, and removable storage, using static random-access memory, dynamic random-access memory, hard disk, and flash memory as the dominant storage technologies, respectively. The storage capacity and speed have been dramatically improved. However, there are still many challenges. Up to now, the density for magnetic storage has been enhanced to near the theoretical limit, due to the physical limit of the ferromagnetic domain size. On the other hand, for CMOS-based flash memory, as the transistor size shrinks to sub-10-nm regime, it becomes easier to leak the charges, leading to possible errors.

The total information of the world has been expanding exponentially, with 2012 marked the beginning of the age of zettabytes (1 ZB $= 10^{21}$ bytes). And in 2018 the world generated 33 ZB of data, which will grow to 175 ZB in 2025 (Reinsel et al., 2018). A new paradigm of "big data" enabled machine learning and AI has already transformed the society. One of recent examples is ChatGPT that could generate logical dialogs at human language level. The transformer model was trained by using a large language data set of hundreds of gigabytes, to form a model with 175 billion parameters. The big data has put greater demands on information storage technologies, regarding data capacity, write/read speeds, and diversity of information mediums. It has become a bottleneck for the machine learning algorithms to transfer large amounts of data between the processor and memory (Zidan et al., 2018).

In this section, we will discuss about applications of low-dimensional materials in information storage. Due to their small size and unique physical properties, low-dimensional materials could improve the performance of current memory devices and importantly could be used to develop information devices with new architectures, such as nonvolatile storage technologies and atomic-level high-density storage technologies.

6.1.2.1 Nonvolatile memory technology

AI, based on deep learning of neural network models, has been improved greatly recently, as represented by AlphaGo and ChatGPT. To train the complex models with billions of parameters, a large amount of matrix calculations is involved to update the weights of the connections between layers of neural networks. To further enhance the performance of AI, it is critical to increase the efficiency and to lower the energy consumption, to transfer the large amount of data between the processor and the memory (Horowitz, 2014). Fundamentally, the problem originates from the von

Neumann architecture that separates the computer's processor from its memory. Thus one of the solutions is to get rid of this architecture and invent a new architecture that performs computations at the memory site. The non-Von Neumann architecture requires nonvolatile memory technologies. Currently, there are mainly four approaches based on following mechanisms: oxide varistor effect, phase change memristive effect, spin transfer torque, and ferroelectric effect.

In metal oxide varistors, the resistance is controlled by an electric field that drives ionic migration and redox reactions to generate and dissipate conductive pathways between the electrodes to form high and low resistive states (Wang, Wu, et al., 2020). Phase-change memristors utilize Joule heating to switch the material between electrodes to undergo an amorphous—crystalline reversible phase transition, accompanied by the switching between high-resistance and low-resistance states (Jo et al., 2009; Wuttig & Yamada, 2007). Spin-transfer torque in a magnetic tunneling junction uses the inverse effect of giant magnetoresistance to change the spin direction of magnetic materials with spin-polarized currents (Chappert et al., 2007). Ferroelectric memories, on the other hand, control the direction of polarization by applying an external voltage in ferroelectric FETs to modulate the resistance of the channel (Polking et al., 2012).

6.1.2.2 High-density atomic storage

The main metrics for information storage devices are the density and capacity, read/write speed, operating voltage and energy consumption, data retention time, and read/write count life. The ultimate storage density and retention time are decided by the ratio of the transition barrier (E_B) between the two states and the thermal energy ($k_B T$): $\tau_0 \exp\left(\frac{E_B}{k_B T}\right)$. Normally, E_B is proportional to the physical size of a memory unit. Therefore there is a fundamental trade-off between the information storage density and the retention time. Possibility of storing information using single atoms, charges, and spins has been explored, by applying nanotechnology of low-dimensional materials.

Scanning probe microscopy (SPM) not only allows atomic-level resolution imaging of material surfaces but also enables manipulation at single-atom level to write information, as well as to read information by electrical, magnetic, and optical measurements (Eigler & Schweizer, 1990). Kalff et al. used scanning tunneling microscopy (STM) to manipulate single atoms to form array of atomic vacancies on chlorine-terminated Cu (1 0 0) surfaces.

8000 bits were stored in the coordinates of atomic positions, with a storage density of up to 502 TB/in.^{2}, three orders of magnitude higher than that of hard disks (Kalff et al., 2016). However, the operating temperature is limited to liquid nitrogen temperature due to thermal perturbation−resulted atomic migration. Eom et al. discovered that surface atoms have different charge states associated with the atomic configurations. Under stimulation of an STM probe, Si atoms on a B-doped Si (1 1 1) $\sqrt{3} \times \sqrt{3}$ reconstructed surface could be transformed between two configurations. Such bi-instability in the atomic configuration and electronic structures could be used to store information. And the local conductivity and density of states could be measured by using scanning tunneling spectroscopy (STS) to read the information. With a memory unit of 2.7 nm^2, a storage density of 37 TB/cm^2 was estimated (Eom et al., 2015).

As a result of exchange interactions between electron spins, materials can form magnetically ordered ferromagnetic and antiferromagnetic phases. Sebastian Loth et al. fabricated iron atomic chains with antiferromagnetically aligned Neel states. By injecting spin-polarized currents with an STM tip to change the spin direction of individual atoms, transitions in the Neel states were achieved (Khajetoorians et al., 2013; Loth et al., 2012). The operating temperature of this memory mechanism depends on the barrier energy of Ising domain-boundary inversion, also requiring cooling of liquid nitrogen.

Diamond has a cubic phase consisting of sp^3 hybridized carbon atoms, with excellent electrical, optical, and spin properties. The wide bandgap in diamond provides a vacuum-like environment for the atomic defects with discrete energy levels. One of the most important defects is the nitrogen-vacancy (NV) color center, consisting of a nitrogen atom and a carbon vacancy. Dhomkar et al. excited an NV defect to a neutral NV^0 or charged NV^1 state by using light of different wavelengths and read the states by observing the photoluminescence, reaching a write and read speed of milliseconds (Dhomkar et al., 2016). Diamond has the highest modulus of elasticity of all materials due to strong carbon−carbon covalent bonding, enabling the NV-based memory devices to work at room temperature. Currently, its storage density is limited by the defect density and lack of precise control of the defect sites.

DNA, which stores genetic information, has a unique helical structure with a diameter of about 2 nm. There are about 3 billion DNA-based pairs in human chromosomes. Understanding the information in the DNA pair sequences is the key to cure genetic diseases. With recent development of synthetic DNA technology, DNA fragments can be synthesized and assembled according to the coding order. And then information can be extracted and decoded by gene sequencing techniques. Unlike other memory technologies based on inorganic materials, DNA memory is based on biomolecules, allowing for the potential storage of information in a living environment. In 2018 Organick et al. (2018) at the University of Washington reported a DNA-based random access reconstruction algorithm that realized storage of a video of 200 MB in 13.4 million gene sequences.

6.1.3 Computation beyond Moore

Since 1970s there has been a clear direction in the development of information technology to reduce the size of transistors to increase the integration scale, thereby increasing the computational power, and to develop general-purpose computers. Software for specific purposes then can be developed and applied to a wide range of problems. In accordance with the famous Moore's law, information industry has grown exponentially over the past half a century. But it has hit its limits, including physical size, manufacturing costs, energy consumption, and heat dissipation.

For complex challenges such as material design, climate simulation, and brain modeling, greater computation power with larger information storage and efficient processing technologies are urgently needed. At the end of Moore's law, it was pointed out in the International Roadmap for Devices and Systems that there are three directions for future development (Khan et al., 2018). The first is the continuation of Moore's law (More Moore), including adopting low-dimensional semiconductor materials to enhance the performance of transistors below 5 nm. The second is a multifunctional integration (More than Moore), which integrates various sensors with the processors and wireless communications to realize the so-called Internet of Things (IoT) (Leiserson et al., 2020). A third approach is the development of new devices based on new principles and architectures (beyond Moore), including neuromorphic computing and quantum computers, which are the topics of this section.

6.1.3.1 Neuromorphic computing

Despite great progress of modern computers based on transistors and integrated circuits, there are still great difficulties in large and complex problems such as brain simulations. One of the greatest challenges is the dramatically increased energy consumption along with the computation scale. In contrast to the low energy efficiency of computers, human brain uses only a few tens of watts, yet it is powerful for pattern recognition, learning, decision-making, and importantly imagination and creation of new ideas. The secret of the brain may lie in its configuration. Human brain consists of a complex neural network structure, where the number of neurons is on the order of 10^{11}, and each neuron is connected to thousands of other neurons by synapses with the number on the order of 10^{14} (Herculano-Houzel, 2009).

Neuromorphic computation is a new information-processing architecture inspired by the brain and neural networks. In a neural network, the strength of connections between neurons is equivalent to weights in the nodes of artificial neural networks and is tuned by ion concentrations at the nodes. Inspired by the onsite-stored memory function, memristors with memory functions are applied in neural networks to store the weights of connections between layers in artificial neural networks (Ambrogio et al., 2018; Wang, Wu, et al., 2020; Yu & Chen, 2016).

In each layer of the neural network, the memristor acts as a synapse consists of input connections, weights, and output connections. Input signals are the voltages from a previous layer. The weights are the conductivity of each node, which is adjusted according to the learning algorithm. And the output signals are the currents into the next layer. Computation in the neural network is carried out by physical principles of Ohm's and Kirchhoff's laws, instead of performing a large number of mathematical operations one by one. The physical results are equivalent to parallel mathematical operations and can therefore dramatically accelerate the computation.

Hybrid nodes composed of CMOS transistors and memristors are capable of embodying the features of neurons and synapses, including pulse time-dependent plasticity. Based on building blocks of nodes, crossbar networks could be constructed, with the CMOS addressing each crossover and the memristor storing the neuron weights. With a 30-nm processing technology, the device density can reach 25 million/cm^2, which is higher than the density of biological neural networks. Minimum linewidth of memristor cross junctions has reached 6 nm, with a fast response speed of $\sim$0.02 Ms and a low energy consumption of $\sim$1 W/cm^2, promising for the energy-efficient deep learning (Pi et al., 2019).

Strukov et al. at the University of California Santa Barbara used an Al_2O_3/TiO_{2-x} double oxide layer to fabricate a memristor with a switching ratio of 10^4, a switching limit of more than 5000 cycles, and a storage time of longer than 10 years. They showed a simple 10×3 neural network structure consisting of 10 inputs and 3 outputs and 10×3 memristor cross-nodes. Nine of the inputs correspond to a 3×3 pixel picture, the remaining input serves as the deviation or error, and the outputs are letters corresponding to the input pictures. Training of the neural network is realized by adjusting the resistance of memristors using CMOS transistors. After 20 training iterations, letters corresponding to input pictures can be recognized (Prezioso et al., 2015).

6.1.3.2 Quantum computers

As the size of transistors scaling to near-atomic level following Moore's law, it has met the physical limit. Future leaps in computing powers will need to be explored in other dimensions. One of the most attractive approaches is to utilize quantum information, including the linear superposition and entanglement states based on principle of quantum mechanics. By replacing N binary bits in the Boolean space with quantum bits in the Hilbert space, an exponential growth will be achieved in the amount of information from N individual states to 2^N states in superposition.

The basic unit of quantum computing is quantum bits or qubits, which can be represented by vectors of $|0\rangle$ and $|1\rangle$. Physically a qubit can be represented by an elementary particle with two quantum states, for example, spin-up and spin-down directions of spin-1/2 particles, or polarization of photons. The difference between classical and quantum bits is that the former can only be in one of the 0 or 1 states, while the latter can be in a superposition of states. N qubits can represent a vector in a 2^N-dimensional vector space. A quantum state is a linear superposition of the 2^N states. Therefore the operations on a quantum state are equivalent to the parallel computations, greatly improving the computational efficiency, especially on certain types of problems.

For large-scale quantum computing, it is still a great challenge to find a suitable quantum system to represent the qubits. In principle, all two-level quantum systems can be used as qubits, the difficulty lies in the trade-off between quantum operations and decoherence, both of which are related with coupling to the external stimuli and environment. Sufficient isolation is needed to preserve their quantum coherence, while strong enough coupling is required to complete manipulation and computation. There are many systems currently being investigated, including superconducting circuits, ion traps, cold atoms, spins on quantum dot, photons, NV centers in diamond, and topological states (de Leon et al., 2021). Apparently, many of the quantum systems are low-dimensional structures, where individual quantum states are confined in nanometer-scale spaces, either naturally or by nano-fabrication.

Quantum computing has grown rapidly in recent years with continued investment from governments and companies. In 2017 Pan group at University of Science and Technology of China reported that photon-based quantum computers have outperformed classical computers on boson sampling problems (Wang et al., 2017). In 2018 IonQ developed a quantum computer based on ion traps, including 11 fully connected quantum bits, 55 independently operated ion pairs, and over 60 quantum gates, with an average single-qubit gate fidelity of 99.5% and an average two-qubit-gate fidelity of 97.5% (Wright et al., 2019). In 2019 a team at Google reported a programmable superconducting qubit-based quantum computer with 53 quantum bits that can perform 1 million times of random samplings in 200 seconds, claiming the so-called quantum supremacy (Arute et al., 2019).

According to the Hefei Declaration on Quantum Technology of the International Conference on Emerging Quantum Technologies in September 2019, development of quantum computers consists of three stages: in the first stage, to surpass classical computers in specific problems; in the second stage, to realize quantum simulations in quantum chemistry and machine learning; and in the third stage, to realize programmable general-purpose quantum computers that can be used in encryption and decryption, analysis of big data and AI. Currently, it is rapidly devloping in the first stage.

6.1.4 Low-dimensional sensors

Sensors are at the interface between virtual digital world and physical world, transducing information from the environment into electrical signals that could be processed, stored, and communicated by information technology. Together with information processing, storage, and communication subsystems, sensors constitute an intelligent system. To some extent, sensors can be viewed as extensions of human senses, to detect optical, acoustic, mechanical, and chemical signals, corresponding to seeing, hearing, touching, and smelling senses. Sensors typically have a much wider range compared with human senses. For example, humans can only see the light within visible range, while optical detectors could cover the whole optical spectrum from deep infrared to deep ultraviolet range.

Sensors could be seen (or hidden) everywhere in daily life. In smartphones, there are GPS and magnetometers for local position, gyroscopes, and accelerometers for direction and speed of movement, CMOS sensors for photos, touchscreens for input by hand, and microphones and speakers for sounds.

Metrics of sensor include sensitivity, selectivity, response speed, and stability. Overall, sensors based on low-dimensional materials have a large specific surface area and therefore a high sensitivity. And because of the short diffusion distance, response speed is fast. Moreover, such sensors are cost-efficient, because of the small mass and size. It can be envisioned that by integration of a variety of low-dimensional sensors, it is possible to build a multifunctional, low-cost, miniaturized intelligent sensing system, becoming crucial infrastructures for the IoT and smart society.

In this section, we will introduce sensors based on low-dimensional materials, including electrical, mechanical sensors, and optical sensors for artificial vision. The discussions will be focused on the fundamental principles and structure-dependent properties.

6.1.4.1 Transistor sensors to detect single molecules

FET sensors are based on charge transfer between the detected object and the semiconductor channel or the contact junction, causing changes of potential and conductance. An FET sensor consists of two main components, a normal FET and receptors that are coupled to the channel. Receptors are critical to determine the sensing selectivity. And coupling efficiency of gate voltage to the channel conductivity is important for the sensitivity. FETs with semiconductor nanowires, carbon nanotubes, and two-dimensional semiconductors as conducting channels allow the gate to regulate the channel conductivity efficiently due to their nanoscale or even atomic-scale thickness. Therefore high sensitivity can be expected for the low-dimensional semiconductor-based FET sensors.

FET sensors based on 1D nanowires and nanotubes have been developed to detect pH, ions, nucleic acids, proteins, and cells (Mu et al., 2015). The first silicon nanowire FET sensor was reported by Lieber group at Harvard University in 2001 (Cui et al., 2001). By using a double gate electrode configuration, the response sensitivity was enhanced to be beyond the Nernst limit of 60 mV/pH (Knopfmacher et al., 2010). A fundamental challenge for nanoscale sensors is the high background noise associated with the high sensitivity. A smart method to improve the signal-to-noise ratio is to analyze the frequency response characteristics to differentiate the signal from background noise (Zheng et al., 2010). An extremely high sensitivity to protein antibody and DNA has been reported for nanowire FET sensors with a detection limit down to fM (10^{-15}M) level (Gao et al., 2011; Stern et al., 2007).

Semiconducting CNTs are promising for FET sensors due to their nanometer scale diameter and near ballistic transport in a micrometer scale. However, it has been challenging to obtain highly pure semiconducting nanotubes and to couple suitable receptors due to the inert surface. By using polymer-assisted method, Peng group in Peking University successfully separated highly pure semiconducting CNTs and fabricated efficient nanotube array transistors. Instead of attaching receptors directly to the CNT channel, DNA receptors were bonded to gold nanoparticles that were loaded to the high-κ Y_2O_3 dielectric layers to avoid direct contact and contamination from the environment to the nanotubes. As a result, the detection limit for target DNA fragments reached as low as 60 aM (10^{-18}M) level (Liang et al., 2020).

Nowadays, individual FET sensors can detect individual molecules and cells. Furthermore, FET arrays can detect electrical signals inside cells and communicating signals between neuron cells (Duan et al., 2012; Tian et al., 2010). Lieber et al. combined nanowire FET arrays with neuronal cells and successfully probed the spatial and temporal characteristics, such as the rate and amplitude of neuronal signal transmission. Reversely, electrical signals could be applied by using FETs to stimulate and control cell activities (Patolsky et al., 2006).

On the basis of transistor sensors, intelligent systems could be created by integrating with computing and memory functional components. Shulaker et al. reported on a 3D integrated system with sensing, computing, and storage on the same chip. Carbon nanotube FETs were used as the sensing input layer, where three different functional groups were combined to respond to different gases selectively. Sensing data was stored in the random access memory units. Then the data was analyzed using a silicon-based integrated circuit processor. Finally, response characteristics to seven gases were categorized using principal component analysis (Fig. 6.4) (Shulaker et al., 2017).

6.1.4.2 Mechanical sensors to weigh a proton

Essentially, interaction between sensor and subject is some kind of force. Mechanical sensors can be used to measure and analyze such forces to determine the type of interaction to get information about the detected substances. One of the ultimate mechanical sensors is nanoprobes of scanning probe microscope (SPM), which can obtain physical and chemical

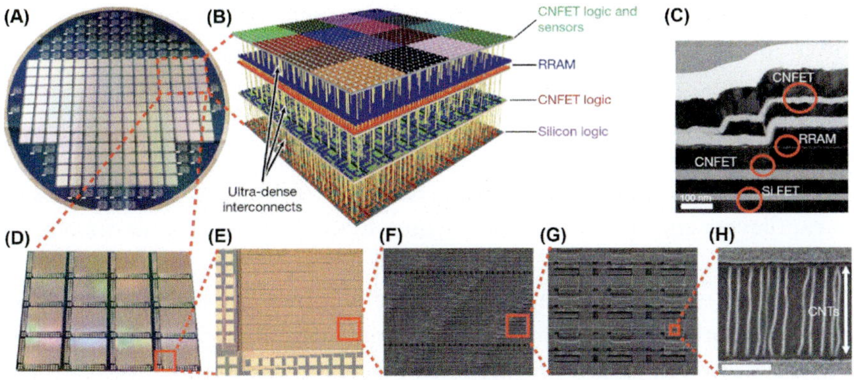

Figure 6.4 ***Carbon nanotube-silicon 3D-integrated gas analyzing system.*** (A) Wafer-level processing; (B) interconnected chip with sensing, storage, and analyzing layers; and (C–H) arrays of carbon nanotube transistors at different length scales (Shulaker et al., 2017).

properties of material surfaces at nanometer or even atomic resolution. By designing suitable functional groups, SPM probe could be used to detect individual molecules and even single chemical bonds. SPM-based mechanical sensors can be categorized into two types: static sensors for measuring the surface force and dynamic sensors to detect the shift of resonance frequency.

The principle of static mechanical sensors is to measure the interaction-induced deformation to the probes. Deformation-associated deflection of the cantilever can be transduced into optical or electrical signals through laser reflection or piezoresistive effect. The accuracy of mechano-optical measurement is very high and can be used for detecting protein antibody (Backmann et al., 2005), DNA (Fritz et al., 2000), and RNA (Zhang et al., 2006). On the other hand, it is easier to integrate mechanical sensors based on piezoresistive effect on chips to fabricate compact intelligent systems (Wee et al., 2005). Piezoresistive effect can be expressed as follows: $\mathrm{d}R/R = \gamma\varepsilon$, where change of resistance is linearly proportional to the strain ε. Piezoresistive effect is an intrinsic property associated with the symmetry of the crystals, while the output signal is also associated with the geometry of probes and cantilevers.

Yoshikawa and Rohrer et al. optimized the geometry of piezoresistive sensors by designing asymmetric Wheatstone bridges for a suspended thin film, where functional component of the sensing part (receptor surface) was separated from the transducer part (piezoresistive beams). The sensitivity was enhanced due to the increase of sensing surface area. In addition, piezoresistive effect was amplified by algebraic combination of the signals from four piezoresistive beams. As a result, the piezoresistive nanomechanical membrane-type surface stress sensor demonstrated a 20-fold improvement compared to ordinary cantilever-based mechanical sensors (Yoshikawa et al., 2011).

Dynamic mechanical sensors are based on mechanical resonators, such as cantilevers and drums. The resonance frequency of a cantilever beam is associated with the device geometry, density, and elastic modulus according to the Euler–Bernoulli theory. When a substance, such as a gas molecule, is adsorbed on the surface, the resonant frequency of the cantilever beam is shifted in relation to the mass of the adsorbed substance. The performance of a resonant mechanical sensor is mainly related to its mass, resonant frequency, and quality factor (Q). Roukes et al. calculated the detection limit of a nano-electromechanical system sensors. It was found that a both-clamped beam with a resonance frequency of 1 GHz and a quality factor of 10^5 is capable of detecting the mass with an accuracy of a single proton, even under thermal perturbations from the environment (Ekinci et al., 2004).

Carbon nanotubes are ideal for resonance sensors due to their 1D structure, low density, small diameter ($\sim$1 nm), and high Young's modulus. Bachtold et al. measured the resonance frequency of a suspended carbon nanotube in an FET configuration. When cooled to 30mK to reduce ambient thermal perturbations, a quality factor of up to 5 million was demonstrated with a resonance frequency of tens of MHz (Moser et al., 2014). At room temperature, the CNT resonators have shown an ultrahigh sensitivity of 25 zg (1E-21 g) because of the small mass and high resonating frequency (Lassagne et al., 2008). As the resonance frequency was further increased to 2 GHz level by reducing the length to 150 nm, the mass detection limit was reduced to 1.7 zg, corresponding to individual naphthalene molecules ($C_{10}H_8$) or xenon atoms (Chaste et al., 2012).

In practical applications, especially for biosensors, one of the biggest challenges comes from the liquid environment and associated large damping of the mechanical vibration. Another big challenge is the long diffusion time of the substance to the sensors due to the low concentration. To overcome the damping effect of the liquid, Favero changed the usual one-dimensional resonant beam structure into a disk shape with higher stiffness, whose "breathing mode" can maintain resonance at GHz level in liquid environment. The resonance frequency is detected by measuring the absorption spectrum of laser light through an optical fiber near the disk, demonstrating a surprisingly high-quality factor of 22,000 in liquid (Gil-Santos et al., 2015).

Instead of setting the sensor in a liquid environment, Burg et al. designed a new type of sensors where the substances to be detected flew through a microfluidic channel inside the cantilever beam. In such case, the environment is confined inside the sensor, largely avoiding the damping and diffusion length problems. The shift of resonance frequency of the cantilever due to mass change of the adsorbed substance could be detected, with a high-quality factor of 15,000. Due to the small mass and high-quality factor, the sensitivity reached fg (10^{-15} g) level, which is six orders of magnitude higher than a normal quartz crystal microbalance (Burg et al., 2007).

6.1.4.3 Optical sensors and machine vision

As the saying goes, "a picture is worth a thousand words." For human beings, more than 80% of information is obtained through seeing. Vision is not only seeing but also an intelligent system including optical sensors (eyes), signal transmission (neurons), and a powerful processor (the brain) that is trained by years of life experience.

Similarly, machine vision is vital in emerging applications like autonomous driving and robotics. Machine vision includes modules of optical signal acquisition, photoelectric conversion, electrical signal transmission, and machine learning modeling. By training machine learning models such as artificial neural networks, the machine can not only see but also recognize what is seen so that decisions could be made for automatic operation. Machine vision based on photodetector arrays and transistor-based neural networks has been investigated intensively in recent years.

Mennel et al. reported a neural network-based image sensor that is fabricated with floating gate transistors of 2D WSe_2 nanosheet, which becomes a direct bandgap semiconductor as the thickness is reduced to monolayer (Mennel et al., 2020). The semiconductor photoelectric detectors form an array with each individual detector corresponding one pixel in the input image. The optical signal is transformed into electrical signals forming a matrix through optical–electrical conversion. The output of photodetector array is connected to the inputs of next layer of the neural network. The responses between the nodes are controlled by gate electrodes, representing the connection weights between neurons in neural network. The neural network model is trained by updating the connection weights based on the loss function calculated from the output as compared with the input image. The weights of the connections are stored onsite as the charge of the floating gate therefore avoiding large amounts of data transfer. They demonstrated supervised machine learning for image classification and unsupervised machine learning for image decoding. Instead of using the same material for photodetection and information processing, Sun et al. fabricated photo-responding transistors with semiconducting carbon nanotubes as the transistor channel and perovskite quantum dots for photodetection (Zhu et al., 2021). The fabricated machine vision sensor showed a high photosensitivity of 5.1E7 A/W, demonstrating reinforced visual learning functions under a weak light pulse of 1 $\mu W/cm^2$, as shown in Fig. 6.5.

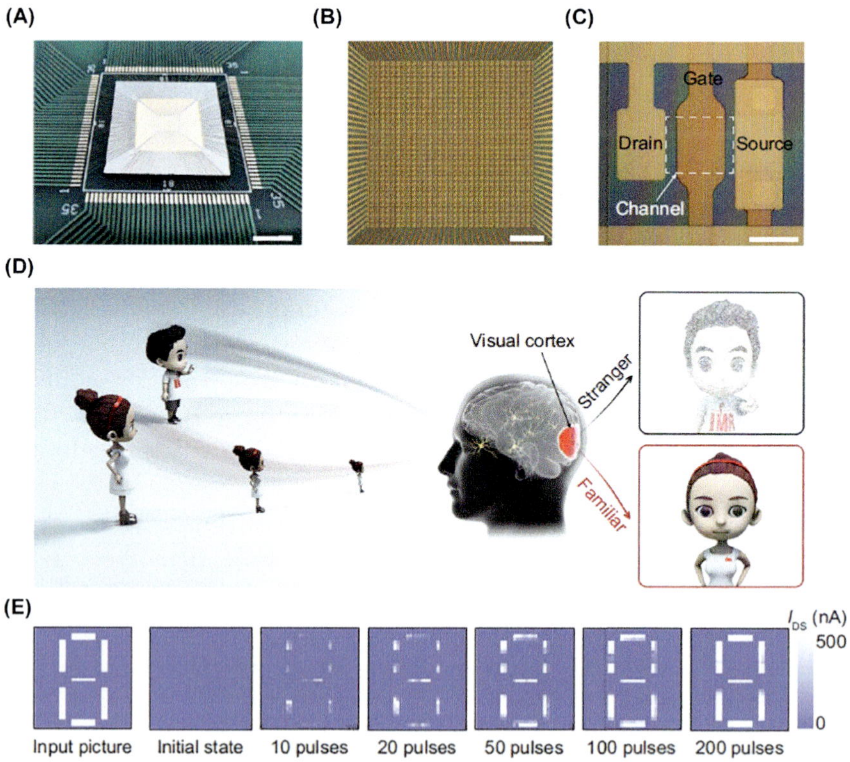

Figure 6.5 ***Neuromorphic machine vision based on carbon nanotube transistors and perovskite quantum dot photodetectors.*** (A–C) A sensor in sensor array; (D) schematics of human vision; and (E) training of machine vision by repeated pulses (Zhu et al., 2021).

6.2 Low-dimensional energy materials

Fundamentally, all activities are dependent on the energy, from food, housing, traveling, clothing, illumination, communication, warming and cooling, and so on. Along with the development of human civilization, especially the deepening understanding of fundamental principles of nature, the forms and scales of energies have changed dramatically.

In the very beginning, sunlight from nuclear fusion brought energy such as light and warmth to all forms of life, including human beings. Then humans learned to use tools, to make fire to light in night, to cook, and to keep warm, by burning firewood. Later, people invented various machines to use mechanical energy from the animal, water, and wind powers.

Over the past 200 years, productivity has skyrocketed in industrial revolutions, driven by enormous amount of fossil energy. In the first industrial revolution, steam engine converted chemical energy in coal into heat and mechanical energy. In the second industrial revolution, power plants converted mechanical energy from heat, water, wind, and later nuclear energy into electricity.

Fossil fuels come from the hydrocarbon compounds accumulated through hundreds of millions of years of photosynthesis in plants to convert solar energy into chemical energy. In a very short time (100 years compared with the 4 billion years of earth history), a massive amount of fossil fuels was burnt to release not only the energy but also greenhouse gases such as carbon dioxide (CO_2), which has changed the energy equilibrium and cycle of carbon elements.

It is not sustainable to keep on burning fossil fuels at the still accelerating pace due to the limited reserves and more practically the environmental problems and already happening climate change. To solve the grand challenges at a global scale, it is critical to find clean and renewable energy sources, to improve the efficiency of energy conversion, and to realize the recycling of materials and regeneration of energy in a sustainable way.

Energy applications of low-dimensional materials can be categorized into four types: (1) energy harvesting devices such as solar cells and artificial photosynthesis; (2) energy storage devices such as lithium-ion batteries and hydrogen storage; (3) energy conversion devices such as fuel cells; and (4) energy recycling devices such as thermoelectric cells.

Due to the principle of energy conservation, essentially all energy-related processes are energy conversion and transfer. In this section, we will discuss the fundamentals shared by various energy devices, the theoretical efficiency limits from the thermodynamics, and practical approaches to enhance the efficiency from a kinetic point of view.

6.2.1 Solar cells

Sunlight is the ultimate source of all energy on earth except geothermal energy, including heat, wind, and hydropower, as well as fossil fuels accumulated from ancient photosynthesis processes. Solar spectrum is consistent with characteristics of blackbody radiation at a temperature of about 5800 K and is predominantly distributed in the visible range, with a peak at around 500 nm (2.48 eV). After atmospheric absorption, the average power density of sunlight at ground level is about 1000 W/m^2. The total solar power at earth's surface amounts to 10^5 TW, far exceeding the annual global energy needs of ~10 TW. In the long run, solar energy is the only reliable energy source for sustainable development of human civilization. In the following sections, we will discuss on solar cells and photosynthetic fuels, which convert solar energy into electrical and chemical energy, respectively.

The basic structure of a solar cell is a semiconductor p–n junction, with a light-absorbing layer, an electron-transporting layer, and a hole-transporting layer. The conversion process from light to electricity mainly includes following steps: (1) Electrons in valance band absorb photons and electron–hole pairs are excited. The electrons at excited state have energy distribution consistent with the solar spectrum and are called hot electrons. (2) After electron–lattice (phonon) interaction and cooling of the hot electrons, electrons and holes are concentrated at the edges of conduction and valence bands, respectively, reaching thermal equilibrium with the environment. (3) Electron and holes are separated by built-in electric field in the p–n junction and supplied to external loads. (4) Through electron–hole recombination, a dynamic equilibrium with light absorption is reached.

The power conversion efficiency (PCE) is the result of the combination of all kinds of loss mechanisms, including light absorption efficiency, separation efficiency, transport mobility, and charge harvesting efficiency. In 1961 Shockley and Queisser considered the equilibrium of band-edge absorption and radiative recombination and proposed the Shockley–Queisser (S–Q) model to predict the theoretical limit of solar cells (Shockley & Queisser, 1961). As bandgap decreases, utilization of

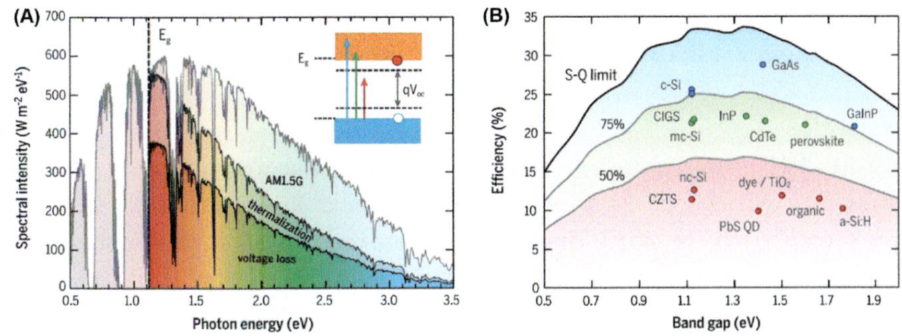

Figure 6.6 ***Theoretical efficiency of solar cells.*** (A) Available solar spectrum for a semiconductor with a certain bandgap (E_g). The inset is a schematic of the photoelectron excitation processes in solar cells; (B) plots of theoretical solar cell conversion efficiency versus the semiconductor bandgap (Polman et al., 2016).

solar spectrum and short-circuit current (J_{SC}) increases; however, the open-circuit voltage (V_{OC}) decreases, resulting an optimal bandgap. For single-junction solar cells, the optimal optical bandgap is 1.34 eV under AM1.5 standard solar spectrum, corresponding to a theoretical efficiency of 33.7% (Fig. 6.6).

6.2.1.1 Nanowire array solar cells

When sunlight hits the surface of a solar cell, part is absorbed, and part is reflected. The absorption efficiency is closely related to the surface morphology. For example, in bulk silicon solar cells, the reflectivity is reduced from 35% to 20%, thanks to the tetrahedral tapered surface by anisotropic etching of ⟨1 0 0⟩ plane. In combination with antireflective coatings, the total reflectivity can be further reduced to less than 10%.

Vertical arrays of nanowires and nanotubes can form effective light traps that fully absorb incident light after multiple reflections (Kayes et al., 2005). It has been reported that carbon nanotube arrays are the world's "blackest" material (Mizuno et al., 2009). Harry Atwater et al. demonstrated the enhancement of optical absorption by forming silicon nanowire arrays (Kelzenberg et al., 2010). With a coverage of only 5%, they were able to absorb 96% of the incident light resulting in an external quantum efficiency of 0.89.

Anttu (2015) calculated the conversion efficiency of III–V group InP nanowire array solar cells. The short-circuit current is lower compared to the bulk solar cells, because of the lower surface coverage. On the other hand, the V_{OC} is enhanced due to reduction of radiative recombination. The theoretical conversion efficiency of the nanowire array solar cells can reach 32.5%, which is similar to the theoretical S–Q efficiency of bulk InP (31%).

In 2010 Yang group reported on the silicon nanowire p–n junction array-based solar cells with a conversion efficiency of about 5% (Garnett & Yang, 2010). In 2013 Bakkers et al. prepared p–n junction solar cells using direct bandgap InP nanowires as an absorber layer, the conversion efficiency was measured to be 11.1% (Cui et al., 2013). Borgström et al. prepared p–i–n junction InP nanowire array solar cells to reduce recombination. With an area coverage of only 12%, a conversion efficiency of 13.8% was demonstrated (Fig. 6.7) (Wallentin et al., 2013).

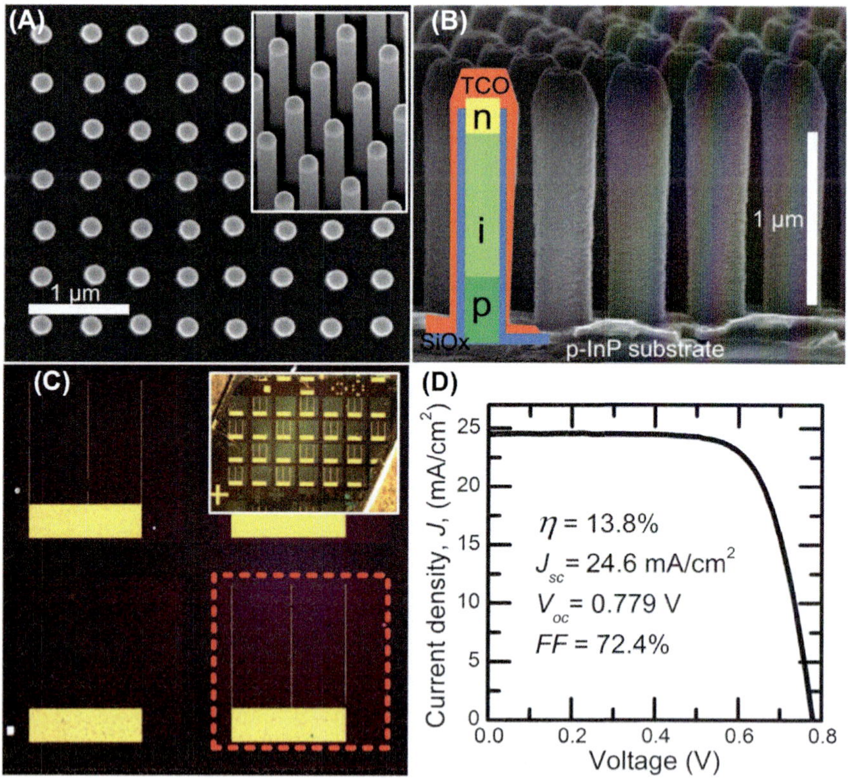

Figure 6.7 ***InP n–i–p nanowire array solar cells.*** (A and B) Top- and tilted-view of SEM images; (C) optical photo of the solar cell with a 1 × 1 mm area; and (D) current density (*J*)–voltage (*V*) plot under simulated solar illumination (Wallentin et al., 2013).

6.2.1.2 Quantum dot solar cells

As predicted by the S–Q model, there is an optimum bandgap for the light absorption layer of solar cells. In conventional semiconductors, the bandgap is controlled by composition. For low-dimensional materials, the bandgap could be further adjusted by size effects. Thanks to the great progresses in fine control of the size, surface, and interface structures of quantum dots, it becomes possible to tune the bandgap of semiconductor nanocrystals to the desired value of 1.34 eV for fabricating ideal single junction solar cells.

In 2005 Sargent group at the University of Toronto reported PbS quantum dot solar cells with a conversion efficiency of ~1% (Mcdonald et al., 2005). The conversion efficiency was improved by an order of magnitude to 9.9% by reducing the density of defects to increase the diffusion length of carriers (Lan et al., 2016). There is still a large space for further improvements, since the V_{OC} is ~0.64 V, which is only 0.56 of the theoretical value; and the short-circuit current is 0.66 of the theoretical value. The performance is limited by the surface defects associated charge hopping transport mechanism (Polman et al., 2016).

In 2020 Wang group of the University of Queensland reported a quantum dot solar cell based on $Cs_{1-x}FA_xPbI_3$ perovskite nanocrystals (Hao et al., 2020). The concentration of defects was reduced by regulating reaction kinetics. The V_{OC}, short-circuit current density, and filling factor reached 1.17 V, 18.3 mA/cm^2, and 78.3%, respectively. The maximum PCE reached 16.6% and maintained at 94% of the initial efficiency under 600 hours of continuous solar illumination (Fig. 6.8), paving the way for practical applications.

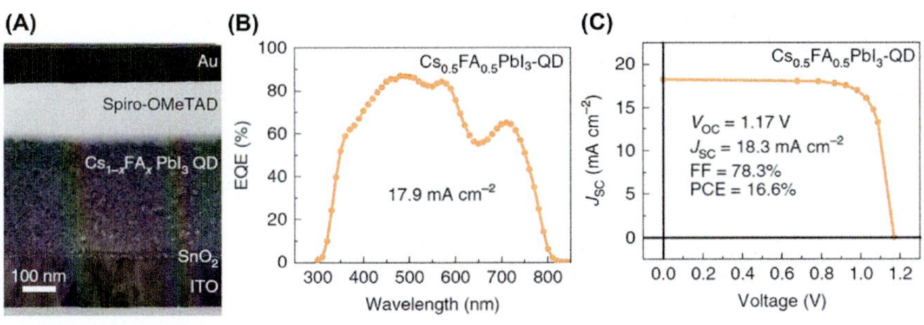

Figure 6.8 ***$Cs_{1-x}FA_xPbI_3$ perovskite nanocrystalline solar cell*** (A) SEM image of cross-sectional structure; (B) external quantum efficiency against wavelength of the incident light; (C) current density (*J*)—voltage (*V*) plot and main indicators of the solar cell performance (Hao et al., 2020).

6.2.1.3 Dye-sensitized solar cells

As the conversion efficiency of a solar cell is determined by a series of processes in multiple materials and junctions, it is challenging to design a high-performance solar cell using a single material. Dye-sensitized solar cells (DSSCs), also known as photoelectrochemical solar cells, have a different structure from the conventional p−n junction solar cells. In DSSC, light absorption, electron transport, and hole transport are realized in different materials. Light is absorbed by a dye molecule (usually ruthenium-based) instead of a semiconductor, that is, the dye sensitization process.

In 1991 Grätzel et al. reported DSSCs with mesoporous TiO_2 films as n-type photoelectrodes (Grätzel, 2001; O'Regan & Grätzel, 1991). Because of the large surface area, a dramatic enhancement of three orders of magnitude in light absorption and photoconversion efficiency was achieved, compared with similar devices with a planar configuration. In DSSC, after dye molecules absorb the light, electrons are excited from highest occupied molecular orbital (HOMO) orbitals to lowest unoccupied molecular orbital (LUMO) orbitals in the dye molecules and then injected into the conduction band of the TiO_2 semiconductor. In the meantime, holes react through an ionic redox pair (I^-/I^{3-}) reaction; and I^-/I^{3-} is then transported to the counter electrode through the liquid phase, where it reacts with electrons transported through the external circuit.

DSSCs have advantages of low cost, flexibility, printability, and semitransparency, and therefore are promising for large-scale applications such as smart windows. A highest conversion efficiency record (11.9%) was reported by Sharp company in 2013 (Chiba et al., 2006). The record was kept for many years until it was broken by Grätzel group in 2022 (Ren et al., 2023). They designed coadsorbed sensitizers that harvest light efficiency across the entire visible range. An independently certified PCE of 15.2% was achieved, with a high operational stability up to 500 hours. To further enhance the conversion efficiency, it is necessary to optimize the multistage photoelectrochemical reactions, including charge transfer at interfaces, electron transport in solid electrodes, and ion transport processes liquid phase.

6.2.1.4 Perovskite solar cells

In recent years, perovskite solar cells have become a hot research frontier because of the surprisingly high efficiency (Green et al., 2019). Organic−inorganic hybrid perovskite materials have unique structures and properties for optoelectronic applications including solar cells. Their bandgaps are determined by the halogen component and can be adjustable from 1.6 to 3.2 eV by replacing iodine with chlorine. In addition, the bandgap could be further adjusted by the organic cation and can be reduced to the optimum bandgap for solar cells with formamidinium (FA, $CH_3(NH_2)_2^+$) to replace methylammonium (MA, $CH_3NH_3^+$). In normal semiconductors, defects states could be traps and scattering centers for charge carriers, which is one of the limiting factors for optoelectronic devices. One special property of the organic–inorganic hybrid perovskite materials is that the defect states are not located within bandgap, and therefore the diffusion length is large, leading to a high carrier mobility (Huang et al., 2017; Yin et al., 2014).

Organic–inorganic hybrid perovskite solar cells were first reported by Miyasaka group at the University of Tokyo in 2006 (Kojima et al., 2007), where $CH_3NH_3PbBr_3$ was introduced as dye molecules in DSSCs, and a modest conversion efficiency of 2.2% was achieved. Since then, there has been great progress in the fabrication techniques, fundamental understanding, and device performance. In 2012 Park and Grätzel et al. introduced a solid-state electrolyte as a hole transport layer, which improved the stability, enhancing the conversion efficiency to 9.7% (Kim et al., 2012). Snaith et al. found that mixed halogens have a higher structural stability and lower electron–hole recombination rates, allowing for larger carrier diffusion distances to micrometer scale (Stranks et al., 2013). It was realized that TiO_2 layer was not necessary; thus the TiO_2 was replaced with a thin insulating layer of Al_2O_3 to increase the V_{OC}, resulting in a conversion efficiency >10% (Lee, Teuscher, et al., 2012). Grätzel et al. developed a stepwise deposition method and got better control of the morphology, reporting a certified conversion efficiency >14% (Burschka et al., 2013). Snaith et al. fabricated perovskite solar cells with a planar heterostructure that was distinct from the porous nanocrystalline architecture, achieving >15% conversion efficiency (Liu et al., 2013).

Despite breakthrough in the conversion efficiency, one bottleneck for practical application of perovskite solar cells is their stability in environments of high temperature, high humidity, and strong light. For practical applications, it is required to maintain 90% of the output power for 10 years and 87% for 25 years. The fundamental origin of the low stability is due to its weak ionic bonding associated low activation energy for defect formation and atomic diffusion (Burschka et al., 2013).

Interface structure and bonding stability are crucial for both the efficiency and stability of solar cells. Grätzel et al. obtained a conversion efficiency >20% using CuSCN as a hole extraction layer. By adding a reduced graphene oxide conductive layer between CuSCN and gold electrodes, the thermal stability was effectively enhanced. Overall, 95% of its initial efficiency was maintained after 1000 hours of operation at full sunlight illumination at 60°C (Arora et al., 2017). Han group at Shanghai Jiao Tong University enhanced the interface strength by using chlorinated graphene oxide to form Pb−Cl and Pb−O bonds with the $[CH(NH_2)_2]_x[CH_3NH_3]_{1-x}Pb_{1+y}I_3$ organic−inorganic perovskite thin film. After 1000 hours under 60°C and AM1.5G simulated solar illumination, 90% of its initial conversion efficiency (21%) was maintained (Wang et al., 2019). On the other hand, Padture et al. at Brown University designed an iodide ion bridging layer to improve the bonding strength between the electron transport layer and the perovskite thin film. Mechanical toughness was increased by 50%, and the conversion efficiency was moderately increased from 20.2% to 21.4%. More importantly, the period to maintain 80% of the initial efficiency under simulated solar illuminations was increased from 700 to 4000 hours (Dai et al., 2021).

6.2.2 Solar fuel

From the perspective of energy and materials, modern civilization is built on the energy from fossil fuels and materials from petrochemicals. Though clean energy, such as solar and wind, has a large capacity; typically the power density is rather low and varies greatly with season, weather, and geography. For large-scale application of clean energy, it is crucial to establish a smart energy and materials system with energy/materials storage and conversion technologies. To achieve the goal of net zero carbon around 2050, it is necessary to develop technologies to produce fuel with zero or negative carbon emission (Shih et al., 2018). One of the most attractive technologies is solar fuels from artificial photosynthesis that convert sunlight into fuel.

6.2.2.1 Natural and artificial photosynthesis

Photosynthesis is a chemical process that water and carbon dioxide are converted into oxygen and hydrocarbon through photocatalysis reactions within plants. It is the main source of oxygen on earth and the ultimate food source for all animals, including human beings. Actually, as indicated by the name, fossil fuels come from the accumulation of hydrocarbon through photosynthesis process over millions of years.

Photosynthesis reaction includes two main steps. In the first step, water is decomposed into oxygen and proton. In the second step, proton and CO_2 react to make hydrocarbon such as glucose. Both steps are two-electron chemical reactions, and the total conversion efficiency is about 1% for the natural photosynthesis (Barber, 2009).

Inspired by natural photosynthesis, it is a dream to synthesize chemicals such as hydrogen, hydrocarbons, and ammonia through photochemical reactions, so that the energy of sunlight is converted and stored in the chemical bonds. To realize the dream, it is important to investigate the fundamental principles of natural photosynthesis, based on which efficient devices could be designed for the light–matter conversion. In such a way, the low-energy-density and time-varying sunlight is transformed into high-energy-density and reliable fuels. By completing the cycling of key elements such as carbon and nitrogen, artificial photosynthesis will be a crucial part of a sustainable future.

In this section, we will examine the thermodynamics for the ultimate conversion efficiency and discuss on the kinetics for approaches to enhance the efficiency. We will focus on application of low-dimensional semiconductors and electrocatalysts in the field of photoelectrochemical conversion in three systems: photocatalytic splitting of water to produce hydrogen, CO_2 reduction for hydrocarbon chemicals, and ammonia synthesis by nitrogen reduction.

Ross and Hsiao calculated the ultimate photoelectrochemical conversion efficiency by using a two-energy-level model, which considered the balance between light excitation and light emission, ignoring nonradiative recombination process (Ross & Hsiao, 1977). The ultimate conversion efficiency was calculated to be $\sim$29%, with the optimum wavelength to be $\sim$1000 nm. It is not surprising that the value is close to the theoretical conversion efficiency of solar cells. Essentially, for both devices, it is the conversion of photon energy into chemical energy, through the photoelectrochemical and photovoltaic conversion, respectively. However, while the conversion efficiency of solar cells is approaching the S–Q limit, the photosynthetic efficiency is still well below the theoretical limit (Barber, 2009).

There is a fundamental difference between photoelectrochemical and photoelectric conversion processes. For solar cells, there is one key parameter determining the conversion efficiency, that is, the bandgap. On the other hand, for photoelectrochemical process, not only the bandgap but importantly the energy levels of the band edges are important for the photoelectrochemical reactions.

From this example, we can see generally the difference between electronic processes in chemistry and physics. While it is desired to construct simple models to understand the principles of electrical devices, chemical processes are typically much more complex. For a chemical reaction, even a "simple" one such as the decomposition of water into hydrogen and oxygen consists of multiple steps, involving multiple electrons for each intermediate state. It is critical to investigate and understand the reaction mechanisms and efficiency loss mechanisms corresponding to each step, to design efficient energy conversion devices.

6.2.2.2 Photocatalytic water splitting: hydrogen cycle

Hydrogen has been considered the ideal energy carrier and ultimate fuel because of the small mass density and high energy density (120 MJ/kg), which is three times that of gasoline. In addition, only water is produced for the combustion, instead of any greenhouse gases pollutants. However, so far hydrogen has not become a major fuel. One reason is that hydrogen is difficult to store, with a low volumetric energy density (8.5 MJ/L), which is only 1/3 that of natural gas and 1/4 that of gasoline. Another important reason is the source of hydrogen, which is mainly derived from natural gas cracking and is still essentially a fossil fuel. In 1972 Japanese scientists Fujishima and Honda (1972) discovered photocatalytic splitting of water by using a TiO_2 semiconductor electrode, opening up a new pathway to hydrogen production.

The change in Gibbs free energy corresponding to water-splitting reaction is $\Delta G_0 = +237.13$ kJ/mol. According to Nernst equation, the equilibrium potential is 1.23 V, thus requiring the semiconductor to have a bandgap greater than 1.23 eV, with the bottom of conduction band and the top of valence band covering the redox energy levels. The reaction mainly includes the following steps: (1) light absorption to excite electron–hole pairs; (2) electrolysis of water into hydrogen and oxygen, so-called hydrogen- and oxygen-evolution reactions (HER and OER), respectively; and (3) separation of hydrogen and oxygen gases, by spatially separated reactions or postproduction purification. Development of low-dimensional materials for photoelectrochemical splitting of water has been focused on (1) modulation of electronic band structure, (2) promotion of electron–hole pair separation, and (3) enhancement of catalytic activities of the HER and OER electrocatalysis.

The ultimate conversion efficiency of a water-splitting system is determined by thermodynamics, which depends on parameters such as bandgap, energy levels of the valence band top, and conduction band bottom, that is the electronic structure. As for many other cases, there are trade-offs for the design. For example, a small bandgap is advantageous to increase the light absorption; however, only when the potential is high enough, an electrochemical reaction occur. Modulation of the semiconductor band structure is a fundamental solution and no less than an art to design an effective photocatalyst. Many approaches have been developed, including size confinement effects, doping, and alloying.

The initial TiO_2 photocatalysts discovered by Fujishima et al. are chemically stable, nontoxic, and low-cost and therefore are highly attractive for large-scale applications. However, bandgaps of anatase and rutile TiO_2 are 3.20 and 3.02 eV, respectively. And the corresponding absorption edges are at 384 and 410 nm, in the UV region, which is only about 4% of the sunlight energy. To enhance light absorption in visible light region, scientists have developed cation-doped TiO_2 with Cr, Fe, and V and anion-doped TiO_2 with N, C, and B (Asahi et al., 2014). However, dopant atoms are usually distributed on the surface of TiO_2 crystals with localized defect energy levels, resulting in trailing in the optical absorption spectrum.

In contrast to the three-dimensional TiO_2 crystals, titanate nanosheets are two-dimensional structures with a thickness of $\sim$1 nm (Sasaki et al., 1996). In such 2D materials, all the atoms are at the surface, including doping atoms. Liu et al. achieved uniformly distributed nitrogen doping in titanate nanosheets, resulting in the enhanced band-edge visible absorption (Liu et al., 2009). Further introduction of hydrogen atoms into the oxygen vacancies of titanate nanosheets has formed a new valence band at the position of 1.72−1.83 eV below the bottom of conduction band, achieving overall light absorption in visible region (Yang et al., 2018).

Alloying is another effective approach to engineer electronic structure and to enhance the light absorption and photocatalytic activities. In 2006 Domen group at the University of Tokyo reported a $(Ga_{1-x}Zn_x)(N_{1-x}O_x)$ solid solution (Maeda et al., 2005, 2006). While both GaN and ZnO are wurtzite structures with bandgaps of 3.4 and 3.2 eV, respectively, the solid solution has a bandgap of 2.58 eV and is active in visible region. With $RhCrO_x$ as a cocatalyst, hydrogen production by photocatalytic splitting using visible light was achieved, with a quantum efficiency of 2.5% at wavelengths from 420 to 440 nm.

In the history of materials science, there have been many times that the breakthrough in technology has come from discovery of an unexpected new system. While most of the explorations of photocatalysts for water splitting have been focused on inorganic semiconductors, Domen group discovered a polymer (C_3N_4) with a bandgap of 2.6 eV and a quantum efficiency of 0.1% in 420−440 nm wavelength range in the presence of a Pt cocatalyst (Wang et al., 2009). Recently, Kang group at Soochow University fabricated a hybrid system of C_3N_4 and carbon quantum dot with a quantum efficiency of 16% at 420 nm and a high light-to-hydrogen conversion efficiency (STH) of 2% (Liu et al., 2015).

After semiconductor absorbs light to excite electron—hole pairs, the separation of electrons—holes pairs is a critical step to determine the conversion efficiency. Various strategies have been developed, such as cocatalysts, sacrificial agents, *Z*-mechanism, and anisotropic crystalline planes. In 2016 Domen group reported the following system: $SrTiO_3$: La, Rh/Au/ $BiVO_4$:Mo, using Ru and RuO_x as HER and OER cocatalysts, respectively. Hydrogen- and oxygen-generating photoelectrodes were spatially spaced apart through the *Z*-mechanism. And electron—hole separation and charge transfer were enhanced through the gold—semiconductor interface. Consequently, the quantum efficiency reached 33% at a wavelength of 419 nm, with an STH efficiency of 1.1%, reaching the level of natural photosynthesis (Wang et al., 2016).

With the advancement of controllable growth of low-dimensional materials, it becomes possible to obtain nanocrystals of uniform size and even crystals composed of specific crystal planes to enhance the chemical activities (Yang et al., 2008). The anisotropic structure and properties of the crystalline surfaces could be used to promote a separation of electron—hole pairs. Li group at Dalian Institute of Chemical Physics, Chinese Academy of Sciences, obtained anisotropic $SrTiO_3$ nanocrystals by controlling the crystal morphology, achieving a fivefold increase in quantum efficiency (Mu et al., 2016). Domen group deposited Rh/Cr_2O_3 and CoOOH as cocatalysts for the HER and OER reactions on the crystalline faces of Al-doped $SrTiO_3$, where built-in electric field generated by the difference in work functions of (1 0 0) and (1 1 0) planes to promote the electron—hole separation. In the wavelength range of 350 and 360 nm, the external quantum efficiency reached 96% and the internal quantum efficiency was close to 100% (Takata et al., 2020). The safety and feasibility for practical applications have been demonstrated at the scale of 100 m^2 (Fig. 6.9) (Nishiyama et al., 2021).

While photocatalyst in powder form has the advantage of large surface area and high catalytic activity, it is challenging to separate the hydrogen and oxygen gases since they are generated simultaneously. On the other hand, in the photovoltaic-electrolysis cell (PV-EC) system, hydrogen and oxygen are produced at the photocathode and photoanode, respectively, naturally separating the gases (Landman et al., 2017). By simply calculating the product of PV conversion efficiency of 20%—30% and electrolysis

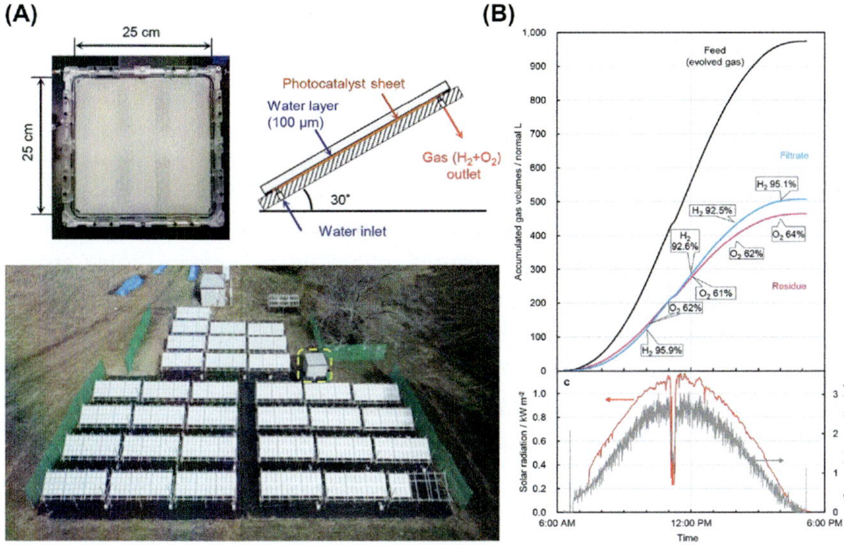

Figure 6.9 ***Demonstration of photocatalytic water splitting at 100-m²*** **scale.** (A) Water-splitting system based on Al-doped $SrTiO_3$ photocatalysts; (B) hydrogen production versus time and illumination intensity (Nishiyama et al., 2021).

conversion efficiency of 60%–70%, the total conversion efficiency of PV-EC can reach 12%–21%, which is much higher than that of natural photosynthesis.

In 2007 Peharz et al. reported an STH conversion efficiency of 18% for a PV-EC system, coupling a III–V semiconductor $Ga_{0.35}In_{0.65}P-Ga_{0.83}In_{0.17}As$ tandem solar cell with a Pt electrocatalyst (Peharz et al., 2007). To fully make use of solar spectrum, in 2016, Jia et al. used an InGaP/GaAs/GaInNAsSb triple-junction tandem solar cell with bandgaps to be 1.9, 1.4, and 1.0 eV for each cell. An average STH efficiency of 30% was demonstrated over 48 hours (Jia et al., 2016).

Despite the high conversion efficiency of the PV-EC devices, the high cost of III–V semiconductor and Pt catalysts is the bottleneck for large-scale applications. In exploration of low-cost solar cells and electrocatalysts, Reece et al. (2011) reported an amorphous silicon-based PV-EC system using NiMoZn as an HER catalyst with Co compounds as an OER catalyst, with an STH efficiency of 4.7% and the efficiency of the whole system to be 2.5%. In 2014 Luo et al. coupled high-efficiency perovskite solar cells with abundant FeNi-layered double hydroxide (LDH) as a bifunctional catalyst for both the OER and HER reactions, with STH conversion efficiency as high as 12.3% (Luo et al., 2014).

6.2.2.3 Photocatalytic reduction of CO_2: carbon cycle

Hydrocarbons have advantages over other energy carriers such as hydrogen in terms of energy density, chemical stability, and established infrastructures (Annual Energy Outlook 2020 with projections to 2050, 2020). However, dramatic rise in atmospheric CO_2 level due to the huge amount burning of fossil fuel has resulted in global environmental problems, including climate change. One solution for the dilemma is a carbon-neutral energy system through recycling of carbon. Since around 2000, liquid solar fuel projects have been launched in many important economic entities, including the United States, China, EU, and Japan. The goal is to reduce CO_2 into hydrocarbon by using clean energy sources such as electricity from solar and wind power.

Photoelectrochemical reduction of CO_2 has similarities with photocatalytic splitting of water, consisting of photoelectrochemical reactions occurring over a semiconductor catalyst. In the meantime, it is important to notice differences in the thermodynamical properties of CO_2 and H_2O, and distinct kinetics of the reactions, which are directly related to the conversion efficiency.

CO_2 reduction reaction (CO_2RR) is a reaction that mimics natural photosynthesis: $CO_2 + H_2O \rightarrow C_xH_yO_z + O_2$. The difficulty of the reaction is the high stability of CO_2, which is thermodynamically the most stable among all carbon-containing molecules. The formation energy of CO_2 is -394 kJ/mol, much higher than that of H_2O (-228 kJ/mol). In addition, CO_2RR is a complex reaction involving gas, liquid, and solid phases. There are many possible reaction paths, involving different numbers of electrons from 2, 4, 6, 8, to 12. And there are up to 16 possible products, such as CO, HCOOH, CH_4, C_2H_4, and C_2H_5OH (Kuhl et al., 2012). Generally, the more electrons involved, the higher the reaction overpotential, thus making electrochemical reduction of CO_2 particularly challenging. Frontiers of current research include (1) development of highly efficient, selective, and stable electrocatalysts to reduce the overpotential and to improve Faraday efficiency, and (2) optimization of mass-transfer processes and device configurations to enhance the current densities and production rates.

Electrocatalysts for electrochemical reduction of CO_2 electrochemical reduction efficiency of CO_2 is most important for the solar-liquid fuel, where the key is the electrochemical catalysts. Catalytic activity is determined by the electronic structure, which can be modulated by size, crystal surface, and doping of the low-dimensional catalysts. The total reaction rate is not only related to the catalytic activity but also dependent on each step in the whole reaction process, including transport of electrons and ions.

Generally, active sites are located at facets and edges on the surface of nanocrystalline catalysts. Therefore as the catalyst size decreases, active sites also increase. In addition to the simple geometric effect, chemical properties of nanocrystals could be fundamentally changed compared with bulk counterparts, due to the changes of electronic properties. In 1985 Hori reported on electrochemical reduction of CO_2 on different metals, where type of products depends strongly on the electronic structure of the catalysts, especially the positions of d-electron orbitals (Hori et al., 1985). Sargent group of the University of Toronto developed a machine learning method combined with a high-throughput computational screening process and identified Cu—Al alloys with optimum CO-binding energy (Zhong et al., 2020). And the Faraday efficiency was enhanced from 66% of pure copper to 80% with a high current density of 400 mA/cm^2 in the reduction reaction of CO_2 to ethylene.

Bao group at Dalian Institute of Chemical Physics, Chinese Academy of Sciences compared the CO_2 reduction electrocatalytic activity of Pd nanocrystals of different sizes and found that the highest reaction rate corresponds to the nanocrystals with an average diameter of 3.7 nm. Current density was 18.4 times higher than that of the nanocrystals with a diameter of 10.3 nm, and Faraday efficiency corresponding to production of CO increased from 5.8% to 91.2%. By analyzing high-resolution transmission electron microscopy (TEM) images, it was found that along with the size reduction, not only active sites were increased, but also the shape was changed. CO_2 and $COOH^*$ tend to be adsorbed at edges of the particles, whereas the adsorption energy of H^* is not strongly dependent on the positions, resulting size-dependent volcano plot of the catalytic activities (Gao et al., 2015).

Electrochemical reduction of CO_2 is a three-phase reaction involving CO_2 gas, electrolyte (liquid phase) to provide protons, and solid-state electrode to transport electrons. One of the challenges is the low solubility of CO_2; thus dissolution and diffusion of gases become the rate-limiting step. Sargent et al. designed a layered electrode of graphite/NP/Cu/PTFE, where carbon nanoparticles and graphite form a network to stabilize the hydrophobic gas diffusion layer, with CO_2 diffusing through the porous electrode, therefore decoupling gas and electrolyte transport processes, resulting an increase in the Faraday efficiency of ethylene to 70% (Dinh et al., 2018). By further introducing a PFSA coating with hydrophilic-hydrophobic bifunctional surfaces, the transport of gas, ion, and electron are decoupled, resulting in close-contacted three-phase interfaces. The diffusion flux of CO_2 is increased by 400 times compared to conventional liquid electrolyte system. The current density was as high as 1 A/cm^2, with a Faraday efficiency of >60% for ethylene and a conversion efficiency of 45% (García de Arquer et al., 2020).

Photoelectrochemical conversion of solar-hydrocarbon fuels includes three types: photocatalysts, photoelectrochemical cells, and photovoltaic-assisted electrolysis. All of them are essentially a combination of light-electron conversion on semiconductors and electrochemical reactions on catalysts. The difference is the coupling manners and transport mechanisms of electrons and ions between light-electron and electron-chemical systems.

Research on reduction of CO_2 in photoelectrochemical cells began in the 1970s. Halmann et al. used p-GaP as the photocathode, carbon and n-TiO_2 as the photoanode to reduce CO_2 to formic acid and methanol (Halmann, 1978). Around the same period, Fujishima et al. reported photoelectrochemical reduction of CO_2 by TiO_2 semiconductor crystals in aqueous solution (Inoue et al., 1979).

Similar strategies to enhance photocatalytic splitting for water have been applied to photocatalytic reduction of CO_2. In 2019 Nakada et al. at Tokyo Institute of Technology designed a *Z*-mechanism photoelectrochemical conversion system in which the photoanode was a TiO_2:Ta/N semiconductor for light-absorbing, the photocathode was an Ru(II)−Re(I) binuclear complex coordination compound, and RuO_x was the oxygen reduction reaction (ORR) catalyst. The Faraday efficiency of solar-driven CO_2 reduction into CO was 71%, and the photochemical energy conversion efficiency is 0.0011% (Fig. 6.10) (Nakada et al., 2019).

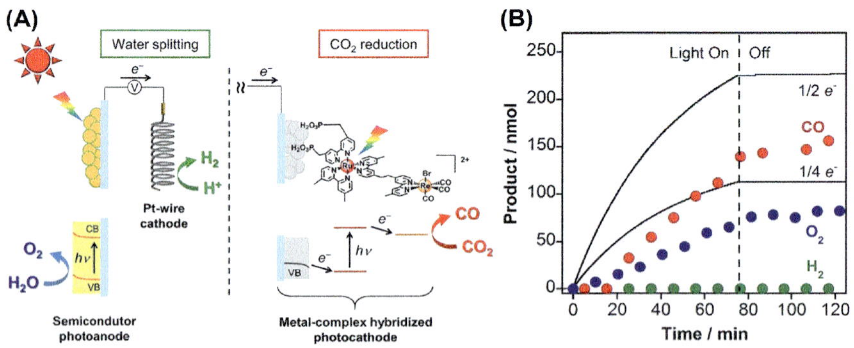

Figure 6.10 ***Solar fuel by photoelectrochemical CO_2*** reduction. (A) Schematic diagram of the working principles; (B) CO and O_2 yields under simulated sunlight illumination with a cell voltage of 0.5 V (Nakada et al., 2019).

In 2020 Wang et al. at University of Cambridge fabricated a *Z*-mechanism photocatalyst consisting of $SrTiO_3$:La, Rh|Au|RuO_2-$BiVO_4$:Mo to enhance light absorption and used a phosphonated cobalt(II) bis(terpyridine) molecular catalyst for CO_2 reduction, with a solar-formic acid conversion efficiency of 0.08% and a Faraday efficiency of ~97% (Wang, Warnan, et al., 2020).

To enhance conversion efficiency, Su-Il In et al. from Daegu Gyeongbuk Institute of Science & Technology, Korea, developed Cu−Pt bimetallic nanocrystalline sensitized titanium oxide for CO_2 reduction to methane and ethane. Under simulated solar illumination (AM1.5), an energy conversion efficiency of ~3% was obtained after an activation process for about 1.5 hours (Sorcar et al., 2019).

With the increasing efficiency and decreasing price of solar cells, it becomes more feasible for photovoltaic-assisted electrolysis to convert solar energy into chemical energy. It is relatively simple to design and optimize because the photoelectric conversion and electrochemical conversion processes are decoupled and can be optimized separately.

Hyo Sang Jeon et al. used cost-effective copper indium gallium selenide (CIGS) thin-film solar cells with an efficiency of 8.58% to couple with Co_3O_4 nanoparticles and gold nanoparticles as electrochemical catalysts for OOR and CO_2RR, respectively (Jeon et al., 2015). A Faraday efficiency of 90% was achieved for the reduction of CO_2 to CO at an operating voltage level of 2.33 V and a current density of 7.32 mA/cm^2. The light-fuel conversion efficiency was 4.23%, exceeding the efficiency of natural photosynthesis (Fig. 6.11). Grätzel group used a triple-junction perovskite solar cell, in combination of gold as CO_2RR catalyst and IrO_2 as ORR catalyst, and obtained a maximum efficiency of 6.5% for CO_2 reduction to CO (Schreier et al., 2015). In an improved work, CuO nanowires with an atomic layer of SnO_2 were used as a bifunctional catalyst for both ORR and CO_2RR in conjunction with a GaInP/GaInAs/Ge triple-junction tandem high-performance solar cell (Schreier et al., 2017). An overall Faraday efficiency of 81% for the reduction of CO_2 to CO and a light-to-fuel conversion efficiency of 13.4% was reported.

The purpose of CO_2 photoelectrochemical reduction is to utilize solar energy to produce liquid fuels that could easily be transported and stored, while reducing CO_2 emission to lower CO_2 concentration in the atmosphere. Only if the process is cost-effective, it can be applied on a large scale. Specifically, the following should be considered: actual cost, net CO_2 reduction, and market value of the products. The energy consumption, costs, and

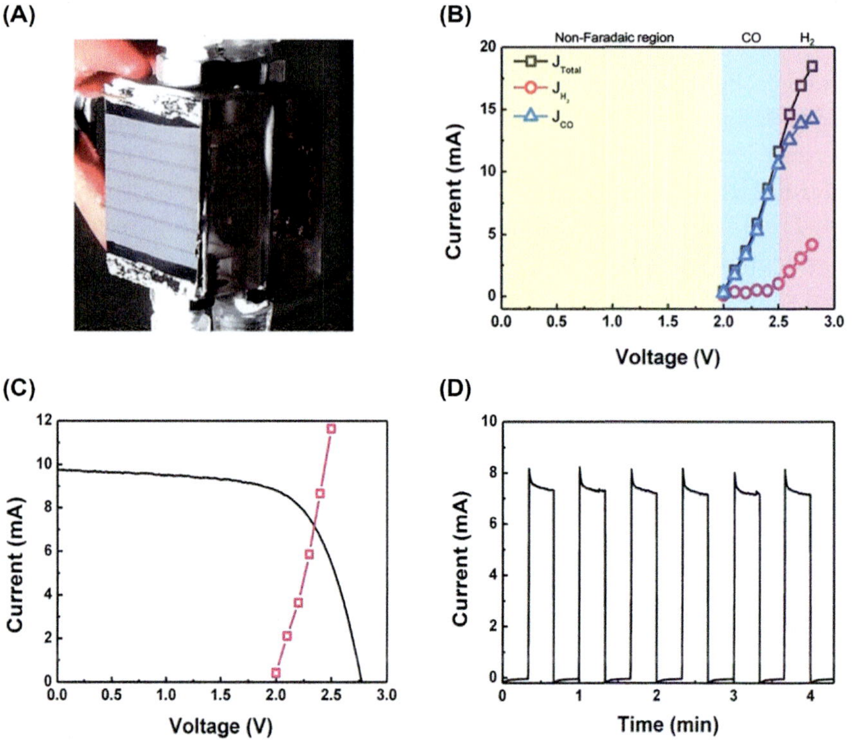

Figure 6.11 ***PV-electrocatalytic reduction of CO_2.*** (A) Photograph of the device, with a CIGS PV module mounted on the outside of the photoelectrochemical (PEC) cell; (B) Steady-state *I*−*V* curves of a Co_3O_4 nanoparticle thin-film anode and a nanostructured gold cathode; (C) *I*−*V* characteristics of the CIGS module, with the intersection point being the operating condition to match working currents; and (D) Working current modulated by a periodically chopped illumination (Jeon et al., 2015). *CIGS*, Copper indium gallium selenide; *PV*, photovoltaic.

outputs in three major steps should be calculated: (1) capture, and separation of CO_2; (2) electrochemical reduction of CO_2; and (3) separation and purification of the products.

Jouny et al. of University of Delaware examined seven possible reaction pathways and products for CO_2 reduction, with the number of reacting electrons ranging from 2, 4, 6, 8, 12, and 18, to form products such as CO, HCOOH, CH_3OH, CH_4, C_2H_4, C_2H_5OH, and C_3H_7OH. Under current technological conditions, conversion rates for CO and formic acid are relatively high, but for both the market size and value are relatively low. To produce chemicals with higher value such as ethanol and *n*-propanol economically, the performance metrics should reach the following targets: overpotential smaller than 0.5 V, current density higher than 300 mA/cm^2, and Faraday efficiency higher than 70% (Jouny et al., 2018).

6.2.2.4 Solar fertilizer: nitrogen cycle

Nitrogen cycle is a major issue to make food to feed the billions of people on this planet. One of the most important goals in the United Nations Sustainable Development Goals (UN SDGS) is to eliminate hunger. In early 20th century, German scientist Fritz Haber invented artificial nitrogen fixation to synthesize ammonia, which is used to produce urea and nitrogen fertilizer, increase food production, and ultimately leading to the "green revolution" (Erisman et al., 2008). Currently, production of ammonia is a $60 billion industry, with an annual scale of about 150 million tons.

As an energy carrier, compared with hydrogen, ammonia has several advantages. Ammonia has a much higher boiling point (−30°C under atmospheric pressure and 20°C under a 0.86 MPa pressure) and is much easier to liquefy. Because of the higher density, its volumetric energy density is 45% higher than that of liquid hydrogen. Ammonia can be easily decomposed to produce hydrogen and nitrogen, involving no greenhouse gas emissions. Because of the higher safety, storage, and transportation infrastructure for ammonia is easier to build. Therefore ammonia is an attractive hydrogen carrier and liquid fuel. In recent years, "ammonia economy" has attracted a lot of attention (MacFarlane et al., 2020).

Synthesis of ammonia seems like a simple chemical reaction: $3H_2 + N_2 \leftrightarrow 2NH_3$, $\Delta H^0 = 46$ kJ/mol. Actually, it is difficult to occur, because the N≡N bond is a nonpolar bond with a high binding energy and therefore difficult to activate. In 1908 Fritz Haber discovered an iron-based catalyst to realize ammonia synthesis (H-B process) under high temperature and high-pressure conditions of 250 atmospheres and 500°C, for which he was awarded the Nobel Prize in Chemistry in 1918.

Because of the harsh reaction conditions and required methane to produce hydrogen as a raw material, the energy consumption of H-B process is very high. Synthesis of ammonia through the H-B process is suitable for centralized production in large industries, resulting in a high cost of delivery of fertilizers to remote areas where most of the farms are located. To replace the thermal catalytic reaction in H-B process, synthesis of ammonia through electrochemical reactions becomes a research frontier because of the mild reaction conditions. It is desired that ammonia can be produced in a small-scale and distributed manner, providing fertilizer for local farming, and reducing the transportation costs.

Research on (photo)electrochemical ammonia synthesis has been reported since the 1960s (Tamelen & Akermark, 1968; Van Tamelen & Seeley, 1969). The fundamental and key reaction in electrochemical ammonia synthesis is the nitrogen reduction reaction (NRR): $N_2 + 6H^+ + 6e^- \rightarrow 2NH_3$. The thermodynamic equilibrium potential is 0.056 V (vs RHE). If the cathode overpotential is 0.3 V, the Faradaic efficiency can reach 35%, and its energy consumption can be comparable with the H-B process. However, this is a complex electrochemical reduction reaction involving six electrons and six protons, resulting in a high kinetic overpotential, which becomes the bottleneck to enhance the conversion efficiency and for practical applications.

Electrochemical synthesis of ammonia can be categorized into aqueous and nonaqueous systems. In aqueous system, due to competition with HER, ammonia yield is typically rather low, around 0.1 $nmol/s/cm^2$. According to Giddey et al., the yield should be at least 100 $nmol/s/cm^2$ for practical applications (Giddey et al., 2013). To avoid competition with HER, nonaqueous electrolytes can be used. McEnaney et al. (2017) obtained an 88.5% Faradaic efficiency and a yield of 1 $nmol/s/cm^2$ in alkaline solution. In an ionic liquid electrolysis system, Zhou et al. reported that the Faradaic efficiency of electrochemical ammonia synthesis reached 60%, but the yield was still only 0.01 $nmol/s/cm^2$.

At present, the overall conversion efficiency of electrochemical ammonia synthesis is relatively low, and there are also problems such as repeatability of the reported experiments. It is very important to establish standard procedures for the best practice. Greenlee et al. (2018) and Douglas MacFarlane (Suryanto et al., 2019) suggested procedures for accurately measuring the efficiency of electrochemical ammonia synthesis, including purifying gas, conducting an open-circuit test, and an argon control test, using $^{15}N_2$ for comparison, testing the long-term stability, repeated testing, normalization of geometric area, and cross-validation of multiple detection methods. Tang and Qiao of the University of Adelaide elaborated on experimental steps and details of electrocatalytic nitrogen reduction to synthesize ammonia, including battery configuration details, electrode treatment process, loading amount of catalyst, details of reference and counter electrodes, ion exchange membrane, and electrolyte (Tang & Qiao, 2019).

6.2.3 Thermoelectric conversion

Overall, 90% of the world's energy is provided by heat engines with a conversion efficiency of about 30%, and the remaining 2/3 of the energy is lost to environment as waste heat. Thermoelectric conversion devices provide a way to reuse the waste heat. They utilize the Seebeck effect to generate power from temperature difference. They are characterized by small size, absence of mechanical vibration, and high reliability. Thermoelectric conversion is becoming an important technology for utilization of waste heat in thermal power plants, automobiles, and many other occasions. They have broad potential applications ranging from wristwatches, pacemakers, sensors, to deep-space probes, and so on.

In thermoelectric conversion devices, it is the migration of charge carriers driven by temperature gradients to generate a thermoelectric electromotive force. Electrons and holes in n- and p-type semiconductors migrate from the hot end to the cold end to generate currents. Multiple n- and p-type thermoelectric material units can be assembled in series to form a thermoelectric conversion module.

The theoretical conversion efficiency of a fully reversible thermoelectric device is equal to the efficiency of Carnot cycle. But in practice, energy will be dissipated in irreversible processes of Joule heating, with a maximum power generation efficiency of $\eta_{\max} = \frac{(T_H - T_C)(\sqrt{1 + ZT_M} - 1)}{T_H(\sqrt{1 + ZT_M} + T_C/T_H)}$, where T_H, T_C, and T_M are temperatures at the hot and cold ends, and the average temperature. The most important factor in determining the conversion efficiency is a dimensionless coefficient: $ZT = S^2\sigma T/(k_{ph} + k_e)$, where T, S, σ, and k are temperature, Seebeck coefficient, electrical conductivity, and thermal conductivity, and $S^2\sigma$ is also called the power factor. Thermal conductivity includes phonon thermal conductivity k_{ph} and electron thermal conductivity k_e.

To enhance the thermoelectric conversion efficiency, it is desired to increase the Seebeck coefficient and electrical conductivity to get a larger power factor and in the meantime to reduce thermal conductivity. An ideal thermoelectric material will be a single crystal for electrons and an amorphous phase for phonons. However, these physical properties are closely related to each other and are functions of the charge carrier's concentration. How to enhance one property, and in the meantime not to compromise the other is the key and an art of designing effective thermoelectric materials (Sales, 1998). Practical thermoelectric conversion devices were developed in the 1950s, with a ZT value of about 1 and a corresponding energy conversion efficiency of 4%–6%. In the 1990s, Mildred Dresselhaus and colleagues proposed new mechanisms such as size confinement, interface, and low-dimensional effects. In recent years, reported

ZT values are higher than 2, where quantum confinement effect and size effect played an important role (Zhao et al., 2014).

6.2.3.1 Low-dimensional effects in thermoelectric conversion

In nano-thermoelectric materials, the intrinsic bandstructure and transport properties of electrons and phonons are strongly influenced by the low dimensionality and small size effects. In 1993 Hicks and Dresselhaus published two theoretical papers about quantum confinement effect on the electronic structures and ultimately the performance of thermoelectric materials (Hicks & Dresselhaus, 1993a, 1993b). Dresselhaus and Saito found that quantum confinement effect has an impact on the power factor related to the de Broglie wavelength. When the feature size is smaller than its de Broglie wavelength, the power factor is significantly enhanced (Hung et al., 2016). Mahan and Sofo (1996) at the University of Tennessee showed that ideal thermoelectric materials have a concentrated distribution of electronic density of states (DOS), with a delta function-like DOS corresponding to the highest power factor. Due to the low-dimensional effect, van Hove singularities appear in the DOS of one-dimensional materials such as carbon nanotubes. In two-dimensional materials such as 2D electron gas of $SrTiO_3$ superlattices, it has been verified that the power factor increases with the decrease of superlattice thickness and corresponding increase of DOS at the bottom of conduction bands (Ohta et al., 2007).

6.2.3.2 Nanostructured thermoelectric materials

In 2007 Yang group (Hochbaum et al., 2008) of the University of California Berkeley, and Heath group (Boukai et al., 2008) of the California Institute of Technology published back-to-back papers in Nature on the thermoelectric properties of silicon nanowires. Yang et al. prepared silicon nanowires with rough surfaces through oxidation-etching methods. By measuring the thermal conductivity, electrical conductivity, and Seebeck coefficient of a single silicon nanowire, it was found that the thermal conductivity was reduced by two orders of magnitude compared with bulk silicon, while the electronic conductivity and Seebeck coefficient are comparable to those of doped silicon, resulting in a *ZT* value of 0.6 at room temperature (Hochbaum et al., 2008). It was reported by Heath et al. that the *ZT* value of silicon nanowires could be as high as 1 at 200K (Boukai et al., 2008). In 2012 Lee et al. of Samsung Electronics fabricated germanium-silicon alloy to reduce thermal conductivity.

At 450K, the thermal conductivity was only 1.2 W/m/K, accordingly, the *ZT* value was calculated to be 0.46 (Lee, Yin, et al., 2012).

Kanatzidis group at Northwestern University introduced a variety of mechanisms at multiple length scales to reduce thermal conductivity and to improve the thermoelectric properties of PbTe materials, of which the *ZT* value reached 2.2 at 915K (Biswas et al., 2012). At atomic scale, 2% sodium alloy atoms were doped to increase the scattering of short-wavelength phonons. At nanometer scale, SrTe nanoparticles with the size of 2−10 nm were included as scattering centers for medium-wavelength phonons. At mesoscopic scale, polycrystalline structure and grain boundaries could effectively scatter the phonons with longer wavelengths.

6.2.3.3 Flexible thermoelectric materials

In recent years, there has been a rapid development of flexible and wearable electronics. It is attractive to use the energy generated by human body for self-powered wearable devices. One possible solution is to generate power by utilizing the temperature difference between body temperature and the environment through thermoelectric conversion (Bahk et al., 2015). Currently, there are still many challenges for flexible thermoelectric conversion devices. On the one hand, it is limited by the conversion efficiency of thermoelectric materials. On the other hand, it is also limited by mechanical flexibility, resulting in poor contact with the skin and a high interface thermal resistance. Traditional thermoelectric materials with reasonably high conversion efficiency, such as Bi_2Te_3, are brittle materials with ionic bonds, while polymer thermoelectric materials have better flexibility but much lower *ZT* values.

Tai and Liu groups of Institute of Metal Research, Chinese Academy of Sciences used flexible networks of single-walled carbon nanotubes as scaffolds to deposit Bi_2Te_3 thin films. In addition to the mechanical flexibility, there is a quasi-epitaxial relation with the Bi_2Te_3 (0 0 0 1) plane aligned to the carbon nanotube bundle direction. The power factor was as high as 1600 $\mu W/m/K^2$ at room temperature, and the *ZT* value reached 0.89. The composite material demonstrated excellent flexibility that the performance did not change after bending for 100 times with a curvature radius of 10 mm (Fig. 6.12) (Jin et al., 2019).

As mentioned earlier, the challenge in thermoelectric design and applications lies in the coupling of electrical and thermal transport properties. Electrons are carriers for both charge and heat, and scattering centers not only scatter phonons but also reduce electron mobility. It is not only a

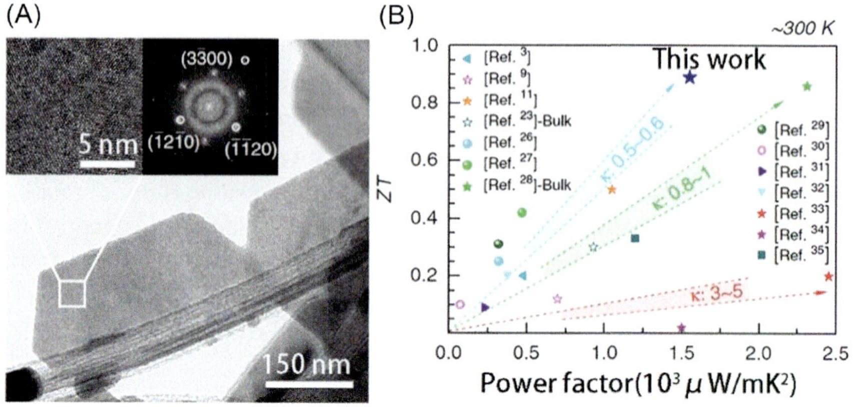

Figure 6.12 ***Flexible thermoelectric materials.*** (A) Single-walled carbon nanotube network-Bi_2Te_3 thin-film composite for flexible thermoelectric conversion; (B) comparison of *ZT* and power factor of flexible thermoelectric materials (Jin et al., 2019).

science but also an art to design efficient thermoelectric materials, to search for the optimal and often compromised solution between structural order and disorder, stable and metastable phases, effective mass and mobility of carriers, and the ratio of covalent to ionic bonds. Many mechanisms have been proposed in recent years, including alloys, strained lattices, modulated electron orbitals, and topological phases, bringing new opportunities for the exploration of high-performance thermoelectric materials (He & Tritt, 2017; Zhao et al., 2014).

6.2.4 Electrochemical energy storage and conversion

Modern economy is dependent on stable power supply, requiring a full-time stability of 99.9%, and energy storage is an important part of a robust energy system (Elliott, 2016). It is getting more challenging to keep the power supply stable for more and more clean energy such as solar energy, wind energy, and hydropower, which fluctuates with the seasons, weather, and time. Currently, the energy storage capacity is equivalent to only 1% of world energy consumption, among which 98% is stored in dams (Larcher & Tarascon, 2015). It is a huge challenge to achieve reliable energy storage to match the global energy demand scale of 14 TW.

"Energy storage" means converting electrical energy into other forms of energy (charging) for storage and transportation, and then releasing it for use (discharging). There are many approaches of energy storage for different applications, including mechanical, thermal, electrical, chemical, and electrochemical methods. Broadly speaking, the "solar fuel" introduced in previous section is also a kind of energy storage, converting low-density and fluctuating sunlight into fuels of high energy density and high reliability.

From the perspective of electron utilization, mechanical conversion does not involve electronic processes, heat storage is accompanied by phase changes, capacitors involve electron and ion transport, and electrochemical energy storage such as fuel cells and batteries involve redox reactions of the valence electrons. Generally, higher energy density could be generated through "inner" electrons with tighter binding. The density of energy stored in the reservoir's gravitational potential energy is about 10^3 J/kg; the energy density of lithium-ion batteries is $\sim 10^5$ J/kg, which is more than 100 times that of the former. And the energy density of fuels could reach the order of $\sim 10^8$ J/kg, because there is no extra weight of electrodes as in the batteries. On the other hand, the total capacity of a water reservoir can be very large. If the length and width are 1 km each and the depth is 100 m, the mass of water in the reservoir is $\sim 10^{11}$ kg that is hard or impossible to build for batteries.

Therefore there is no single "perfect" energy storage solution for all the scenarios. The choice of energy storage should not be based on which technology is the most "advanced," but rather based on specific application requirements for energy and power density, scale, charging and discharging rates, costs, environmental impacts, and so on. It is crucial to construct smart energy storage architectures to adjust the ratio and output for each storage subsystem.

6.2.4.1 A comparison of information and energy devices

This section focuses on application of low-dimensional materials in electrochemical energy storage and conversion, including secondary batteries, fuel cells, and supercapacitors (Winter & Brodd, 2004). It is interesting to make a comparison between energy and information devices. Over the past half century, the density and performance of semiconductor transistors have followed Moore's law, doubling every 18 months, with an annual growth rate of $\sim 60\%$. On the other hand, secondary batteries have evolved from Ni–Cd in the 1970s, Ni–MH batteries in the 1980s, to commercialized lithium-ion batteries in the 1990s. Despite the great progress, the energy density is less than 400 W h/kg, and the annual growth rate is only $\sim 5\%$. From smartphones to electric vehicles, there is a modern CPU chip with an area the size of a fingernail integrating billions of transistors, while the largest component is often a battery, forming a great contrast. Such difference is a result of the different basic principles and process complexity of the two types of electrical devices of switches and reservoirs, respectively. A transistor is to control the "on" and "off" states of electrons or currents. The size of a single transistor has been reduced from micron scale to 10-nm scale. The physical limit is a single-electron on a single atom. On the other hand, for a certain energy capacity, the number of electrons and therefore the amount of mass is fixed for the batteries. In addition, the battery includes much more complex processes that not only include electron transfer at the interface through redox reactions, but also mass and ion transport in multiscale porous structures and multiple phases. Therefore it is much more challenging to achieve theoretical energy capacity.

6.2.4.2 Metal ion batteries

A secondary battery is based on a pair of reversible oxidation–reduction reactions. In the discharging process, oxidation and reduction reactions take place at the negative (anode) and positive (cathode) electrodes, respectively. In the meantime, ions in electrolyte pass through the separator and electrons passing through the external circuit, driving the load. Charging is the inverse of the above process.

The voltage of a battery is determined by the change of Gibbs free energy of the redox reaction following Nernst equation: $\Delta G = -nFE$, where n is number of electrons, F is Faraday's constant, and E is reaction electromotive force. The capacity of a battery is determined by the number of electrons involved in the reaction: $Q = nF/M$ in unit of A h/kg, where F is Faraday's constant, and M is the equivalent molecular weight. Therefore an "ideal" battery should have high voltage, small atomic weight, and low density. Among the metallic elements, lithium has the smallest atomic number, the lowest density, and the highest electrochemical potential, making it an ideal anode material for electrochemical batteries.

The 2019 Nobel Prize in Chemistry was awarded to Goodenough of University of Texas at Austin, Whittingham of State University of New York, and Yoshino of Asahi Kasei Corporation in recognition of their contributions to the invention, development, and industrial application of lithium-ion batteries. At present, lithium-ion batteries are close to the theoretical capacity, meaning that almost every atom of the active electrode has been fully utilized with the ratio of lithium ions to the host electrode according to the stoichiometric ratio. To further increase the storage capacity, new material systems are actively being explored. High-capacity anodes include silicon and lithium, and new cathode materials include sulfur and oxygen.

6.2.4.2.1 High-capacity anode materials silicon-based anode

Silicon can form alloy with lithium such as $Li_{15}Si_4$, which has 10 times the theoretical energy density of graphite. However, accompanying lithiation and delithiation, volume expansion and contraction up to 320% is generated, leading to rupture and detachment of the active material. In addition, a solid electrolyte interface (SEI) film grows repeatedly to consume electrolyte, severely restricting its cycling performance. Size effects of low-dimensional materials and composite interfaces are effective means to inhibit the volume expansion and surface reactions.

Volume expansion and pulverization due to lithiation is fundamentally a mechanical problem. Whether a crack could propagate leading to rupture depends on the competition between elastic energy and fracture strength. Elastic energy is proportional to volume, while fracture strength is inversely proportional to the featured size. Generally, a higher elastic strain could be tolerated for a particle with a smaller size (Liu et al., 2012). Cui group at Stanford University reported silicon nanowire arrays as anode for lithium-ion batteries. The nanowire arrays were grown directly on the current collector to maintain good mechanical and electrical contacts. Void between nanowires in the array could accommodate volume expansion. As a result, the capacity reached 3500 mA h/g at 0.2 C charge/discharge rate (Chan et al., 2008).

Nanosilicon can solve the problem of cyclic rupture, but its low density leads to a low volumetric energy density. And the large specific surface area results in side reactions and a reduced Coulombic efficiency. On the other hand, microsilicon anodes are much cheaper and have a lower surface. However, they cannot withstand the large strain during lithiation process. Yang group at Tianjin University used graphene and microsilicon particle to form a composite electrode, where graphene network could tolerate the volume expansion of silicon particles. The specific volume capacity reached 1200 mA h/L at a surface capacity of 3 mA h/cm^2, and a capacity retention rate of 76% after 100 times of cycling (Han et al., 2020). Such nanocarbon−microsilicon composite structure was further optimized. A double-layer carbon framework with an inner carbon layer by CVD deposition and an outer graphene hydrogel was designed to provide space to accommodate volume expansion and to maintain electrical connectivity of the overall electrode. The Coulombic efficiency was higher than 99.5% in 1000 cycles. A full battery was fabricated, combined with an NCM811 cathode, and a high volumetric energy density of 1048 W h/L was achieved (Chen et al., 2021).

6.2.4.2.2 Lithium metal anode

The ultimate anode of lithium battery is lithium metal. Its theoretical energy density is 11,586 W h/kg, close to that of gasoline (13,200 W h/kg). If combined with oxygen to form a lithium-air battery, its energy density is 3505 W h/kg. And lithium can be combined with sulfur to form a lithium–sulfur battery, with an energy density of 2567 W h/kg, much higher than that of a lithium-ion battery (387 W h/kg). However, there are still great challenges and concerns over the safety issues due to growth of lithium dendrites and low Coulombic efficiency due to the instable SEI.

Dendrite growth is a classical metallurgical problem of inhomogeneous growth, which often occurs during solidification and crystallization due to overcooling associated with redistribution of composition. In electroplating, one reason for dendrite growth is the locally enhanced electric field due to the curvature of a sharp tip (Monroe & Newman, 2003). Zhang et al. of Pacific Northwest National Laboratory added cations such as Cs^+ and Rb^+ to the electrolyte, which were preferentially deposited at protruded positions during lithium deposition process, thus shielding the local tip-enhancement effect to avoid the growth of lithium dendrites (Ding et al., 2013). It is also possible to modify the substrate structure for a similar effect. Zhang group of Tsinghua University covered a layer of glass fiber network on a copper collector, to which lithium ions are strongly absorbed due to the polar functional groups on its surface; therefore redistribution of lithium ions is inhibited and so as the dendrite formation (Cheng et al., 2016).

In addition to the crystal growth dynamics, there are approaches to inhibit the dendrite growth by controlling the nucleation and mass transport processes. In terms of nucleation control, Tour et al. used copper substrate with graphene or carbon nanotubes coating as a template for lithium metal deposition to increase the nucleation sites and to inhibit nonuniform dendrite growth (Raji et al., 2017). In terms of ion diffusion kinetics, Lu et al. added halides such as LiF at the interface between lithium metal and electrolyte to improve wettability, to increase the ionic conductivity to reduce the composition inhomogeneity, resulting in a smooth surface of the deposited lithium (Lu et al., 2014).

6.2.4.2.3 Lithium–sulfur batteries

The theoretical energy density of lithium–sulfur batteries can reach 2567 W h/kg, which is about 10 times that of a classical layered oxide cathode. The main challenges of sulfur cathodes are (1) sulfur and reaction products Li_2S and Li_2S_2 are insulators, limiting their rate performance; (2) volume expansion of sulfur during charge and discharge; (3) slow reaction kinetics; and (4) shuttle effect of soluble polysulfides (Mikhaylik & Akridge, 2004).

Loading sulfur into porous conductive electrodes can serve multiple purposes to improve conductivity, to limit volume expansion of sulfur during charging, and to limit diffusion of polysulfide compounds, thereby improving rate and cycle performance. Li group at Institute of Metal Research, Chinese Academy of Sciences used a network of individually dispersed single-walled carbon nanotubes as a porous electrode to obtain a sulfur content of 95%, corresponding to a sulfur loading of 7.2 mg/cm^2. Accordingly, an area capacity of 8.63 mA h/cm^2 was obtained, which is twice that of lithium-ion batteries. Due to the high conductivity and excellent mechanical properties of carbon nanotubes, the cycling performance was effectively improved, resulting in a Coulombic efficiency of higher than 99% during 300 times of cycles (Fang et al., 2017). Fang et al. designed a composite graphene electrode with a highly conductive graphene film as current collector and another sulfur-loaded porous graphene layer as cathode, achieving a sulfur loading of 5 mg/cm^2. The assembled lithium–sulfur battery showed an initial capacity of 1500 mA h/g, an area capacity of 7.5 mA h/cm^2, and a Coulombic efficiency exceeding 99.5% over 400 cycles (Fig. 6.13) (Fang et al., 2016).

In terms of accelerating lithium–sulfur reaction kinetics, Song et al. from the University of Science and Technology of China and Niu et al. from Nankai University designed a nitrogen-doped graphene-supported Ni single-atom catalyst. $Ni-N_4$ active sites effectively combine with lithium polysulfide to reduce its decomposition energy barrier. The capacity after 500 cycles at 1 C was 826 mA h/g, maintaining 78% of the initial capacity, and the corresponding Coulombic efficiency was as high as 99.94% (Zhang et al., 2019). Similarly, Huang et al. from Beijing Institute of Technology reported a single-atomic Co–N–C electrocatalyst, which has a significant catalytic activity for the deposition and decomposition of polysulfides, resulting in a high surface capacity (10.9 mA h/cm^2) and a Coulomb efficiency of 99.9% after 300 charging cycles (Li, Kong, et al., 2019).

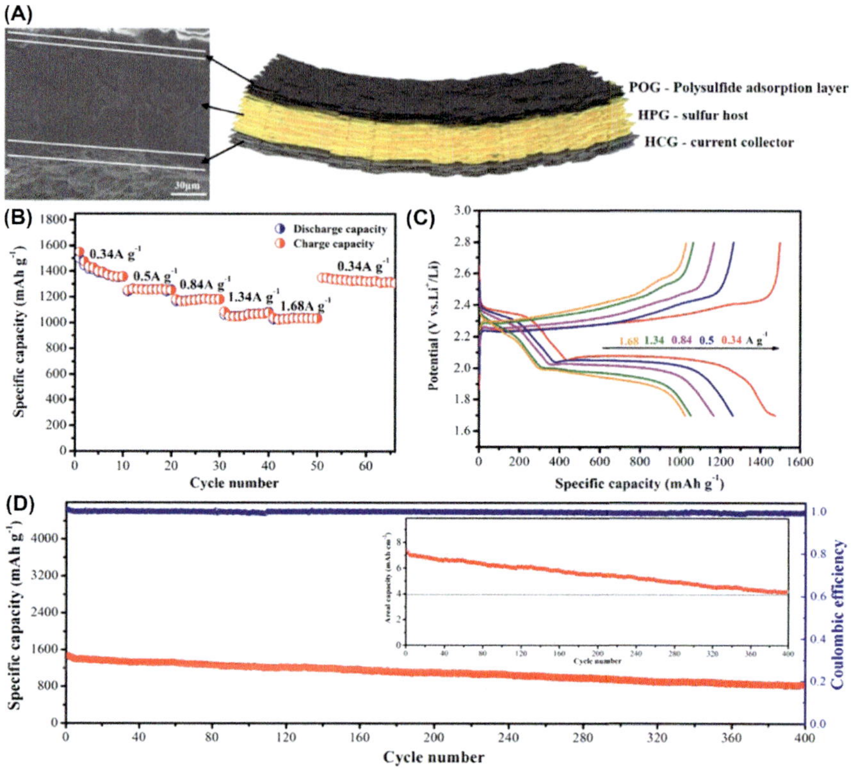

Figure 6.13 ***Graphene-based lithium–sulfur batteries.*** (A) Graphene sulfur composite cathode; (B) rate performance; (C) cyclic voltammetry (CV) curve at a scanning speed of 0.1 mV/s; and (D) cycling performance at a current density of 0.34 A/g (Fang et al., 2016).

6.2.4.2.4 Metal–air batteries

If metallic lithium is the ultimate negative electrode, oxygen could be said to be the ideal positive electrode. The theoretical capacity of a lithium–air battery is as high as 3842 mA h/g. Lithium–air batteries have something in common with fuel cells in the oxygen-involved electrochemical reactions and catalysts. The difference is that in lithium–air batteries, lithium ions participate in the reaction to produce solid compounds such as Li_2O_2. And only ORR occurs in the fuel cell while, in the lithium–air battery, undergoes both ORR and OER reactions during the discharging and charging process, respectively, requiring the catalyst to have both ORR and OER activities (Wang et al., 2014). Because the reaction product Li_2O_2 is insoluble and electrically insulating, it is a great challenge to design a suitable cathode electrode for lithium–air batteries to transport gas, electrolyte, and electrons for the catalytic reactions (Mirzaeian & Hall, 2009).

The practical application of lithium-air batteries is facing many challenges, including high overpotential, poor cycle performance, and low energy efficiency. To solve the problem of overpotential, it is crucial to search for efficient catalysts. The poor cycle performance is because Li_2O_2 blocks the pores and prevents the diffusion of oxygen and electrolyte. The stability of the electrolyte and electrode at a high potential is also an important factor affecting the performance, especially the cycling life.

Peter Bruce of University of St. Andrews reported a high-rate reversible lithium–oxygen battery (Peng et al., 2012), where dimethyl sulfoxide (DMSO)–based electrolyte and porous gold electrodes were used, and the capacity was maintained at 95% after 100 cycles. However, DMSO is unstable and therefore the cycle times were limited. In addition, the use of gold electrodes also makes its cost unacceptable.

Researchers from the University of Illinois at Chicago and Argonne National Laboratory reported a lithium–oxygen battery with a long cycle life of more than 700 cycles (Asadi et al., 2018). In this battery, surface of lithium metal was coated with a Li_2CO_3/C protective layer. By using MoS_2 nanosheets as cathode, ionic liquids, and DMSO as electrolytes, side reactions of CO_2 and H_2O were largely inhibited. A reasonably high capacity of 500 mA h/g was obtained along with improved cycling performance.

6.2.4.3 Supercapacitors

Electrochemical capacitors, also known as supercapacitors, are a kind of electrochemical energy storage device based on formation of an electrochemical double layer between a porous electrode and an electrolyte. The surface storage mechanism of supercapacitors contrasts with the batteries regarding the electrochemical mechanism and performance characteristics. Supercapacitors can charge and discharge faster because the electrochemical process only occurs at the surface. They have excellent cycling performance since the structure of the electrodes is not changed. In many ways, supercapacitors are complementary to volume-based batteries. Supercapacitors have a high-power density (~10 kW/kg), a long cycle life (millions of times), and a short charge and discharge time (seconds).

Main performance indicators of supercapacitors include capacitance, energy density, and power density. The capacitance is defined as $C = \frac{\varepsilon_0 \varepsilon_r A}{d}$, where d is the thickness of the electrochemical double layer, and A is the specific surface area. The origin of the superhigh capacitance of supercapacitors is the largespecific surface area.

The energy density is defined as $E = \frac{1}{2}CU^2$, where C is the capacitance, and U is the operating voltage. Therefore there are two approaches to increase the energy density, either to increase the capacitance or the operating voltage. The capacitance could be increased by increasing the surface area, increasing the dielectric coefficient, or shortening the distances between the interfaces. On the other hand, energy density is proportional to the square of operating voltage, which is limited by the stability of the electrolyte. Therefore the supercapacitors based on ionic liquid and organic electrolytes typically have higher energy density than those based on aqueous electrolytes.

The power density $\left(P = \frac{E}{t}\right)$ depends on the kinetics of charging and discharging, including the transport of ions and electrons, ion adsorption on the electrode surface, and the rate of the Faradaic process in pseudocapacitors.

Among the other structural features, the porous structure plays a critical role for both the capacitance and the transport dynamics. It has been generally believed that the pores must be larger than the size of the ions in the electrolyte to transport and store ions. In the mesoporous range, when the pore size is reduced to twice the ion diameter, the capacitance will decrease due to the screening effect. Gogotsi, Simon et al. discovered an abnormal pore size effect (Chmiola et al., 2006). The capacitance continues to increase as the pore size decreases, until reaching the ion size, due to a confinement effect from the pores (Largeot et al., 2008).

High-rate supercapacitors porous carbon materials are commonly used for supercapacitors, because of their stable physicochemical properties, high conductivity, and low cost. However, porous-carbon-based ECs suffer from the poor kinetics related to the slow ion transport. To optimize the ion transport and storage, Wang et al. from Institute of Metal Research, Chinese Academy of Sciences designed and fabricated a hierarchical porous graphitic carbon (HPGC) material for the high-rate electrochemical capacitors (Wang et al., 2008).

The hierarchical porous structures consist of pores on macro—meso—micro scales. Macropores provide a large ion buffer space and reduce resistance for ion transport. Mesopores provide efficient ion diffusion paths. And micropores provide a large number of ion storage sites. In addition, locally graphitized carbon shells enhanced electrical conductivity. As a result, both energy and power density have been greatly improved. The energy density and power density reached 22.9 W h/kg and 23 kW/kg, respectively (Fig. 6.14), exceeding the power target of the PNGV (Partnership for a New Generation of Vehicles).

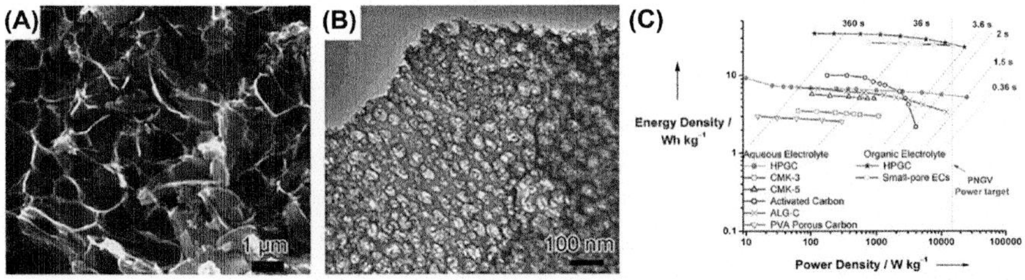

Figure 6.14 ***Hierarchical porous carbon based supercapacitors.*** (A and B) SEM and TEM images of the hierarchical porous carbon structures, (C) Ragone plot showing the energy and power density of the HPGC material in comparison with those of commercially available porous carbon materials (Wang et al., 2008). *HPGC*, Hierarchical porous graphitic carbon.

High volumetric capacitance is crucial for mobile electronics applications, such as smartphones and electric vehicles . The challenge in improving volumetric performance is the compromise between capacity and density. High capacitance requires a pore structure with a large specific surface area; however, porous materials tend to have relatively low densities, limiting the volumetric capacitance.

Yang group of Tianjin University reported highly dense but porous carbons assembled from graphene sheets (Li, Tao, et al., 2016). Obtained by an evaporation-induced drying of a graphene hydrogel, graphene sheets are tightly interlinked into a three-dimensional connected nanoporous network, combining two seemingly incompatible characteristics: a porous microstructure and a high density, with a density of 1.58 g/cm^3, 70% that of graphite. Volume capacitance of the assembled electrochemical capacitor reached 376 F/cm^3, which was the highest value reported for carbon-based electrode materials in aqueous electrolytes (Tao et al., 2013).

Li group of Institute of Metal Research, Chinese Academy of Sciences found that by adjusting the content of graphene oxide (GO) and exfoliated graphene (EG), the microstructure, including the graphitic spacing and wedge-shaped pore size in the graphene-assembled self-supporting film, can be precisely tuned (Li, Gadipelli, et al., 2020). The pore size is optimized according to the size of ions, and the porous structure is fully utilized to obtain a best combination of specific surface area and density. Volume-specific capacitance of the graphene electrode assembled from GO and EG in equal proportions reached 216 F/cm^3. And volumetric energy density of the symmetrical electrochemical capacitor reached 88.1 W h/L in an ionic liquid electrolyte (Fig. 6.15). The optimized graphene film electrode was further assembled into an all-solid-state, flexible

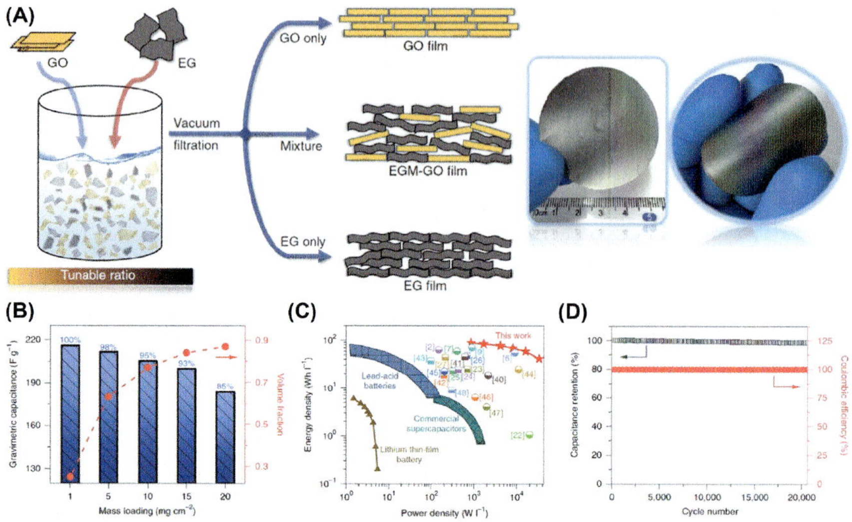

Figure 6.15 ***Graphene-based high volumetric capacitance.*** (A) Assembly of graphene membrane with tunable interlayer spacing by adjusting the ratio of GO and EG; (B) specific capacitance retention along with increasing areal mass loading; (C) Ragone plots of the 50% EG film-based supercapacitor in comparison with other energy storage devices; (D) cycling performance at a current density 10 A/g (Li, Gadipelli, et al., 2020). *EG*, Exfoliated graphene; *GO*, graphene oxide.

smart device, demonstrating potential applications as a portable power supply for flexible and mobile electronics.

6.2.4.4 Fuel cells

In a fuel cell, chemical energy is converted into electricity through electrochemical oxidation–reduction reactions between fuels such as hydrogen and oxygen. Theoretical specific capacity of hydrogen fuel cells is about 3300 W h/kg, which is much higher than that of lithium-ion batteries. Therefore fuel cells have been considered an important solution of clean energy, especially suitable for long-distance and heavy-duty vehicles (Cano et al., 2018). Toyota and other manufacturers have launched fuel-cell vehicles (Yoshida & Kojima, 2015). At 2022 Beijing Winter Olympics, hydrogen fuel cell–powered buses made by FOTON provided services for the athletes and showed satisfied performance regarding carbon emission, safety, and particularly cold start of the engines in extremely cold weather.

A bottleneck for large-scale application of fuel cells is the precious platinum catalysts (Kodama et al., 2021). How to reduce platinum loading without compromising catalytic activity and long-term durability is one of the most important topics of fuel cells. Therefore research has been focused on designing new electrocatalysts with high activity, low cost, and resistance to poisoning.

One of the most important candidate solutions is nanostructured electrocatalysts, for the large specific surface area and large number of active sites. In addition, nanocrystals with specific crystalline facets can effectively enhance catalytic activity. Typical examples include single-crystalline Pt_3Ni nanocrystals with (1 1 1) facets and Pt_3Ni octahedrons (Huang et al., 2015).

Through adjusting the composition, interface, and strain, alloy catalysts could simultaneously improve the catalytic activity and structural stability to reduce the use of precious metal catalysts (Stamenkovic et al., 2007; Strasser et al., 2010; Wu, Qi, et al., 2012). Pt-skin catalysts with precious platinum distributed in the outer layer and earth-abundant transition metals in the core have shown an optimum combination of catalytic activity, stability, and cost (Chen et al., 2014).

Nanosized catalysts have been applied in commercial fuel-cell electric vehicles. In next-generation fuel-cell vehicles, nanocrystalline catalysts with controlled composition distribution, shape, and planes are expected to be incorporated (Fig. 6.16).

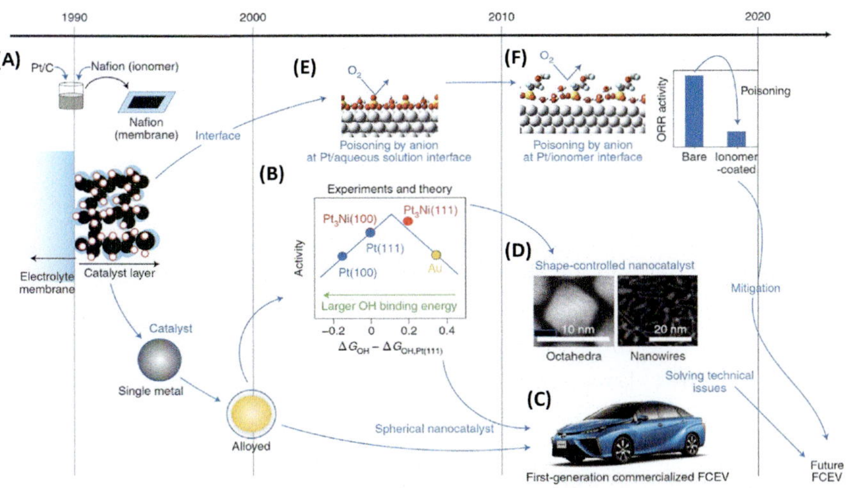

Figure 6.16 ***Development of catalysts for PEFC.*** (A) Catalyst preparation process; (B) volcano plot of ORR activity and OH binding energy; (C) first-generation of commercial fuel-cell electric vehicle; (D) low-dimensional catalysts with controlled morphology; (E and F) catalyst poisoning mechanism (Kodama et al., 2021). *PEFC*, polymer electrolyte fuel cell.

6.3 Low-dimensional medical materials

Needless to emphasize, health care is the most basic and important need for people. Among the 17 Sustainable Development Goals of United Nations, at least 4 are directly related to health: no poverty, zero hunger, good health and well-being, clean water, and so on. Human health has improved greatly over the past 200 years. The world's average life expectancy has increased from ~40 years in the 1800s to ~70 years nowadays. Many diseases that used to be deadly have been largely eradicated, including smallpox and polio. This is mainly due to the advancement of biology and medical science, development of effective drugs and vaccines (Malik, 2014).

However, there are still great challenges for health care (Luo & Gao, 2020), as fully demonstrated during the pandemic in the 3 years since 2020. COVID-19 has caused 774 million infections and 7 million deaths worldwide (record up to February 11 2024). The most fundamental and effective approach to fight against COVID-19 is to understand virus structure, and the mechanism of infection and transmission routes between the people. Due to the nanoscale size of viruses, naturally, nanoscience and low-dimensional materials played an important role.

This section will briefly introduce applications of low-dimensional materials in health care, including preventive health care, health monitoring, disease diagnosis, and treatment, as well as biological safety.

6.3.1 Preventive health care

Prevention is one of the most important and effective approaches to deal with infectious diseases. To prevent spread of the virus from infected people to uninfected people, the key is isolation, including maintaining social distance, wearing masks, and sanitizing public space. On the other hand, vaccination is the most fundamental approach to establish herd immunity against the virus. Low-dimensional materials have important applications in both isolation and vaccination.

6.3.1.1 Disinfection and sterilization

In terms of disinfection and sterilization, one of the most effective ways in daily life is to wash hands with soap. Viral envelope of a virus can be regarded as a self-assembled double-layer membrane structure composed of lipid molecules with a hydrophilic end and a hydrophobic end.

Sodium stearate in soap is also composed of hydrophilic and hydrophobic ends. Therefore washing hands with soap can dissolve and destroy the viral envelopes for effective disinfecting.

In addition to viruses, bacteria is another type of microorganism to cause infections and diseases. It is desirable to have material itself to be antibacterial, to avoid breeding and transmission of bacteria, especially for the surface of medical facilities. It has been well known that metals, such as silver and copper, have antibacterial functions. Metal nanoparticles are more active and have stronger antibacterial capabilities. On the one hand, positively charged nanoparticles could damage the negatively charged bacteria and viruses through electrostatic force. On the other hand, metal ions can enter bacterial cells to produce oxidative substances that disturb the cell functions. Antimicrobial coatings and additives based on metal nanoparticles have been applied in skin care products, metal surfaces, inks, and food packaging (Ermini & Voliani, 2021; Lemire et al., 2013; Sánchez-López et al., 2020).

TiO_2 coating is another kind of inorganic nano-antibacterial material, initially developed by Fujishima et al. based on photocatalysis. Under ultraviolet light, electron—hole pairs are excited to generate highly active hydroxyl radicals and hydrogen peroxide to destroy microorganisms (Fujishima et al., 2008; Hashimoto et al., 2005). Based on this technology, TOTO has developed a variety of products, which have been widely applied to hospital walls and floors, as well as the disinfection of complex medical instruments such as catheters used in surgery (Sekiguchi et al., 2007). Photocatalytic antibacterial materials combined with recently developed ultraviolet LED chips have been applied for water purification (Chen et al., 2017), which is crucial in remote areas and especially during disasters when normal electric power and water supply is cutoff.

6.3.1.2 Air filtering

The main transmission route of respiratory infections such as coronavirus disease 2019 (COVID-19) is through virus-containing droplets and aerosols exhaled by infected people when breathing, talking, especially coughing, and sneezing. Aerosol particles are nanoparticles floating in air, and their stability and residence time are related to particle size, humidity, and temperature of the environment. Micro- and nanoscale particles can stay in air for up to 12 hours (Tellier et al., 2019). To avoid inhaling virus-containing droplets and aerosol particles, it is crucial to filter the air by wearing masks, which has been strongly recommended by WHO and requested by governments to reduce the infection rate of coronavirus (Chu et al., 2020; Feng et al., 2020).

Filtration mechanism of masks is related to the particle size. Large particles are settled down by gravity and elastic collision, medium-sized particles are intercepted by pores between the fibers, and small particles are adsorbed through electrostatic attraction. Filtering effect is related to the structure and properties of the fibers, including hydrophobicity, density, size, and electron affinity. The materials mainly include natural cotton, silk, wool, and artificial fibers. One of the most commonly used and most effective one is the "N95" mask, which can block 95% of nonoily particles with a diameter of 0.3 microns or larger. The key material is polymer nanofibers fabricated by electrospinning. Due to its small size and large specific surface area, it can filter and absorb small particles more effectively. In addition, static charges on the surface can ionize the air and capture viruses, bacteria, and dust through electrostatic attraction.

6.3.1.3 Nanovaccines

To completely eradicate a disease, the most powerful tool is an effective vaccine. The basic principle of vaccines is to introduce into the human body antigens that have certain viral characteristics but are harmless or less harmful than the virus, allowing the immune system to "recognize" and "remember" the features of the virus and produce antibodies to damage the virus when infected, thereby getting immunity. After the outbreak of COVID-19 epidemic, China, United States, United Kingdom, and other countries developed vaccines at an unprecedented speed. Within a year, a variety of vaccines have passed clinical trials, including inactivated vaccines, messenger RNA (mRNA) technology, and protein subunit vaccines (Florindo et al., 2020; Nel & Miller, 2021; Shin et al., 2020).

As indicated by the name, for producing traditional inactivated vaccines, the actual virus was modified and inactivated to lose the ability to cause disease and used to induce an immune response. The advantage is that the vaccine is relatively stable and can be stored at 2–8°C in a conventional refrigerator. Therefore vaccines could be produced in a large scale and conveniently distributed to developing countries and regions with relatively poor infrastructures.

Basic principle of messenger RNA technology is to inject RNA sequences with virus characteristicss into human body to induce the body to synthesize fighting proteins to cause immune responses (Pardi et al., 2018). One of the challenges for mRNA technology is the low stability of RNA in human body. To improve its stability and delivery efficiency, RNA sequences are delivered through lipid nanoparticles (Nel & Miller, 2021).

A third type is the protein subunit vaccine, which includes a characteristic spike protein of SARS-CoV-2, allowing the immune system to recognize this protein and generate immunity to the actual coronavirus. One of the keys to enhance the delivery efficiency is also the protection from lipid nanoparticles (Chung et al., 2020).

In 2023 Nobel Prize in Physiology or Medicine was awarded jointly to Katalin Karikó and Drew Weissman "for their discoveries concerning nucleoside base modifications that enabled the development of effective mRNA vaccines against COVID-19."

6.3.2 Health monitoring

Convenient, fast, accurate, and noninvasive health monitoring is important for health management, personalized medicine, and especially for health care for the elderly of fast aging societies. For athletes, monitoring their health conditions has become a routine for preventing injuries, optimizing their performance to challenge the physical limits. Thanks to the application of new technologies, world records are continuously broken in Olympic Games. Traditionally, large medical equipments can only be found in professional hospitals. With the development of wearable electronic devices, many of which are based low-dimensional materials, people can monitor their personal health information more conveniently in daily life.

Because of the small size and large specific surface area, low-dimensional materials have great potential for applications as biosensors. Biological signals of human body, including body fluids, saliva, sweat, and tears, are converted to chemical, mechanical, and optical information through receptor materials. Biosensor is a new field that combines low-dimensional materials, flexible electronics, electrochemical sensing, microfluidics, AI, and so on (Fig. 6.17). There are various types of biosensors under development, for the use in eyes, teeth, chests, wrists, arms, and soles, for the monitoring of nucleic acids, proteins, bacteria, viruses, drug molecules, metabolites, and so on (Kim et al., 2019).

In addition to monitoring biochemical signals, another research frontier is to detect neural signals to understand the mechanisms of nerves and brains, ultimately the intelligence, by using low-dimensional materials-based probes. Low-dimensional materials have many advantages for

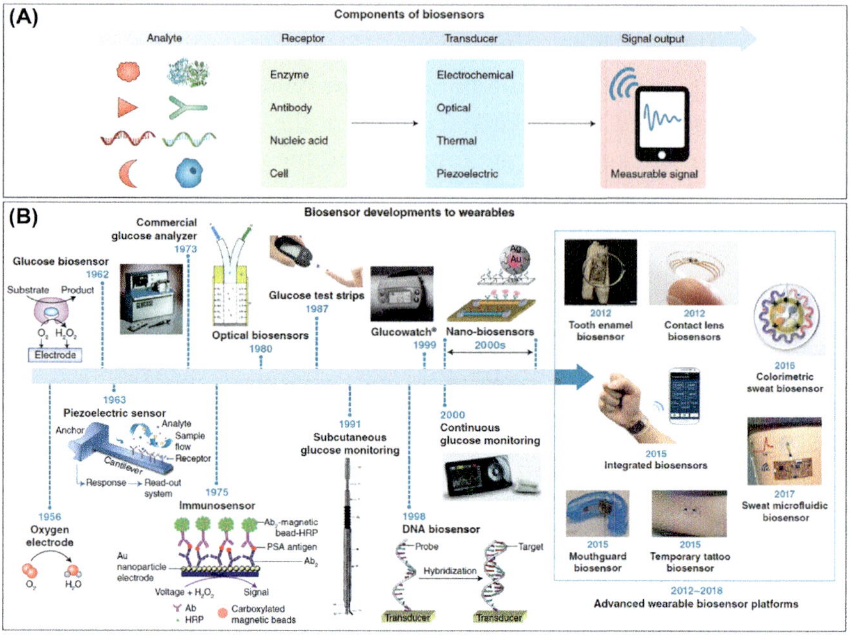

Figure 6.17 ***Low-dimensional materials-based biosensors.*** Principles (A) and development history (B) of biosensors (Kim et al., 2019).

applications in neural interfaces, including high-resolution sensing due to the nanoscale size, reduced damage to the neural tissue due to excellent flexibility (Acarón Ledesma et al., 2019). Lieber group at Harvard University developed a three-dimensional porous sensing device where nanowires are used as a scaffold for cell growth. Therefore high-density electrodes are naturally implanted during the growth of the cells. It can detect three-dimensional bioelectric signals with a detection limit of 50 nanosiemens, a time resolution of milliseconds, and a spatial resolution of submicrometers (Tian et al., 2012). Rogers group of University of Illinois at Urbana-Champaign reported a biodegradable detector for brain activities. The device was based on piezoresistive effect of silicon thin films. With specific functionalization, it can realize multifunctional measurements of electrical signals, pressure, motion, temperature, flow rate, and pH value (Kang et al., 2016).

6.3.3 Disease diagnosis and treatment

Early and accurate diagnosis is the first and probably the most important step for curing diseases. Especially for serious diseases such as cancer, early diagnosis is the key to effective treatment and to improve the survival rate. In ancient times, doctors depended on human senses to get information by "looking, hearing, asking, and touching." In modern times, advanced techniques have been developed to get various, deeper, and more accurate information, including thermometers, stethoscopes, sphygmomanometers, electrocardiograms, X-rays, nuclear magnetic resonance, and ultrasound imaging. Such methods are similar to the methods for material characterization, with the purpose of understanding the structure and more importantly to understand the relation between the behaviors and the structures at multiple length scales. Taking tumors as an example, it is necessary to figure out the cell structure, location and distribution, size and growth dynamics, and their interface with surrounding tissues.

There are two main approaches of disease diagnosis based on low-dimensional materials. One is the detection at molecular level to see if there are abnormal proteins and other molecules. For example, diagnosis of COVID-19 mainly depends on polymerase chain reaction (PCR) test to amplify and to check nucleic acid of the virus. The other approach of disease diagnosis is to directly image at the tissue level to check if there are any abnormalities, such as damage or tumors.

6.3.3.1 Breath diagnosis

Exhaled gases from breath contain a variety of volatile organic compounds that include indications for diseases; therefore breath analysis could be used for diagnosis. Compared with other approaches, breath diagnosis is noninvasive, simple to perform, and can be monitored in real time. Traditional breath analysis is typically based on mass spectrometry and optical spectroscopy, which are highly accurate, but specialized so that only professional medical institutions can do. Gas sensors based on low-dimensional materials have advantages of high sensitivity and small size and therefore have a great potential to be applied for more convenient breath analysis.

Haick group at Israel Institute of Technology reported an exhalation diagnostic method based on silicon nanowire FET gas sensors (Shehada et al., 2016). Instead of targeting a gas using a single receptor, several types of receptors are combined to functionalize the nanowires. Combined response from the nanowire FET arrays demonstrated a much higher selectivity to the mixed gases. Exhaled gases from patients with gastric cancer, lung cancer, and asthma were used to train the sensor. Assisted by a neural network–based machine learning model, the detection sensitivity reached 1 ppm level, and the accuracy was 80%.

6.3.3.2 Diagnostic imaging

While molecular diagnosis can indicate diseases, it is often necessary to take images of the tissues to confirm and to check how serious the disease is. Because of the small size, low-dimensional materials can enter tissues and even cells, serving as markers to show the characteristics and possibly abnormal features. Subcellular spatial resolution could be realized by using infrared light detection, and greater penetration depth can be obtained by using magnetic resonance methods (Shin et al., 2015).

Optical imaging diagnosis can be divided into two mechanisms: surface plasmon resonance of metal nanoparticles and fluorescence luminescence of semiconductor nanoparticles. El-Sayed of Georgia Institute of Technology reported on infrared imaging of cancer cells by gold nanorods (Huang et al., 2006). Resonant energy of surface plasmon is closely related to size and shape of the nanorods. As the aspect ratio increases, longitudinal resonance peak shifts from 650 to 950 nm. In near-infrared region, ordinary cells and water environment are transparent. Therefore in the dark field imaging mode, gold nanorod-enriched area can show a high-contrast image of cancer cells.

Nie et al. at Emory University reported cancer-targeted imaging based on core–shell structured CdSe–ZnS semiconductor nanocrystals (Gao et al., 2004). The surface of the nanocrystals is covered with a block copolymer, including hydrophilic and hydrophobic parts. The nanocrystal and surface ligands are encapsulated by the hydrophobic end, while the hydrophilic end ensures stable dispersion in body fluid environment. In addition to passively accumulating to the tumor site during delivery process, antibodies have been used to modify the surface so that nanocrystals are preferably absorbed to the targeted tumors. By imaging with fluorescent luminescence from the nanocrystals, prostate tumors in mice were identified.

Yan group at Peking University reported diagnosis and treatment of tumors using rare earth nanocrystals. Nd^{3+}-sensitized erbium nanocrystals emit fluorescence through upconversion under near-infrared laser illumination and can display the location of tumors in mice through fluorescence bioimaging (Wang et al., 2013). It was found that lanthanide nanoparticles have two emission modes under red laser. The first is near-infrared light with a wavelength of 980 nm to locate the position of tumors. The second mode is green light through upconversion, which can be used for inhibiting tumor growth by exciting the rose bengal molecules covalently bonded to the surface of nanocrystals (Li, Tang, et al., 2016). An effective tumor inhibition rate of 67% was reported after photodynamic therapy for 2 weeks.

6.3.3.3 Magnetic resonance imaging

Magnetic resonance imaging (MRI) is based on nuclear magnetic resonance (NMR), where ^1H resonantly absorbs electromagnetic waves at Larmor frequency under the action of a magnetic field. Generally, the contrast in MRI is small. Magnetic nanoparticles have been used to enhance the contrast, because of the acceleration effects from local magnetic field to the spin relaxation (Shin et al., 2015).

Chouly et al. (1996) reported that by using superparamagnetic iron oxide nanoparticles as an imaging contrast agent, they were able to detect early liver cancer cells from patients with cirrhosis (Bellin et al., 1994; Denys et al., 1994). MRI contrast was enhanced and cancerous lymph node tumors of about 1 mm could be detected (Mouli et al., 2010; Wunderbaldinger et al., 2002). Viola et al. (2015) from Northwestern University constructed a composite "probe" of antibodies and superparamagnetic nanoparticles, which can be used for early diagnosis of Alzheimer's disease and evaluation of drug effectiveness.

6.3.3.4 Nanomedicines

Nanomedicine based on interactions of low-dimensional materials with the human body is an attractive research field. For nanomedicines, it is critical to guide low-dimensional materials to enter a specific place (organ), stay for a desired time, and be discharged according to the designed path. This determines whether the medicines are beneficial or toxic to the human body. For cancer-treating drugs, it is necessary to increase their concentration in cancer cells and reduce the concentration in normal tissue cells. And it is necessary to increase the circulating time to improve its effect, and to shorten the time to stay in the liver to reduce toxicity (Tsoi et al., 2016). Therefore it is important to investigate the interactions not only between the medicines with the targeted tissues but also with the whole body as a system.

The interactions between low-dimensional materials and living organisms have chemical, mechanical, magnetic, and thermal effects (Shin & Cheon, 2017). Chemical interactions include binding of drug molecules to tumors, as well as the incorporation of drug carriers with cell membranes driven by surface hydrophilicity and charge. Under a magnetic field, the spin direction of magnetic nanoparticles tends to be aligned, converting magnetic interaction into a mechanical force. In an alternating magnetic field, Brownian motion coupled with the magnetic force causes magnetic particles to rotate or vibrate rapidly, producing a thermal effect for magnetothermal therapy (Yamaguchi et al., 2014).

Interaction between low-dimensional materials and organs depends on size, shape, hardness, surface charge, and so on (Du et al., 2018). For example, hard fibrous materials, such as asbestos, can induce lung cancer, while soft spherical lipid nanoparticles can enter cells and are used as drug carriers. Generally, particles with a small hydrodynamic diameter (<6 nm) can pass through the filtering membrane in kidney and to be excreted in urine (Soo Choi et al., 2007). The shape also plays an important role. It was reported that flexible carbon nanotubes with a diameter of 1 nm and a length of 100−1000 nm tend to be oriented in the flow direction, making it easier to be excreted into urine (Ruggiero et al., 2010).

From 1995 to 2017 the United States Food and Drug Administration (FDA) approved about 50 types of nanomedicines, including liposomes, polymers, micelles, nanocrystals, inorganic nanoparticles, and proteins (Ventola, 2017). Among them, photothermal therapy based on nanoparticles has successfully advanced from fundamental research to practical clinical application. For example, for the treatment of acne, a company called "Sebacia" developed a nanomedicine based on SiO_x−Au core−shell nanoparticles. Such particles can effectively absorb near-infrared light and convert into local heat to damages the pilosebaceous glands without affecting surrounding skin, thereby inhibiting excess sebum and reducing side effects (Paithankar et al., 2015). In another example, Rastinehad et al. reported a clinical trial of photothermal therapy using SiO_x−Au nanoparticles for treating prostate tumors, which is the most commonly diagnosed cancer and the second leading cause of cancer-related death for men in the United States. Among 16 patients diagnosed with low- or intermediate-risk prostate cancer, 15 of them showed tumor remission after 3 months of treatment, and 14 patients were tested negative of tumors after 1 year (Rastinehad et al., 2019).

6.3.4 Biological safety of low-dimensional materials

Biosafety of low-dimensional materials is a crucial issue, especially for their applications in health care. In November 2019 there was a landmark event that aroused widespread attention and heated discussion, when carbon nanotubes were included in a Substitute It Now (SIN) list by the

International Chemistry Secretariat (ChemSec). In response, 39 researchers of carbon nanotubes from the United States, the United Kingdom, Spain, Italy, France, Switzerland, Belgium, Australia, Ireland, Poland, Germany, China, Canada, and Japan, coauthored an article "banning carbon nanotubes would be scientifically unjustified and damaging to innovation" (Heller et al., 2020).

It was pointed out that properties of carbon nanotubes are determined by many structural characteristics and processing parameters, resulting in diverse physical, chemical, mechanical, and biological properties. The diameter ranges from 1 nm to tens of nanometers, and its length spans a range from nanometers to meters. They have been assembled with macroscopic forms of powder, sponge, film on substrate, and dispersion in solution. And the surface can be modified by various covalent and noncovalent functionalizations. Therefore it is not scientific to simply include all kinds of carbon nanotubes with diverse structures and properties as a whole on the ban list, practically hindering safe and effective applications of carbon nanotubes.

From this incident, we can see that on the safety of new materials, scientists and nonprofit organizations may hold different positions. Researchers can obtain relatively accurate and comprehensive information from their own research and communications with peers and understand the complexity of the safety issue. On the other hand, it is the intension of nonprofit organizations to provide clear and simple guidelines for the general public. In the future, it is crucial to conduct comprehensive and systematic research on the safety of low-dimensional materials to obtain accurate correlations between their structural characteristics and safety. It is necessary to establish standards for structural characterization, experimental procedures, and safety assessment methods. Equally important, the scientific community needs to explain about research results and talk with general citizens in a timely manner to gain public trust.

6.4 Low-dimensional composite materials

In Chapter 5 we discussed properties of low-dimensional materials associated with the dimensions and length scales. As an example, nanocarbon materials, including carbon nanotubes and graphene, have combined attractive properties, such as low density, ultrahigh elastic modulus, tensile strength, current carrying capacity, and thermal conductivity (Geim & Novoselov, 2007). Grand projects, such as space elevators, have been envisioned, based on the low density and high strength of carbon nanotubes (Edwards, 2000; Pugno, 2006). Is it possible to grow carbon nanotubes of unlimited lengths? Is it possible to achieve the properties at nanoscale (10^{-9} m) of low-dimensional materials at macroscopic or even space scale of up to 10^{8} m? There are still many grand challenges and important fundamental questions to be answered.

There are many lessons from the history of materials science. In the 1950s there were many reports on crystal whiskers. At micrometer scale, single crystalline whiskers of metals and nonmetallic materials have demonstrated tensile strength approaching the theoretical limit (Brenner, 1956). However, due to the brittle nature, as the size becomes bigger, probability of fracture exponentially increases. The large gap between the properties of individual nanostructures and macroscopic components is a common challenge for the large-scale applications of low-dimensional materials in many fields, such as mechanical structural materials, and electrically and thermally conductive materials.

Composite material is an important solution for solving the dilemma between small-scale structural feature and large-scale applications. On the one hand, by dispersing and embedding nanostructures into the matrix, unique properties at small scale could be maintained. On the other hand, by "gluing" the nanostructure with the matrix, larger components could be fabricated for applications at macroscale lengths. In addition, interface between matrix and the second phase could be designed to strengthen and to enhance the transport properties to achieve an effect of $1 + 1 > 2$.

Traditional composite materials, based on the matrixes of polymer, metal, and ceramics, have been widely applied in aerospace, aviation, automotive, and many other fields. In addition to the mechanical strengthening composite, broadly speaking, catalysts, vaccines, and medicines also have composite structures. In this section, we will introduce the principles and applications of low-dimensional materials in composite materials for mechanically strengthening, electrical conducting, thermal managements, and multifunctional materials.

6.4.1 Theoretical models of composite materials

There are three main elements for the composite materials, matrix, secondary phase, and interface. For properties that are not sensitive to the microstructure, such as density and modulus, properties of composite structure could be calculated by linear combination. For other properties that are sensitive to microstructure and morphology, such as strength and toughness, the interface plays a decisive role. Depending on the distribution of the secondary phase, composite materials could be modeled as fiber array and random networks, respectively.

Considering a simple fiber-array model, where secondary phase is distributed parallelly. Along the fiber direction, the matrix and secondary phase are equal in mechanical strain, temperature gradient, or potential difference. In this case, Young's modulus, thermal conductivity, and electrical conductivity could be simply calculated following the mixing rule: $P_c = \sum_{t=1}^{n} V_i P_i$, $n = 2$.

For a typical polymer matrix, the modulus, thermal conductivity, and electrical conductivity are on the order of 1 GPa, 0.1 W/(m K), and 0.01 S/cm. As a secondary phase, the values of the modulus, thermal conductivity, and electrical conductivity of carbon nanotubes are about 10^3 GPa, 10^3 W/(m K), and 10^6 S/cm. With a concentration of 10%, the modulus, thermal conductivity, and electrical conductivity of polymer-based carbon nanotube composite materials are expected to be 100 GPa, 100 W/(m K), and 10^5 S/cm, which are 100, 1000, and 10^7 folds those of the matrix. To achieve the theoretically predicted enhancement, it is necessary to optimize the structures at multiple length scales, from intrinsic structure, and properties of individual nanostructures, dispersion, alignment, and length, to the interface.

For the random network model, fibers of finite lengths are randomly positioned in the matrix. Since there are orders of magnitude differences in electrical and thermal conductivity between the secondary phase and the matrix, the electrical and thermal transport relies on the formation of a three-dimensional continuous network of the secondary phase following a percolation model (Foygel et al., 2005): $\sigma(\phi; a) = \sigma_0 [\phi - \phi_c(a)]^{t(a)}$, where $\phi_c(a)$ is the percolation threshold. In this model, the associated physical properties increase exponentially with content (ϕ) of the secondary phase.

6.4.2 Mechanically reinforced composite materials

The purpose of composite materials is to strengthen the weaknesses of matrix phase, for example, to improve the stiffness and strength of polymer, to enhance the strength and reduce the weight of metals, and to increase the toughness of ceramics.

Tensile strength of fiber-reinforced composites can be described by an interface shearing model. When matrix is under tensile loading, tensile stress is transferred to the fiber by a shear stress at the interface between the phases. A critical condition of failure is when the interface shear stress reaches the shear strength of matrix, and the maximum tensile stress at the fiber reaches tensile strength of the fiber. Accordingly, a critical length could be calculated: $l_c = \sigma_r \frac{Al}{\tau_m S}$, where A is cross-sectional area, S is interface area, τ_m is shear strength of interface, and σ_r is tensile strength of fiber. When the length of fiber is greater than the critical length, tensile stresses of fiber in distance $l - l_c$ can reach the tensile strength. Tensile strength of the composite can be calculated as $\sigma_c = \sigma_r V_r \left(\frac{l}{2l_c}\right) + \sigma_m V_m$, where V_r is filling ratio, σ_r is reinforcing phase strength, and l is fiber length. To achieve a high reinforcement effect, it is necessary to increase the filling ratio, tensile strength of the fiber, and importantly the shear strength of the interface.

Carbon fibers have been widely used in mechanically reinforced composite materials because of the high tensile strength, practically infinite length, and strong coupling between the fiber and matrix. Carbon nanotubes can be regarded as carbon nanofibers with a diameter in nanometer range; therefore they have been considered an ideal fiber-strengthening material. Compared with carbon fibers, carbon nanotubes have smaller diameters, therefore larger interface surface areas, so that the critical length is shorter, and the theoretical strength of the composite is higher. In carbon nanotube-reinforced polymer composites with low volume fractions, the mechanical properties were indeed significantly improved. Salvetat et al. at the University of Orléans reported that inclusion of multiwalled carbon nanotubes with a concentration of 8.3 wt% can increase Young's modulus of natural rubber by 10 times and in the meantime double the tensile strength (Bhattacharyya et al., 2008).

Fracture toughness is the ability of a material to resist crack propagation and therefore is one of the most important mechanical properties for a reliable mechanical component. The challenge to design a tough material is the compromise among toughness, modulus, and strength. In nature, there are many ingenious designs to achieve the toughening effect of $1 + 1 > 2$, through formation of composite materials. An example is the seashell, which consists of hard minerals and soft organic matter forming a layered structure (Tang et al., 2003). Inspired by this, Tang group from the National Center for Nanoscience and Technology of China designed a layered composite material consisting of graphene oxide and a conjugated polymer (PCDO). Its strength reached 106 MPa, and the toughness was 2.5 MJ/m^3, which is 40% higher than that of the seashells (Cheng et al., 2013).

In addition to static mechanical properties, low-dimensional strengthening phase can also improve the dynamic mechanical properties. Ajayan group reported that carbon nanotube could increase the damping coefficient of resin composites by 15 times (Suhr et al., 2005). Koratkar et al. reported that with an addition of 2% weight of oxidized single-walled nanotube fillers, up to 1000% increase in modulus of polycarbonate could be achieved. The increase in damping is due to frictional sliding at the nanotube—polymer interfaces, which were dramatically increased due to the nanoscale dimensions (Koratkar et al., 2005). CNT—polymer composite materials with enhanced energy dissipation efficiency have been applied in golf clubs, baseball bats, bicycle frames, and other sporting instruments. Since 2005 Swiss BMC bicycle racing team has started to use carbon nanotube-reinforced composite materials to reduce the frame weight of the bicycles in Tour de France (De Volder et al., 2013).

Overall, a small addition of low-dimensional materials could show a significant improvement of the mechanical properties of composite materials. However, as the addition continues to increase, mechanical properties of the composite materials typically reach saturation or even decrease (Blighe et al., 2011). The main reason is that low-dimensional materials, such as carbon nanotubes, are difficult to disperse and have a nonideal interface with the matrix. The high modulus and tensile strength of carbon nanotubes originate from the strong covalent bonds and absence of dangling bonds on surface, which also is the reason for the difficulty to couple with the matrix. Another challenge is the entanglement and agglomeration of carbon nanotubes, due to the van der Waals forces.

To counter such challenges, carbon nanotubes have been functionalized to improve dispersion and bonding with the matrix. In addition, carbon nanotubes have been assembled into highly aligned fibers to avoid agglomeration. Pasquali et al. at Rice University fabricated carbon nanotube fibers through a liquid crystal transition. The nanotube fibers demonstrated impressive comprehensive properties with an electrical conductivity of 5 Ms/m, a thermal conductivity of 380 W/m K, a density of 1.0—1.5 g/cm^3, a tensile strength of 1—3 GPa, and Young's modulus of $\sim$120 GPa (Behabtu et al., 2013).

Graphene has also been assembled into a liquid crystal phase and fabricated into well-aligned fibers. Lian et al. Rensselaer Polytechnic Institute obtained graphene fibers by directional assembly of large and small sheets densely packed with graphene oxide. After heat treatment and annealing processing, thermal conductivity reached 1290 W/m K, and tensile strength was 1.08 GPa (Xin et al., 2015). Gao group of Zhejiang University has been working on fabrication and applications of liquid crystal assembly graphene fibers (Chang et al., 2021; Li, Yang, et al., 2020; Xu & Gao, 2011, 2014). A highly oriented near-crystalline graphene flexible paper is obtained by removing the intralayer wrinkles during the liquid crystal-assembly process. A tensile strength of 1.1 GPa and Young's modulus of 62.8 GPa was achieved. The assembled graphene fibers formed a composite material with a resin matrix, where the tensile strength was increased from 190 to 634 MPa. The composite material exhibits excellent electromagnetic shielding performance, with a shielding efficiency of 30−40 dB, in the frequency range of 2−18 GHz (Li, Yang, et al., 2020).

6.4.3 Electrically conductive composites

Conductivity of polymer is typically at the level of 10^{-4} S/cm, while the conductivity of carbon nanotubes is as high as 10^{5} S/cm (Seliuta et al., 2013), so the percolation threshold theory is applicable for describing the conductive properties of carbon nanotube network-polymer composites.

The percolation threshold of multiwalled carbon nanotubes in thermoplastic polyurethane composites was reported to be 0.5% (Koerner et al., 2005). With a filler ratio of 0.5%−10%, the electrical conductivity could reach 1.0−10 S/cm. Single-walled carbon nanotubes are superior to multiwalled carbon nanotubes due to their small diameters and larger contact areas. A lower percolation threshold of 0.3% was reported for the single-walled carbon nanotube/polyaniline composites, where a conductivity higher than 10 S/cm was achieved with 10% carbon nanotube filler (Blanchet et al., 2003).

Du et al. comparatively investigated the conductive properties of carbon nanotubes and graphene-strengthened high density polyethylene (HDPE) composites and found that the percolation threshold of carbon nanotubes is 0.15 vol.%, which is an order of magnitude smaller than that of graphene (1 vol.%). It is attributed to the one-dimensional structure of carbon nanotubes that is more effective to form a three-dimensional conductive network (Fig. 6.18) (Du et al., 2011).

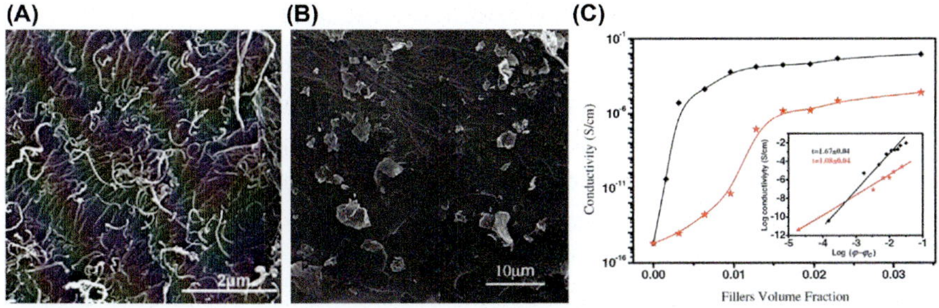

Figure 6.18 ***Carbon nanotube and graphene-based composite materials.*** SEM images of composites of carbon nanotube (A) and graphene (B) with HDPE; (C) plots of conductivity and volume filling ratio of CNT/HDPE (*black*) and graphene/HDPE (*red*) composites (Du et al., 2011). *HDPE*, High density polyethylene.

An important application of electrically conductive composites is electromagnetic shielding, based on reflection and absorption mechanisms. The reflection mechanism is related to the carrier conductivity. And the absorption mechanism is related to polarization, including electric and magnetic dipoles. Due to the skin effect of alternating electric current (AC), interaction between high-frequency electromagnetic waves and the material is concentrated mainly at the surface within a submicron depth. Compared to bulk metals, low-dimensional materials have a larger specific surface area, thus providing an advantage in electromagnetic shielding at high-frequency range.

Conventional electromagnetic shielding materials, such as nickel wires, have a typical shielding efficiency of 87 dB. Chung et al. of State University of New York reported a shielding efficiency of 130 dB for flexible graphite, attributed to its high electrical conductivity and a large specific surface area (Luo & Chung, 1996).

Low-dimensional materials, such as carbon nanotubes and graphene, are highly expected in high-frequency electromagnetic shielding and wave-absorbing materials due to their high conductivity and nanometer size (Li, Zhang, et al., 2019). Composites of highly conductive carbon nanomaterials and magnetic nanoparticles can further enhance the wave-absorbing properties by exploiting both electrode polarization and magnetic polarization mechanisms. Carbon nanotube–polymer composite films developed by Nanocomp Technologies have been used by NASA for electromagnetic shielding of satellites (De Volder et al., 2013).

6.4.4 Thermal interface composite materials

Electrical properties of semiconductors are very sensitive to temperature changes; therefore thermal management is crucial to semiconductor devices such as solar cells, LEDs, and microprocessors. As transistor sizes get smaller and smaller, the power density gets higher and higher. For CPUs at 14nm nodes, power density has reached 100 W/cm^2, and heat dissipation has become a bottleneck limiting the performance (Chen et al., 2016).

Currently, metal heat sinks, such as aluminum, are commonly used for thermal packaging of semiconductor devices. One of the challenges is the hard contact and high interface thermal resistance between metal and silicon-based semiconductors. And thermal interface materials are required to reduce the interface thermal resistance. Low-dimensional materials are more flexible than bulk materials due to the small size and large aspect ratio. In addition, thermal conductivity is expected to be enhanced because of the reduced phonon scattering of the dangling bonds-free surface. Indeed, in-plane thermal conductivity as high as 1000 W/m K has been reported for carbon nanotubes, graphene, and h-BN, making them ideal for applications as directional thermal interface materials (Balandin et al., 2008; Berber et al., 2000; Cai et al., 2019).

Thermal conductivity of carbon nanotube composites can be calculated using the following equation: $\frac{K_e}{K_m} = 1 + \frac{fp}{3}\frac{K_c/K_m}{p + \frac{2a_K}{d}\frac{K_c}{K_m}}$, where K_c, K_m, and K_e are the thermal conductivities of the carbon nanotubes, matrix phase, and composite. f is carbon nanotube content, p is carbon nanotube aspect ratio, d is the diameter of the carbon nanotubes, and $a_K = R_K K_m$ is a parameter related to the interface thermal resistance R_K (Nan et al., 2004). Apparently, the higher the intrinsic thermal conductivity of carbon nanotubes and the higher the carbon nanotube content, the higher the thermal conductivity of the composite is. Under an ideal condition of infinitely long carbon nanotubes, the formula can be simplified as $\frac{K_e}{K_m} = 1 + \frac{f}{3}K_c/K_m$, where the enhancement of thermal conductivity of the composite material is proportional to carbon nanotube content.

Based on above discussions, there are mainly three approaches to improve the thermal conductivity of carbon nanotube composites: (1) to improve the crystal quality and intrinsic thermal conductivity of carbon nanotubes (K_c); (2) to form a close contact with the matrix phase as well as the heat source, to reduce interfacial thermal resistance (R_K); and (3) to increase the length of the carbon nanotubes, to increase the aspect ratio

(p). An ideal structure is an array of compactly aligned carbon nanotubes that forms close contact with the matrix and the heat source and sinks.

For the carbon nanotube–polymer composite materials, Chauvet et al. reported that addition of 7% single-walled carbon nanotubes into poly(methyl methacrylate) could increase the electrical conductivity by orders of magnitude and the thermal conductivity by about 55%, from 0.23 W/m K to 0.35 W/m K (Bonnet et al., 2007). To enhance thermal transport at the interface between carbon nanotubes and matrix, Fan group at Tsinghua University injected polymers into vertically aligned carbon nanotubes to fabricate a composite. With a carbon nanotube content of only 0.4 vol %, the thermal conductivity was increased to 1.2 W/m K (Huang et al., 2005). Similarly, to fully make use of the high in-plane thermal conductivity of graphene, Liu et al. of Suzhou Institute of Nano-tech and Nano-bionics, Chinese Academy of Sciences, fabricated parallelly aligned graphene/resin composites. In the range between 40°C and 90°C, the thermal conductivity increases with temperature from 16.75 to 33.54 W/m K (Li, Guo, et al., 2014).

As mentioned before, the function of thermal interface materials is to enhance thermal transport between silicon chips and metal sinks, where thermal contact at the interface is critical. To reduce thermal resistance at the interfaces, Cola et al. at Purdue University directly grew carbon nanotube arrays on copper foil (Cola et al., 2007). The height was 10 μm and the density was $\sim 10^8/\text{mm}^2$. Under a pressure of 100 kPa, the total thermal resistance of the Si–Ag–CNT–Cu interface was about 10 mm^2 K/W.

To further enhance the interface transport at the top surface, Liu group of Institute of Metal Research, Chinese Academy of Sciences, covered a thin carbon layer on top of carbon nanotube arrays grown on a copper foil (Ping et al., 2017). The thermal conductivity of the carbon nanotube arrays was 2.7 W/m K, and the thermal resistivity of the Cu–CNT array-Cu interface was reduced to 0.8 mm^2 K/W (Fig. 6.19).

6.4.5 Multifunctional low-dimensional composite materials

In addition to composite materials targeting a specific physical property to perform a single function, multifunctional composite materials couple multiple physical properties, such as mechanical, electrical, thermal, and optical properties, to achieve multiple functions, including sensing, actuating, display, and energy conversion. Such multifunctional composite materials will be important for smart devices that are the infrastructure of the coming smart society.

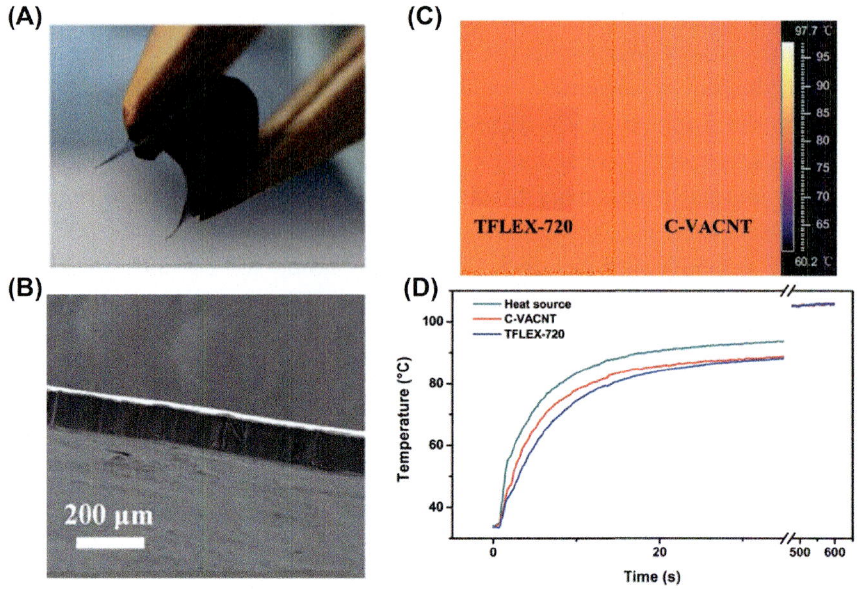

Figure 6.19 ***Carbon nanotube array-based thermal interface materials.*** (A and B) Optical and SEM images of the carbon nanotube array; (C) thermography comparison of a commercially available thermal interface material (TFLEX 720) with the vertically aligned carbon nanotubes (VACNT); (D) plots of temperature again heating time for the heat source and thermal interface material-covered surfaces (Ping et al., 2017).

Inspired by multifunctional skins coupling protecting, sensing, and self-healing functions, Bao group of Stanford University fabricated electronic skins that can sense pressure and bending strain, based on a nickel nanoparticles/supramolecular organic polymer composite (Tee et al., 2012). The addition of nickel nanoparticles improved the mechanical and thermal properties of the organic polymer, and its glass transition temperature increases from −20°C to 10°C. In the meantime, the electrical conductivity was also increased and can be modulated by compression or bending stress; therefore they can be applied as mechanical sensors. By using the enhanced glass transition temperature, an interesting self-healing function was achieved. Simply by heating the cut polymer composite at 50°C for a short time, both the mechanical and electrical properties were recovered.

To mimic the animal muscles that transform the electrochemical power into mechanical movements, Baughman group at University of Texas at Dallas invented artificial muscles based on carbon nanotubes (Foroughi et al., 2011). Carbon nanotube fibers are twisted into a spiral rope. During electrochemical cycles, the volume expands due to intercalation of ions, resulting in rotation, and stretching of the rope. A maximum expansion of 16.5% was achieved by interlacing two strands of carbon nanotube ropes to amplify the volume expansion effect (Lee et al., 2017). Movements of pure carbon nanotube yarns are bipolar in response to the potential. Such bipolarity was overcome by fabricating a carbon nanotube–polymer composite spiral structure, which demonstrated unipolar stretching behavior. By adjusting the potential scan rate to control the effective size of the ions, a stroke as high as 21% was reached for the artificial muscles (Chu et al., 2021).

By combining sensing and display functions, Peng group of Fudan University reported on a large-area textile-based multifunctional low-dimensional composite device (Shi et al., 2021). It is interwoven with warp and weft threads. The weft threads consist of transparent conductive polymer fibers, and the warp threads are silver-plated conductive composite fibers. There is a thin layer of zinc sulfide phosphor at the weft–warp contact for display. This flexible display has excellent bending and stretching properties. Integration of flexible display, flexible energy storage, and pressure sensing on clothing was demonstrated. Combined with brainwave signal detection and analysis, it was possible to monitor and display the psychological states in real time, such as being relaxed or anxious (Fig. 6.20).

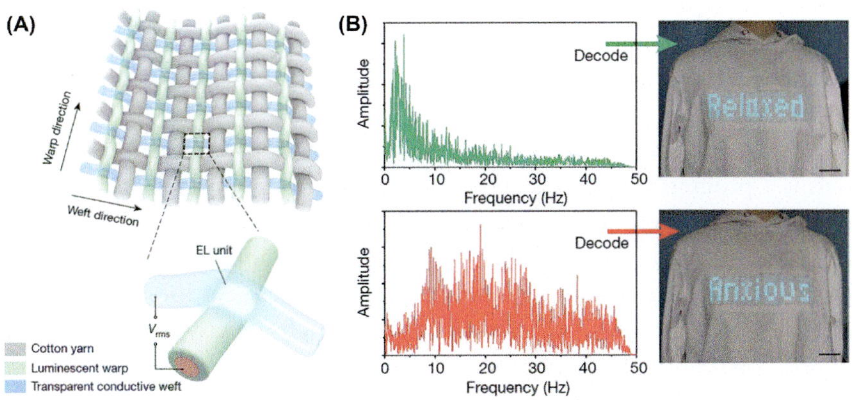

Figure 6.20 ***Multifunctional display textiles.*** (A) Silver-plated polymer fiber and ZnS phosphor to form an electroluminescent network; (B) combined with biosensing, psychological states are analyzed and displayed on clothes in real time (Shi et al., 2021).

6.5 Summary

After years of exploration, low-dimensional materials are being started to be applied to a wide range of fields, ranging from information electronic, clean energy, sustainable environment, and health care.

In information electronics, nanotransistors have continued Moore's law to scale the size to below 10 nm. Low-dimensional sensors and thin-film transistors form the basis of wearable electronic devices and IoT. Furthermore, breakthroughs have been made in the exploration of devices based on new principles such as quantum computers and neuromorphic computing.

For clean energy, perovskite solar cells are promising for low-cost and high-efficiency solar power conversion. Artificial photosynthesis-based solar fuels are expected to realize the cycling of hydrogen, carbon, and nitrogen. And low-dimensional materials have been widely used as electrode materials in secondary batteries and supercapacitors and as catalysts in fuel cells.

In health care, nano-vaccines have been successfully developed by using lipid bilayers to protect the antigen cargo to enhance the immunogenicity, and to end the COVID-19 pandemic. Biosensors greatly improve health management by monitoring the health information from skin and tears to brain and nerves. Magnetic nanoparticles and nanocrystal-based drugs are being used in clinical diagnosis and treatment.

In composite materials, low-dimensional materials have shown strengthening effect and improvements in thermal and electrical conductivity. Multifunctional low-dimensional composite materials are forming new research frontiers in flexible skin electronics, artificial muscles, and wearable displays.

It should be mentioned that the application of low-dimensional materials is a progressing process and cannot be achieved overnight. There are still many different fundamentals and practical challenges for different applications. Taking solar fuel as an example, the bottleneck for hydrogen production by photosplitting water is the efficiency. And selectivity is still a great challenge for CO_2 reduction, while high-activity catalysts are still under exploration for ammonia synthesis. To fully realize the potential applications of low-dimensional materials, it is important to further strengthening the basic research to explore "killer" applications based on the intrinsic properties. A bright future depends on continued support, investment, and collaboration of society, government, and enterprises.

References

Acarón Ledesma, H., Li, X., Carvalho-de-Souza, J. L., Wei, W., Bezanilla, F., & Tian, B. (2019). An atlas of nano-enabled neural interfaces. *Nature Nanotechnology.*, *14*(7), 645–657. Available from https://doi.org/10.1038/s41565-019-0487-x, http://www.nature.com/nnano/index.html.

Ambrogio, S., Narayanan, P., Tsai, H., Shelby, R. M., Boybat, I., Di Nolfo, C., Sidler, S., Giordano, M., Bodini, M., Farinha, N. C. P., Killeen, B., Cheng, C., Jaoudi, Y., & Burr, G. W. (2018). Equivalent-accuracy accelerated neural-network training using analogue memory. *Nature*, *558*(7708), 60–67. Available from https://doi.org/10.1038/s41586-018-0180-5, http://www.nature.com/nature/index.html.

Anttu, N. (2015). Shockley–Queisser detailed balance efficiency limit for nanowire solar cells. *ACS Photonics.*, *2*(3), 446–453. Available from https://doi.org/10.1021/ph5004835, http://pubs.acs.org/journal/apchd5.

Arora, N., Dar, M. I., Hinderhofer, A., Pellet, N., Schreiber, F., Zakeeruddin, S. M., & Grätzel, M. (2017). Perovskite solar cells with CuSCN hole extraction layers yield stabilized efficiencies greater than 20%. *Science (New York, N.Y.)*, *358*(6364), 768–771. Available from https://doi.org/10.1126/science.aam5655, http://science.sciencemag.org/content/sci/358/6364/768.full.pdf.

Arute, F., Arya, K., Babbush, R., Bacon, D., Bardin, J. C., Barends, R., Biswas, R., Boixo, S., Brandao, F. G. S. L., Buell, D. A., Burkett, B., Chen, Y., Chen, Z., Chiaro, B., Collins, R., Courtney, W., Dunsworth, A., Farhi, E., Foxen, B., ... Martinis, J. M. (2019). Quantum supremacy using a programmable superconducting processor. *Nature*, *574*(7779), 505–510. Available from https://doi.org/10.1038/s41586-019-1666-5, http://www.nature.com/nature/index.html.

Asadi, M., Sayahpour, B., Abbasi, P., Ngo, A. T., Karis, K., Jokisaari, J. R., Liu, C., Narayanan, B., Gerard, M., Yasaei, P., Hu, X., Mukherjee, A., Lau, K. C., Assary, R. S., Khalili-Araghi, F., Klie, R. F., Curtiss, L. A., & Salehi-Khojin, A. (2018). A lithium–oxygen battery with a long cycle life in an air-like atmosphere. *Nature*, *555* (7697), 502–506. Available from https://doi.org/10.1038/nature25984, http://www.nature.com/nature/index.html.

Asahi, R., Morikawa, T., Irie, H., & Ohwaki, T. (2014). Nitrogen-doped titanium dioxide as visible-light-sensitive photocatalyst: Designs, developments, and prospects. *Chemical Reviews*, *114*(19), 9824–9852. Available from https://doi.org/10.1021/cr5000738, http://pubs.acs.org/journal/chreay.

Backmann, N., Zahnd, C., Huber, F., Bietsch, A., Plückthun, A., Lang, H. P., Güntherodt, H. J., Hegner, M., & Gerber, C. (2005). A label-free immunosensor array using single-chain antibody fragments. *Proceedings of the National Academy of Sciences of the United States of America*, *102*(41), 14587–14592. Available from https://doi.org/10.1073/pnas.0504917102.

Bahk, J. H., Fang, H., Yazawa, K., & Shakouri, A. (2015). Flexible thermoelectric materials and device optimization for wearable energy harvesting. *Journal of Materials Chemistry C.*, *3*(40), 10362–10374. Available from https://doi.org/10.1039/c5tc01644d, http://pubs.rsc.org/en/journals/journalissues/tc.

Balandin, A. A., Ghosh, S., Bao, W., Calizo, I., Teweldebrhan, D., Miao, F., & Lau, C. N. (2008). Superior thermal conductivity of single-layer graphene. *Nano Letters*, *8* (3), 902–907. Available from https://doi.org/10.1021/nl0731872.

Barber, J. (2009). Photosynthetic energy conversion: Natural and artificial. *Chemical Society Reviews*, *38*(1), 185–196. Available from https://doi.org/10.1039/b802262n.

Behabtu, N., Young, C. C., Tsentalovich, D. E., Kleinerman, O., Wang, X., Ma, A. W. K., Bengio, E. A., Ter Waarbeek, R. F., De Jong, J. J., Hoogerwerf, R. E., Fairchild, S. B., Ferguson, J. B., Maruyama, B., Kono, J., Talmon, Y., Cohen, Y.,

Otto, M. J., & Pasquali, M. (2013). Strong, light, multifunctional fibers of carbon nanotubes with ultrahigh conductivity. *Science (New York, N.Y.)*, *339*(6116), 182–186. Available from https://doi.org/10.1126/science.1228061, http://www.sciencemag.org/content/339/6116/182.full.pdf.

Bellin, M. F., Zaim, S., Auberton, E., Sarfati, G., Duron, J. J., Khayat, D., & Grellet, J. (1994). Liver metastases: Safety and efficacy of detection with superparamagnetic iron oxide in MR imaging. *Radiology*, *193*(3), 657–663. Available from https://doi.org/10.1148/radiology.193.3.7972804.

Berber, S., Kwon, Y. K., & Tománek, D. (2000). Unusually high thermal conductivity of carbon nanotubes. *Physical Review Letters*, *84*(20), 4613–4616. Available from https://doi.org/10.1103/PhysRevLett.84.4613.

Bhattacharyya, S., Sinturel, C., Bahloul, O., Saboungi, M. L., Thomas, S., & Salvetat, J. P. (2008). Improving reinforcement of natural rubber by networking of activated carbon nanotubes. *Carbon*, *46*(7), 1037–1045. Available from https://doi.org/10.1016/j.carbon.2008.03.011.

Bishop, M. D., Hills, G., Srimani, T., Lau, C., Murphy, D., Fuller, S., Humes, J., Ratkovich, A., Nelson, M., & Shulaker, M. M. (2020). Fabrication of carbon nanotube field-effect transistors in commercial silicon manufacturing facilities. *Nature Electronics.*, *3*(8), 492–501. Available from https://doi.org/10.1038/s41928-020-0419-7, https://www.nature.com/natelectron/.

Biswas, K., He, J., Blum, I. D., Wu, C. I., Hogan, T. P., Seidman, D. N., Dravid, V. P., & Kanatzidis, M. G. (2012). High-performance bulk thermoelectrics with all-scale hierarchical architectures. *Nature*, *489*(7416), 414–418. Available from https://doi.org/10.1038/nature11439, http://www.nature.com/nature/index.html.

Blanchet, G. B., Fincher, C. R., & Gao, F. (2003). Polyaniline nanotube composites: A high-resolution printable conductor. *Applied Physics Letters*, *82*(8), 1290–1292. Available from https://doi.org/10.1063/1.1553991.

Blighe, F. M., Young, K., Vilatela, J. J., Windle, A. H., Kinloch, I. A., Deng, L., Young, R. J., & Coleman, J. N. (2011). The effect of nanotube content and orientation on the mechanical properties of polymer-nanotube composite fibers: Separating intrinsic reinforcement from orientational effects. *Advanced Functional Materials.*, *21*(2), 364–371. Available from https://doi.org/10.1002/adfm.201000940, http://onlinelibrary.wiley.com/journal/10.1002/(ISSN)1616-3028.

Bonnet, P., Sireude, D., Garnier, B., & Chauvet, O. (2007). Thermal properties and percolation in carbon nanotube-polymer composites. *Applied Physics Letters*, *91*(20). Available from https://doi.org/10.1063/1.2813625.

Boukai, A. I., Bunimovich, Y., Tahir-Kheli, J., Yu, J. K., Goddard, W. A., & Heath, J. R. (2008). Silicon nanowires as efficient thermoelectric materials. *Nature*, *451*(7175), 168–171. Available from https://doi.org/10.1038/nature06458, http://www.nature.com/nature/index.html.

Brenner, S. S. (1956). Tensile strength of whiskers. *Journal of Applied Physics*, *27*(12), 1484–1491. Available from https://doi.org/10.1063/1.1722294.

Burg, T. P., Godin, M., Knudsen, S. M., Shen, W., Carlson, G., Foster, J. S., Babcock, K., & Manalis, S. R. (2007). Weighing of biomolecules, single cells and single nanoparticles in fluid. *Nature*, *446*(7139), 1066–1069. Available from https://doi.org/10.1038/nature05741, http://www.nature.com/nature/index.html.

Burschka, J., Pellet, N., Moon, S. J., Humphry-Baker, R., Gao, P., Nazeeruddin, M. K., & Grätzel, M. (2013). Sequential deposition as a route to high-performance perovskite-sensitized solar cells. *Nature*, *499*(7458), 316–319. Available from https://doi.org/10.1038/nature12340.

Cai, Q., Scullion, D., Gan, W., Falin, A., Zhang, S., Watanabe, K., Taniguchi, T., Chen, Y., Santos, E. J. G., & Li, L. H. (2019). High thermal conductivity of high-quality monolayer boron nitride and its thermal expansion. *Science Advances.*, *5*(6). Available from https://doi.org/10.1126/sciadv.aav0129, https://advances.sciencemag.org/content/5/6/eaav0129/tab-pdf.

Cano, Z. P., Banham, D., Ye, S., Hintennach, A., Lu, J., Fowler, M., & Chen, Z. (2018). Batteries and fuel cells for emerging electric vehicle markets. *Nature Energy.*, *3*(4), 279–289. Available from https://doi.org/10.1038/s41560-018-0108-1, http://www.nature.com/nenergy/.

Cao, Q., Kim, H. S., Pimparkar, N., Kulkarni, J. P., Wang, C., Shim, M., Roy, K., Alam, M. A., & Rogers, J. A. (2008). Medium-scale carbon nanotube thin-film integrated circuits on flexible plastic substrates. *Nature*, *454*(7203), 495–500. Available from https://doi.org/10.1038/nature07110, http://www.nature.com/nature/index.html.

Cao, Q., Tersoff, J., Farmer, D. B., Zhu, Y., & Han, S. J. (2017). Carbon nanotube transistors scaled to a 40-nanometer footprint. *Science (New York, N.Y.)*, *356*(6345), 1369–1372. Available from https://doi.org/10.1126/science.aan2476, http://science.sciencemag.org/content/sci/356/6345/1369.full.pdf.

Cao, Y., Brady, G. J., Gui, H., Rutherglen, C., Arnold, M. S., & Zhou, C. (2016). Radio frequency transistors using aligned semiconducting carbon nanotubes with current-gain cutoff frequency and maximum oscillation frequency simultaneously greater than 70 GHz. *ACS Nano.*, *10*(7), 6782–6790. Available from https://doi.org/10.1021/acsnano.6b02395, http://pubs.acs.org/journal/ancac3.

Chan, C. K., Peng, H., Liu, G., McIlwrath, K., Zhang, X. F., Huggins, R. A., & Cui, Y. (2008). High-performance lithium battery anodes using silicon nanowires. *Nature Nanotechnology.*, *3*(1), 31–35. Available from https://doi.org/10.1038/nnano.2007.411.

Chang, D., Liu, J., Fang, B., Xu, Z., Li, Z., Liu, Y., Brassart, L., Guo, F., Gao, W., & Gao, C. (2021). Reversible fusion and fission of graphene oxide-based fibers. *Science (New York, N.Y.)*, *372*(6542), 614–617. Available from https://doi.org/10.1126/science.abb6640, https://science.sciencemag.org/content/372/6542/614.full.

Chang, E. Y., Kuo, C. I., Hsu, H. T., Chiang, C. Y., & Miyamoto, Y. (2013). InAs thin-channel high-electron-mobility transistors with very high current-gain cutoff frequency for emerging submillimeter-wave applications. *Applied Physics Express*, *6*(3). Available from https://doi.org/10.7567/APEX.6.034001Taiwan, http://apex.jsap.jp/link?APEX/6/034001/pdf.

Chappert, C., Fert, A., & Van Dau, F. N. (2007). The emergence of spin electronics in data storage. *Nature Materials*, *6*(11), 813–823. Available from https://doi.org/10.1038/nmat2024, http://www.nature.com/nmat/.

Chaste, J., Eichler, A., Moser, J., Ceballos, G., Rurali, R., & Bachtold, A. (2012). A nanomechanical mass sensor with yoctogram resolution. *Nature Nanotechnology.*, 7(5), 301–304. Available from https://doi.org/10.1038/nnano.2012.42.

Chen, C., Kang, Y., Huo, Z., Zhu, Z., Huang, W., Xin, H. L., Snyder, J. D., Li, D., Herron, J. A., Mavrikakis, M., Chi, M., More, K. L., Li, Y., Markovic, N. M., Somorjai, G. A., Yang, P., & Stamenkovic, V. R. (2014). Highly crystalline multimetallic nanoframes with three-dimensional electrocatalytic surfaces. *Science (New York, N.Y.)*, *343*(6177), 1339–1343. Available from https://doi.org/10.1126/science.1249061, http://www.sciencemag.org/content/343/6177/1339.full.pdf.

Chen, F., Han, J., Kong, D., Yuan, Y., Xiao, J., Wu, S., Tang, D.-M., Deng, Y., Lv, W., Lu, J., Kang, F., & Yang, Q.-H. (2021). 1000WhL^{-1} lithium-ion batteries enabled by crosslink-shrunk tough carbon encapsulated silicon microparticle anodes. *National Science Review*, *8*(9). Available from https://doi.org/10.1093/nsr/nwab012.

Chen, H., Ginzburg, V. V., Yang, J., Yang, Y., Liu, W., Huang, Y., Du, L., & Chen, B. (2016). Thermal conductivity of polymer-based composites: Fundamentals and applications. *Progress in Polymer Science.*, *59*, 41–85. Available from https://doi.org/10.1016/j.progpolymsci.2016.03.001, http://www.sciencedirect.com/science/journal/00796700.

Chen, J., Loeb, S., & Kim, J. H. (2017). LED revolution: Fundamentals and prospects for UV disinfection applications. *Environmental Science: Water Research and Technology*, *3*(2), 188–202. Available from https://doi.org/10.1039/c6ew00241b, http://pubs.rsc.org/en/journals/journalissues/ew#!recentarticles&adv.

Cheng, Q., Wu, M., Li, M., Jiang, L., & Tang, Z. (2013). Ultratough artificial nacre based on conjugated cross-linked graphene oxide. *Angewandte Chemie—International Edition.*, *52*(13), 3750–3755. Available from https://doi.org/10.1002/anie.201210166.

Cheng, R., Bai, J., Liao, L., Zhou, H., Chen, Y., Liu, L., Lin, Y.-C., Jiang, S., Huang, Y., & Duan, X. (2012). High-frequency self-aligned graphene transistors with transferred gate stacks. *Proceedings of the National Academy of Sciences*, *109*(29), 11588–11592. Available from https://doi.org/10.1073/pnas.1205696109.

Cheng, X. B., Hou, T. Z., Zhang, R., Peng, H. J., Zhao, C. Z., Huang, J. Q., & Zhang, Q. (2016). Dendrite-free lithium deposition induced by uniformly distributed lithium ions for efficient lithium metal batteries. *Advanced Materials.*, *28*(15), 2888–2895. Available from https://doi.org/10.1002/adma.201506124, http://www3.interscience.wiley.com/journal/119030556/issue.

Chhowalla, M., Jena, D., & Zhang, H. (2016). Two-dimensional semiconductors for transistors. *Nature Reviews Materials.*, *1*(11). Available from https://doi.org/10.1038/natrevmats.2016.52, http://www.nature.com/natrevmats/.

Chiba, Y., Islam, A., Watanabe, Y., Komiya, R., Koide, N., & Han, L. (2006). Dye-sensitized solar cells with conversion efficiency of 11.1%. *Japanese Journal of Applied Physics*, *45*(7L). Available from https://doi.org/10.1143/jjap.45.l638.

Chmiola, J., Yushin, G., Gogotsi, Y., Portet, C., Simon, P., & Taberna, P. L. (2006). Anomalous increase in carbon capacitance at pore sizes less than 1 nanometer. *Science (New York, N.Y.)*, *313*(5794), 1760–1763. Available from https://doi.org/10.1126/science.1132195.

Chouly, C., Pouliquen, D., Lucet, I., Jeune, J. J., & Jallet, P. (1996). Development of superparamagnetic nanoparticles for MRI: Effect of particle size, charge and surface nature on biodistribution. *Journal of Microencapsulation*, *13*(3), 245–255. Available from https://doi.org/10.3109/02652049609026013.

Chu, D. K., Akl, E. A., Duda, S., Solo, K., Yaacoub, S., Schünemann, H. J., El-harakeh, A., Bognanni, A., Lotfi, T., Loeb, M., Hajizadeh, A., Bak, A., Izcovich, A., Cuello-Garcia, C. A., Chen, C., Harris, D. J., Borowiack, E., Chamseddine, F., Schünemann, F., … Reinap, M. (2020). Physical distancing, face masks, and eye protection to prevent person-to-person transmission of SARS-CoV-2 and COVID-19: A systematic review and meta-analysis. *The Lancet*, *395*(10242), 1973–1987. Available from https://doi.org/10.1016/S0140-6736(20)31142-9, http://www.journals.elsevier.com/the-lancet/.

Chu, H., Hu, X., Wang, Z., Mu, J., Li, N., Zhou, X., Fang, S., Haines, C. S., Park, J. W., Qin, S., Yuan, N., Xu, J., Tawfick, S., Kim, H., Conlin, P., Cho, M., Cho, K., Oh, J., Nielsen, S., … Baughman, R. H. (2021). Unipolar stroke, electroosmotic pump carbon nanotube yarn muscles. *Science (New York, N.Y.)*, *371*(6528), 494–498. Available from https://doi.org/10.1126/science.abc4538.

Chung, Y. H., Beiss, V., Fiering, S. N., & Steinmetz, N. F. (2020). Covid-19 vaccine frontrunners and their nanotechnology design. *ACS Nano.*, *14*(10), 12522–12537. Available from https://doi.org/10.1021/acsnano.0c07197, http://pubs.acs.org/journal/ancac3.

Cola, B. A., Xu, X., & Fisher, T. S. (2007). Increased real contact in thermal interfaces: A carbon nanotube/foil material. *Applied Physics Letters*, *90*(9). Available from https://doi.org/10.1063/1.2644018.

Colinge, J. P., Lee, C. W., Afzalian, A., Akhavan, N. D., Yan, R., Ferain, I., Razavi, P., O'Neill, B., Blake, A., White, M., Kelleher, A. M., McCarthy, B., & Murphy, R. (2010). Nanowire transistors without junctions. *Nature Nanotechnology.*, *5*(3),

225–229. Available from https://doi.org/10.1038/nnano.2010.15, http://www.nature.com/nnano/index.html.

Cui, Y., Wang, J., Plissard, S. R., Cavalli, A., Vu, T. T. T., Van Veldhoven, R. P. J., Gao, L., Trainor, M., Verheijen, M. A., Haverkort, J. E. M., & Bakkers, E. P. A. M. (2013). Efficiency enhancement of InP nanowire solar cells by surface cleaning. *Nano Letters, 13*(9), 4113–4117. Available from https://doi.org/10.1021/nl4016182.

Cui, Y., Wei, Q., Park, H., & Lieber, C. M. (2001). Nanowire nanosensors for highly sensitive and selective detection of biological and chemical species. *Science (New York, N.Y.), 293*(5533), 1289–1292. Available from https://doi.org/10.1126/science.1062711.

Dai, Z., Yadavalli, S. K., Chen, M., Abbaspourtamijani, A., Qi, Y., & Padture, N. P. (2021). Interfacial toughening with self-assembled monolayers enhances perovskite solar cell reliability. *Science (New York, N.Y.), 372*(6542), 618–622. Available from https://doi.org/10.1126/science.abf5602.

Dang, S., Amin, O., Shihada, B., & Alouini, M. S. (2020). What should 6G be? *Nature Electronics., 3*(1), 20–29. Available from https://doi.org/10.1038/s41928-019-0355-6, https://www.nature.com/natelectron/.

Das, S., Zhang, W., Demarteau, M., Hoffmann, A., Dubey, M., & Roelofs, A. (2014). Tunable transport gap in phosphorene. *Nano Letters, 14*(10), 5733–5739. Available from https://doi.org/10.1021/nl5025535, http://pubs.acs.org/journal/nalefd.

Denys, A., Arrive, L., Servois, V., Dubray, B., Najmark, D., Sibert, A., & Menu, Y. (1994). Hepatic tumors: Detection and characterization at 1-T MR imaging enhanced with AMI-25. *Radiology, 193*(3), 665–669. Available from https://doi.org/10.1148/radiology.193.3.7972805.

Desai, S. B., Madhvapathy, S. R., Sachid, A. B., Llinas, J. P., Wang, Q., Ahn, G. H., Pitner, G., Kim, M. J., Bokor, J., Hu, C., Wong, H. S. P., & Javey, A. (2016). MoS_2 transistors with 1-nanometer gate lengths. *Science (New York, N.Y.), 354*(6308), 99–102. Available from https://doi.org/10.1126/science.aah4698, http://science.sciencemag.org/content/sci/354/6308/99.full.pdf.

Dhomkar, S., Henshaw, J., Jayakumar, H., & Meriles, C. A. (2016). Long-term data storage in diamond. *Science Advances., 2*(10). Available from https://doi.org/10.1126/sciadv.1600911, http://advances.sciencemag.org/content/advances/2/10/e1600911.full.pdf.

Ding, F., Xu, W., Graff, G. L., Zhang, J., Sushko, M. L., Chen, X., Shao, Y., Engelhard, M. H., Nie, Z., Xiao, J., Liu, X., Sushko, P. V., Liu, J., & Zhang, J. G. (2013). Dendrite-free lithium deposition via self-healing electrostatic shield mechanism. *Journal of the American Chemical Society, 135*(11), 4450–4456. Available from https://doi.org/10.1021/ja312241y.

Dinh, C.-T., Burdyny, T., Kibria, M. G., Seifitokaldani, A., Gabardo, C. M., de Arquer, F. P. G., Kiani, A., Edwards, J. P., De Luna, P., Bushuyev, O. S., Zou, C., Quintero-Bermudez, R., Pang, Y., Sinton, D., & Sargent, E. H. (2018). CO_2 electroreduction to ethylene via hydroxide-mediated copper catalysis at an abrupt interface. *Science (New York, N.Y.), 360*(6390), 783–787. Available from https://doi.org/10.1126/science.aas9100.

Du, B., Yu, M., & Zheng, J. (2018). Transport and interactions of nanoparticles in the kidneys. *Nature Reviews Materials., 3*(10), 358–374. Available from https://doi.org/10.1038/s41578-018-0038-3, http://www.nature.com/natrevmats/.

Du, J., Zhao, L., Zeng, Y., Zhang, L., Li, F., Liu, P., & Liu, C. (2011). Comparison of electrical properties between multi-walled carbon nanotube and graphene nanosheet/high density polyethylene composites with a segregated network structure. *Carbon, 49*(4), 1094–1100. Available from https://doi.org/10.1016/j.carbon.2010.11.013.

Duan, X., Gao, R., Xie, P., Cohen-Karni, T., Qing, Q., Choe, H. S., Tian, B., Jiang, X., & Lieber, C. M. (2012). Intracellular recordings of action potentials by an extracellular nanoscale field-effect transistor. *Nature Nanotechnology.*, *7*(3), 174–179. Available from https://doi.org/10.1038/nnano.2011.223, http://www.nature.com/nnano/index.htm, l.

Edwards, B. C. (2000). Design and deployment of a space elevator. *Acta Astronautica*, *47* (10), 735–744. Available from https://doi.org/10.1016/S0094-5765(00)00111-9.

Eigler, D. M., & Schweizer, E. K. (1990). Positioning single atoms with a scanning tunnelling microscope. *Nature*, *344*(6266), 524–526. Available from https://doi.org/10.1038/344524a0.

Ekinci, K. L., Yang, Y. T., & Roukes, M. L. (2004). Ultimate limits to inertial mass sensing based upon nanoelectromechanical systems. *Journal of Applied Physics*, *95*(5), 2682–2689. Available from https://doi.org/10.1063/1.1642738.

Elliott, D. (2016). A balancing act for renewables. *Nature Energy.*, *1*(1). Available from https://doi.org/10.1038/nenergy.2015.3, http://www.nature.com/nenergy/.

Eom, D., Moon, C. Y., & Koo, J. Y. (2015). Switching the charge state of individual surface atoms at Si(111)-$\sqrt{3} \times \sqrt{3}$:B surfaces. *Nano Letters*, *15*(1), 398–402. Available from https://doi.org/10.1021/nl503724x, http://pubs.acs.org/journal/nalefd.

Erisman, J. W., Sutton, M. A., Galloway, J., Klimont, Z., & Winiwarter, W. (2008). How a century of ammonia synthesis changed the world. *Nature Geoscience.*, *1*(10), 636–639. Available from https://doi.org/10.1038/ngeo325.

Ermini, M. L., & Voliani, V. (2021). Antimicrobial nano-agents: The copper age. *ACS Nano.*, *15*(4), 6008–6029. Available from https://doi.org/10.1021/acsnano.0c10756, http://pubs.acs.org/journal/ancac3.

Fang, R., Li, G., Zhao, S., Yin, L., Du, K., Hou, P., Wang, S., Cheng, H. M., Liu, C., & Li, F. (2017). Single-wall carbon nanotube network enabled ultrahigh sulfur-content electrodes for high-performance lithium-sulfur batteries. *Nano Energy*, *42*, 205–214. Available from https://doi.org/10.1016/j.nanoen.2017.10.053, http://www.journals.elsevier.com/nano-energy/.

Fang, R., Zhao, S., Pei, S., Qian, X., Hou, P. X., Cheng, H. M., Liu, C., & Li, F. (2016). Toward more reliable lithium-sulfur batteries: An all-graphene cathode structure. *ACS Nano*, *10*(9), 8676–8682. Available from https://doi.org/10.1021/acsnano.6b04019, http://pubs.acs.org/journal/ancac3.

Feng, S., Shen, C., Xia, N., Song, W., Fan, M., & Cowling, B. J. (2020). Rational use of face masks in the COVID-19 pandemic. *The Lancet Respiratory Medicine.*, *8*(5), 434–436. Available from https://doi.org/10.1016/S2213-2600(20)30134-X, http://www.elsevier.com/journals/the-lancet-respiratory-medicine/2213-2600.

Florindo, H. F., Kleiner, R., Vaskovich-Koubi, D., Acúrcio, R. C., Carreira, B., Yeini, E., Tiram, G., Liubomirski, Y., & Satchi-Fainaro, R. (2020). Immune-mediated approaches against COVID-19. *Nature Nanotechnology.*, *15*(8), 630–645. Available from https://doi.org/10.1038/s41565-020-0732-3, http://www.nature.com/nnano/index.html.

Foroughi, J., Spinks, G. M., Wallace, G. G., Oh, J., Kozlov, M. E., Fang, S., Mirfakhrai, T., Madden, J. D. W., Shin, M. K., Kim, S. J., & Baughman, R. H. (2011). Torsional carbon nanotube artificial muscles. *Science (New York, N.Y.)*, *334*(6055), 494–497. Available from https://doi.org/10.1126/science.1211220, http://www.sciencemag.org/content/334/6055/494.full.pdf.

Foygel, M., Morris, R. D., Anez, D., French, S., & Sobolev, V. L. (2005). Theoretical and computational studies of carbon nanotube composites and suspensions: Electrical and thermal conductivity. *Physical Review B.*, *71*(10). Available from https://doi.org/10.1103/physrevb.71.104201.

Fritz, J., Baller, M. K., Lang, H. P., Rothuizen, H., Vettiger, P., Meyer, E., Güntherodt, H.-J., Gerber, C., & Gimzewski, J. K. (2000). Translating biomolecular recognition into nanomechanics. *Science (New York, N.Y.)*, *288*(5464), 316–318. Available from https://doi.org/10.1126/science.288.5464.316.

Fujishima, A., & Honda, K. (1972). Electrochemical photolysis of water at a semiconductor electrode. *Nature, 238*(5358), 37–38. Available from https://doi.org/10.1038/238037a0.

Fujishima, A., Zhang, X., & Tryk, D. A. (2008). TiO_2 photocatalysis and related surface phenomena. *Surface Science Reports., 63*(12), 515–582. Available from https://doi.org/10.1016/j.surfrep.2008.10.001.

Gao, A., Lu, N., Dai, P., Li, T., Pei, H., Gao, X., Gong, Y., Wang, Y., & Fan, C. (2011). Silicon-nanowire-based CMOS-compatible field-effect transistor nanosensors for ultrasensitive electrical detection of nucleic acids. *Nano Letters, 11*(9), 3974–3978. Available from https://doi.org/10.1021/nl202303y.

Gao, D., Zhou, H., Wang, J., Miao, S., Yang, F., Wang, G., Wang, J., & Bao, X. (2015). Size-dependent electrocatalytic reduction of CO_2 over Pd nanoparticles. *Journal of the American Chemical Society, 137*(13), 4288–4291. Available from https://doi.org/10.1021/jacs.5b00046, http://pubs.acs.org/journal/jacsat.

Gao, X., Cui, Y., Levenson, R. M., Chung, L. W. K., & Nie, S. (2004). In vivo cancer targeting and imaging with semiconductor quantum dots. *Nature Biotechnology, 22*(8), 969–976. Available from https://doi.org/10.1038/nbt994.

García de Arquer, F. P., Dinh, C. T., Ozden, A., Wicks, J., McCallum, C., Kirmani, A. R., Nam, D. H., Gabardo, C., Seifitokaldani, A., Wang, X., Li, Y. C., Li, F., Edwards, J., Richter, L. J., Thorpe, S. J., Sinton, D., & Sargent, E. H. (2020). CO_2 electrolysis to multicarbon products at activities greater than $1 Acm^{-2}$. *Science (New York, N.Y.), 367*(6478), 661–666. Available from https://doi.org/10.1126/science.aay4217, https://science.sciencemag.org/content/367/6478/661/tab-pdf.

Garnett, E., & Yang, P. (2010). Light trapping in silicon nanowire solar cells. *Nano Letters, 10*(3), 1082–1087. Available from https://doi.org/10.1021/nl100161z.

Geim, A. K., & Novoselov, K. S. (2007). The rise of graphene. *Nature Materials, 6*(3), 183–191. Available from https://doi.org/10.1038/nmat1849.

Giddey, S., Badwal, S. P. S., & Kulkarni, A. (2013). Review of electrochemical ammonia production technologies and materials. *International Journal of Hydrogen Energy, 38*(34), 14576–14594. Available from https://doi.org/10.1016/j.ijhydene.2013.09.054.

Gil-Santos, E., Baker, C., Nguyen, D. T., Hease, W., Gomez, C., Lemaître, A., Ducci, S., Leo, G., & Favero, I. (2015). High-frequency nano-optomechanical disk resonators in liquids. *Nature Nanotechnology., 10*(9), 810–816. Available from https://doi.org/10.1038/nnano.2015.160.

Grätzel, M. (2001). Photoelectrochemical cells. *Nature, 414*(6861), 338–344. Available from https://doi.org/10.1038/35104607.

Green, M. A., Hishikawa, Y., Dunlop, E. D., Levi, D. H., Hohl-Ebinger, J., Yoshita, M., & Ho-Baillie, A. W. Y. (2019). Solar cell efficiency tables (Version 53). *Progress in Photovoltaics: Research and Applications., 27*(1), 3–12. Available from https://doi.org/10.1002/pip.3102, http://onlinelibrary.wiley.com/journal/10.1002/(ISSN)1099-159X.

Greenlee, L. F., Renner, J. N., & Foster, S. L. (2018). The use of controls for consistent and accurate measurements of electrocatalytic ammonia synthesis from dinitrogen. *ACS Catalysis., 8*(9), 7820–7827. Available from https://doi.org/10.1021/acscatal.8b02120, http://pubs.acs.org/page/accacs/about.html.

Halmann, M. (1978). Photoelectrochemical reduction of aqueous carbon dioxide on p-type gallium phosphide in liquid junction solar cells. *Nature, 275*(5676), 115–116. Available from https://doi.org/10.1038/275115a0.

Han, J., Tang, D. M., Kong, D., Chen, F., Xiao, J., Zhao, Z., Pan, S., Wu, S., & Yang, Q. H. (2020). A thick yet dense silicon anode with enhanced interface stability in lithium storage evidenced by in situ TEM observations. *Science Bulletin., 65*(18), 1563–1569. Available from https://doi.org/10.1016/j.scib.2020.05.018, http://link.springer.com/journal/11434.

Han, S. J., Tang, J., Kumar, B., Falk, A., Farmer, D., Tulevski, G., Jenkins, K., Afzali, A., Oida, S., Ott, J., Hannon, J., & Haensch, W. (2017). High-speed logic integrated circuits with solution-processed self-assembled carbon nanotubes. *Nature Nanotechnology.*, *12*(9), 861–865. Available from https://doi.org/10.1038/nnano.2017.115, http://www.nature.com/nnano/index.html.

Hao, M., Bai, Y., Zeiske, S., Ren, L., Liu, J., Yuan, Y., Zarrabi, N., Cheng, N., Ghasemi, M., Chen, P., Lyu, M., He, D., Yun, J. H., Du, Y., Wang, Y., Ding, S., Armin, A., Meredith, P., Liu, G., ... Wang, L. (2020). Ligand-assisted cation-exchange engineering for high-efficiency colloidal Cs1 − *x*FA*x*PbI3 quantum dot solar cells with reduced phase segregation. *Nature Energy.*, *5*(1), 79–88. Available from https://doi.org/10.1038/s41560-019-0535-7, http://www.nature.com/nenergy/.

Hasan, S., Salahuddin, S., Vaidyanathan, M., & Alam, M. A. (2006). High-frequency performance projections for ballistic carbon-nanotube transistors. *IEEE Transactions On Nanotechnology.*, *5*(1), 14–22. Available from https://doi.org/10.1109/TNANO.2005.858594.

Hashimoto, K., Irie, H., & Fujishima, A. (2005). TiO_2 photocatalysis: A historical overview and future prospects. *Japanese Journal of Applied Physics, Part 1: Regular Papers and Short Notes and Review Papers*, *44*(12), 8269–8285. Available from https://doi.org/10.1143/JJAP.44.8269Japan, http://jjap.ipap.jp/link?JJAP/44/8269/pdf.

He, J., & Tritt, T. M. (2017). Advances in thermoelectric materials research: Looking back and moving forward. *Science (New York, N.Y.)*, *357*(6358). Available from https://doi.org/10.1126/science.aak9997, http://science.sciencemag.org/content/sci/357/6358/eaak9997.full.pdf.

Heller, D. A., Jena, P. V., Pasquali, M., Kostarelos, K., Delogu, L. G., Meidl, R. E., Rotkin, S. V., Scheinberg, D. A., Schwartz, R. E., Terrones, M., Wang, Y. H., Bianco, A., Boghossian, A. A., Cambré, S., Cognet, L., Corrie, S. R., Demokritou, P., Giordani, S., Hertel, T., ... Yudasaka, M. (2020). Banning carbon nanotubes would be scientifically unjustified and damaging to innovation. *Nature Nanotechnology.*, *15*(3), 164–166. Available from https://doi.org/10.1038/s41565-020-0656-y, http://www.nature.com/nnano/index.html.

Herculano-Houzel, S. (2009). The human brain in numbers: A linearly scaled-up primate brain. *Frontiers in Human Neuroscience*, *3*. Available from https://doi.org/10.3389/neuro.09.031.2009, http://www.frontiersin.org/Journal/DownloadFile.ashx?pdf = 1&FileId = 25104&articleId = 857&Version = 1&ContentTypeId = 15&FileName = 031_2009.pdf.

Hicks, L. D., & Dresselhaus, M. S. (1993a). Effect of quantum-well structures on the thermoelectric figure of merit. *Physical Review B*, *47*(19), 12727–12731. Available from https://doi.org/10.1103/physrevb.47.12727.

Hicks, L. D., & Dresselhaus, M. S. (1993b). Thermoelectric figure of merit of a one-dimensional conductor. *Physical Review B.*, *47*(24), 16631–16634. Available from https://doi.org/10.1103/physrevb.47.16631.

Hills, G., Lau, C., Wright, A., Fuller, S., Bishop, M. D., Srimani, T., Kanhaiya, P., Ho, R., Amer, A., Stein, Y., Murphy, D., Arvind., Chandrakasan, A., & Shulaker, M. M. (2019). Modern microprocessor built from complementary carbon nanotube transistors. *Nature*, *572*(7771), 595–602. Available from https://doi.org/10.1038/s41586-019-1493-8, http://www.nature.com/nature/index.html.

Hochbaum, A. I., Chen, R., Delgado, R. D., Liang, W., Garnett, E. C., Najarian, M., Majumdar, A., & Yang, P. (2008). Enhanced thermoelectric performance of rough silicon nanowires. *Nature*, *451*(7175), 163–167. Available from https://doi.org/10.1038/nature06381.

Hori, Y., Kikuchi, K., & Suzuki, S. (1985). Production of CO and CH_4 in electrochemical reduction of CO_2 at metal electrodes in aqueous hydrogencarbonate solution.

Chemistry Letters., *14*(11), 1695–1698. Available from https://doi.org/10.1246/cl.1985.1695.

Annual Energy Outlook 2020 with projections to 2050. (2020). U.S. Energy Information Administration Office of Energy Analysis U.S. Department of Energy Washington, DC 20585.

Horowitz, M. (2014). 1.1 Computing's energy problem (and what we can do about it). *IEEE international solid-state circuits conference digest of technical papers*. IEEE. 57, 10–14, 10.1109/ISSCC.2014.6757323.

Huang, H., Liu, C. H., Wu, Y., & Fan, S. (2005). Aligned carbon nanotube composite films for thermal management. *Advanced Materials.*, *17*(13), 1652–1656. Available from https://doi.org/10.1002/adma.200500467.

Huang, J., Yuan, Y., Shao, Y., & Yan, Y. (2017). Understanding the physical properties of hybrid perovskites for photovoltaic applications. *Nature Reviews Materials.*, *2*. Available from https://doi.org/10.1038/natrevmats.2017.42, http://www.nature.com/natrevmats/.

Huang, X., El-Sayed, I. H., Qian, W., & El-Sayed, M. A. (2006). Cancer cell imaging and photothermal therapy in the near-infrared region by using gold nanorods. *Journal of the American Chemical Society*, *128*(6), 2115–2120. Available from https://doi.org/10.1021/ja057254a.

Huang, X., Zhao, Z., Cao, L., Chen, Y., Zhu, E., Lin, Z., Li, M., Yan, A., Zettl, A., Wang, Y. M., Duan, X., Mueller, T., & Huang, Y. (2015). High-performance transition metal-doped Pt_3Ni octahedra for oxygen reduction reaction. *Science (New York, N.Y.)*, *348*(6240), 1230–1234. Available from https://doi.org/10.1126/science.aaa8765, http://www.sciencemag.org/content/348/6240/1230.full.pdf.

Hung, N. T., Hasdeo, E. H., Nugraha, A. R. T., Dresselhaus, M. S., & Saito, R. (2016). Quantum effects in the thermoelectric power factor of low-dimensional semiconductors. *Physical Review Letters*, *117*(3). Available from https://doi.org/10.1103/PhysRevLett.117.036602, http://harvest.aps.org/bagit/articles/10.1103/PhysRevLett.117.036602/apsxml.

Inoue, T., Fujishima, A., Konishi, S., & Honda, K. (1979). Photoelectrocatalytic reduction of carbon dioxide in aqueous suspensions of semiconductor powders. *Nature*, *277* (5698), 637–638. Available from https://doi.org/10.1038/277637a0.

Javey, A., Guo, J., Wang, Q., Lundstrom, M., & Dai, H. (2003). Ballistic carbon nanotube field-effect transistors. *Nature*, *424*(6949), 654–657. Available from https://doi.org/10.1038/nature01797.

Jeon, H. S., Koh, J. H., Park, S. J., Jee, M. S., Ko, D. H., Hwang, Y. J., & Min, B. K. (2015). A monolithic and standalone solar-fuel device having comparable efficiency to photosynthesis in nature. *Journal of Materials Chemistry A.*, *3*(11), 5835–5842. Available from https://doi.org/10.1039/c4ta06495j, http://pubs.rsc.org/en/journals/journalissues/ta.

Jia, J., Seitz, L. C., Benck, J. D., Huo, Y., Chen, Y., Ng, J. W. D., Bilir, T., Harris, J. S., & Jaramillo, T. F. (2016). Solar water splitting by photovoltaic-electrolysis with a solar-to-hydrogen efficiency over 30%. *Nature Communications.*, 7. Available from https://doi.org/10.1038/ncomms13237, http://www.nature.com/ncomms/index.html.

Jin, Q., Jiang, S., Zhao, Y., Wang, D., Qiu, J., Tang, D. M., Tan, J., Sun, D. M., Hou, P. X., Chen, X. Q., Tai, K., Gao, N., Liu, C., Cheng, H. M., & Jiang, X. (2019). Flexible layer-structured Bi_2Te_3 thermoelectric on a carbon nanotube scaffold. *Nature Materials*, *18*(1), 62–68. Available from https://doi.org/10.1038/s41563-018-0217-z, http://www.nature.com/nmat/.

Jo, A., Joo, W., Jin, W. H., Nam, H., & Kim, J. K. (2009). Ultrahigh-density phase-change data storage without the use of heating. *Nature Nanotechnology.*, *4*(11), 727–731. Available from https://doi.org/10.1038/nnano.2009.260, http://www.nature.com/nnano/index.html.

Joo, Y., Brady, G. J., Arnold, M. S., & Gopalan, P. (2014). Dose-controlled, floating evaporative self-assembly and alignment of semiconducting carbon nanotubes from organic solvents. *Langmuir: The ACS Journal of Surfaces and Colloids*, *30*(12), 3460–3466. Available from https://doi.org/10.1021/la500162x, http://pubs.acs.org/journal/langd5.

Jouny, M., Luc, W., & Jiao, F. (2018). General techno-economic analysis of CO_2 electrolysis systems. *Industrial and Engineering Chemistry Research.*, *57*(6), 2165–2177. Available from https://doi.org/10.1021/acs.iecr.7b03514, http://pubs.acs.org/journal/iecred.

Kalff, F. E., Rebergen, M. P., Fahrenfort, E., Girovsky, J., Toskovic, R., Lado, J. L., Fernández-Rossier, J., & Otte, A. F. (2016). A kilobyte rewritable atomic memory. *Nature Nanotechnology.*, *11*(11), 926–929. Available from https://doi.org/10.1038/nnano.2016.131.

Kang, S. K., Murphy, R. K. J., Hwang, S. W., Lee, S. M., Harburg, D. V., Krueger, N. A., Shin, J., Gamble, P., Cheng, H., Yu, S., Liu, Z., McCall, J. G., Stephen, M., Ying, H., Kim, J., Park, G., Webb, R. C., Lee, C. H., Chung, S., ... Rogers, J. A. (2016). Bioresorbable silicon electronic sensors for the brain. *Nature*, *530*(7588), 71–76. Available from https://doi.org/10.1038/nature16492, http://www.nature.com/nature/index.html.

Kayes, B. M., Atwater, H. A., & Lewis, N. S. (2005). Comparison of the device physics principles of planar and radial p–n junction nanorod solar cells. *Journal of Applied Physics*, *97*(11). Available from https://doi.org/10.1063/1.1901835.

Kelzenberg, M. D., Boettcher, S. W., Petykiewicz, J. A., Turner-Evans, D. B., Putnam, M. C., Warren, E. L., Spurgeon, J. M., Briggs, R. M., Lewis, N. S., & Atwater, H. A. (2010). Enhanced absorption and carrier collection in Si wire arrays for photovoltaic applications. *Nature Materials*, *9*(3), 239–244. Available from https://doi.org/10.1038/nmat2635, http://www.nature.com/nmat/.

Khajetoorians, A. A., Baxevanis, B., Hübner, C., Schlenk, T., Krause, S., Wehling, T. O., Lounis, S., Lichtenstein, A., Pfannkuche, D., Wiebe, J., & Wiesendanger, R. (2013). Current-driven spin dynamics of artificially constructed quantum magnets. *Science (New York, N.Y.)*, *339*(6115), 55–59. Available from https://doi.org/10.1126/science.1228519, http://www.sciencemag.org/content/339/6115/55.full.pdf.

Khan, H. N., Hounshell, D. A., & Fuchs, E. R. H. (2018). Science and research policy at the end of Moore's law. *Nature Electronics.*, *1*(1), 14–21. Available from https://doi.org/10.1038/s41928-017-0005-9, https://www.nature.com/natelectron/.

Kim, H. S., Lee, C. R., Im, J. H., Lee, K. B., Moehl, T., Marchioro, A., Moon, S. J., Humphry-Baker, R., Yum, J. H., Moser, J. E., Grätzel, M., & Park, N. G. (2012). Lead iodide perovskite sensitized all-solid-state submicron thin film mesoscopic solar cell with efficiency exceeding 9%. *Scientific Reports.*, *2*. Available from https://doi.org/10.1038/srep00591.

Kim, J., Campbell, A. S., de Ávila, B. E. F., & Wang, J. (2019). Wearable biosensors for healthcare monitoring. *Nature Biotechnology*, *37*(4), 389–406. Available from https://doi.org/10.1038/s41587-019-0045-y, http://www.nature.com/nbt/index.html.

Knopfmacher, O., Tarasov, A., Fu, W., Wipf, M., Niesen, B., Calame, M., & Schönenberger, C. (2010). Nernst limit in dual-gated Si-nanowire FET sensors. *Nano Letters*, *10*(6), 2268–2274. Available from https://doi.org/10.1021/nl100892y.

Kodama, K., Nagai, T., Kuwaki, A., Jinnouchi, R., & Morimoto, Y. (2021). Challenges in applying highly active Pt-based nanostructured catalysts for oxygen reduction reactions to fuel cell vehicles. *Nature Nanotechnology.*, *16*(2), 140–147. Available from https://doi.org/10.1038/s41565-020-00824-w, http://www.nature.com/nnano/index.html.

Koerner, H., Liu, W., Alexander, M., Mirau, P., Dowty, H., & Vaia, R. A. (2005). Deformation-morphology correlations in electrically conductive carbon nanotube—

Thermoplastic polyurethane nanocomposites. *Polymer*, *46*(12), 4405–4420. Available from https://doi.org/10.1016/j.polymer.2005.02.025, http://www.journals.elsevier.com/polymer/.

Kojima, A., Teshima, K., Shirai, Y., & Miyasaka, T. (2007). Novel photoelectrochemical cell with mesoscopic electrodes sensitized by lead-halide compounds (5). *ECS Meeting Abstracts*, *MA2007-02*(8), 352. Available from https://doi.org/10.1149/ma2007-02/8/352.

Koratkar, N. A., Suhr, J., Joshi, A., Kane, R. S., Schadler, L. S., Ajayan, P. M., & Bartolucci, S. (2005). Characterizing energy dissipation in single-walled carbon nanotube polycarbonate composites. *Applied Physics Letters*, *87*(6). Available from https://doi.org/10.1063/1.2007867.

Kuhl, K. P., Cave, E. R., Abram, D. N., & Jaramillo, T. F. (2012). New insights into the electrochemical reduction of carbon dioxide on metallic copper surfaces. *Energy and Environmental Science*, *5*(5), 7050–7059. Available from https://doi.org/10.1039/c2ee21234j.

Lachner, R. (2014). Towards 0.7 terahertz silicon germanium heterojunction bipolar technology—The DOTSEVEN project. *ECS Transactions*, *64*(6), 21–37. Available from https://doi.org/10.1149/06406.0021ecst, http://ecst.ecsdl.org/.

Lam, K.-T., Dong, Z., & Guo, J. (2014). Performance limits projection of black phosphorous field-effect transistors. *IEEE Electron Device Letters*, *35*(9), 963–965. Available from https://doi.org/10.1109/led.2014.2333368.

Lan, X., Voznyy, O., Kiani, A., García De Arquer, F. P., Abbas, A. S., Kim, G. H., Liu, M., Yang, Z., Walters, G., Xu, J., Yuan, M., Ning, Z., Fan, F., Kanjanaboos, P., Kramer, I., Zhitomirsky, D., Lee, P., Perelgut, A., Hoogland, S., & Sargent, E. H. (2016). Passivation using molecular halides increases quantum dot solar cell performance. *Advanced Materials.*, *28*(2), 299–304. Available from https://doi.org/10.1002/adma.201503657, http://www3.interscience.wiley.com/journal/119030556/issue.

Landman, A., Dotan, H., Shter, G. E., Wullenkord, M., Houaijia, A., Maljusch, A., Grader, G. S., & Rothschild, A. (2017). Photoelectrochemical water splitting in separate oxygen and hydrogen cells. *Nature Materials*, *16*(6), 646–651. Available from https://doi.org/10.1038/nmat4876, http://www.nature.com/nmat/.

Larcher, D., & Tarascon, J.-M. (2015). Towards greener and more sustainable batteries for electrical energy storage. *Nature Chemistry.*, 7(1), 19–29. Available from https://doi.org/10.1038/nchem.2085.

Largeot, C., Portet, C., Chmiola, J., Taberna, P. L., Gogotsi, Y., & Simon, P. (2008). Relation between the ion size and pore size for an electric double-layer capacitor. *Journal of the American Chemical Society*, *130*(9), 2730–2731. Available from https://doi.org/10.1021/ja7106178.

Lassagne, B., Garcia-Sanchez, D., Aguasca, A., & Bachtold, A. (2008). Ultrasensitive mass sensing with a nanotube electromechanical resonator. *Nano Letters*, *8*(11), 3735–3738. Available from https://doi.org/10.1021/nl801982v.

Lee, E. K., Yin, L., Lee, Y., Lee, J. W., Lee, S. J., Lee, J., Cha, S. N., Whang, D., Hwang, G. S., Hippalgaonkar, K., Majumdar, A., Yu, C., Choi, B. L., Kim, J. M., & Kim, K. (2012). Large thermoelectric figure-of-merits from SiGe nanowires by simultaneously measuring electrical and thermal transport properties. *Nano Letters*, *12*(6), 2918–2923. Available from https://doi.org/10.1021/nl300587u.

Lee, G. H., Yu, Y. J., Cui, X., Petrone, N., Lee, C. H., Choi, M. S., Lee, D. Y., Lee, C., Yoo, W. J., Watanabe, K., Taniguchi, T., Nuckolls, C., Kim, P., & Hone, J. (2013). Flexible and transparent MoS_2 field-effect transistors on hexagonal boron nitride-graphene heterostructures. *ACS Nano*, 7(9), 7931–7936. Available from https://doi.org/10.1021/nn402954e.

Lee, J. A., Li, N., Haines, C. S., Kim, K. J., Lepró, X., Ovalle-Robles, R., Kim, S. J., & Baughman, R. H. (2017). Electrochemically powered, energy-conserving carbon nanotube artificial muscles. *Advanced Materials.*, *29*(31). Available from https://doi.

org/10.1002/adma.201700870, http://www3.interscience.wiley.com/journal/119030556/issue.

Lee, M. M., Teuscher, J., Miyasaka, T., Murakami, T. N., & Snaith, H. J. (2012). Efficient hybrid solar cells based on meso-superstructured organometal halide perovskites. *Science (New York, N.Y.)*, *338*(6107), 643–647. Available from https://doi.org/10.1126/science.1228604, http://www.sciencemag.org/content/338/6107/643.full.pdf.

Lee, S., Jagannathan, B., Narasimha, S., Chou, A., Zamdmer, N., Johnson, J., Williams, R., Wagner, L., Kim, J., Plouchart, J.O., Pekarik, J., Springer, S., & Freeman G. (2007). Record RF performance of 45-nm SOI CMOS technology. *Technical Digest—International Electron Devices Meeting, IEDM*. IEEE. 255–258, Available from https://doi.org/10.1109/IEDM.2007.4418916.

Leiserson, C. E., Thompson, N. C., Emer, J. S., Kuszmaul, B. C., Lampson, B. W., Sanchez, D., & Schardl, T. B. (2020). There's plenty of room at the top: What will drive computer performance after Moore's law? *Science (New York, N.Y.)*, *368*(6495). Available from https://doi.org/10.1126/science.aam9744.

Lemire, J. A., Harrison, J. J., & Turner, R. J. (2013). Antimicrobial activity of metals: Mechanisms, molecular targets and applications. *Nature Reviews. Microbiology*, *11*(6), 371–384. Available from https://doi.org/10.1038/nrmicro3028.

de Leon, N. P., Itoh, K. M., Kim, D., Mehta, K. K., Northup, T. E., Paik, H., Palmer, B. S., Samarth, N., Sangtawesin, S., & Steuerman, D. W. (2021). Materials challenges and opportunities for quantum computing hardware. *Science (New York, N.Y.)*, *372* (6539). Available from https://doi.org/10.1126/science.abb2823, https://science.sciencemag.org/content/sci/372/6539/eabb2823.full.pdf.

Li, B. Q., Kong, L., Zhao, C. X., Jin, Q., Chen, X., Peng, H. J., Qin, J. L., Chen, J. X., Yuan, H., Zhang, Q., & Huang, J. Q. (2019). Expediting redox kinetics of sulfur species by atomic-scale electrocatalysts in lithium–sulfur batteries. *China InfoMat.*, *1*(4), 533–541. Available from https://doi.org/10.1002/inf2.12056, onlinelibrary.wiley.com/journal/25673165.

Li, H., Tao, Y., Zheng, X., Luo, J., Kang, F., Cheng, H. M., & Yang, Q. H. (2016). Ultra-thick graphene bulk supercapacitor electrodes for compact energy storage. *Energy and Environmental Science*, *9*(10), 3135–3142. Available from https://doi.org/10.1039/c6ee00941g, http://www.rsc.org/Publishing/Journals/EE/About.asp.

Li, L., Yu, Y., Ye, G. J., Ge, Q., Ou, X., Wu, H., Feng, D., Chen, X. H., & Zhang, Y. (2014). Black phosphorus field-effect transistors. *Nature Nanotechnology.*, *9*(5), 372–377. Available from https://doi.org/10.1038/nnano.2014.35, http://www.nature.com/nnano/index.html.

Li, P., Yang, M., Liu, Y., Qin, H., Liu, J., Xu, Z., Liu, Y., Meng, F., Lin, J., Wang, F., & Gao, C. (2020). Continuous crystalline graphene papers with gigapascal strength by intercalation modulated plasticization. *Nature Communications.*, *11*(1). Available from https://doi.org/10.1038/s41467-020-16494-0, http://www.nature.com/ncomms/index.html.

Li, Q., Guo, Y., Li, W., Qiu, S., Zhu, C., Wei, X., Chen, M., Liu, C., Liao, S., Gong, Y., Mishra, A. K., & Liu, L. (2014). Ultrahigh thermal conductivity of assembled aligned multilayer graphene/epoxy composite. *Chemistry of Materials.*, *26*(15), 4459–4465. Available from https://doi.org/10.1021/cm501473t, http://pubs.acs.org/journal/cmatex.

Li, Q., Zhang, Z., Qi, L., Liao, Q., Kang, Z., & Zhang, Y. (2019). Toward the application of high frequency electromagnetic wave absorption by carbon nanostructures. *Advanced Science.*, *6*(8). Available from https://doi.org/10.1002/advs.201801057, http://onlinelibrary.wiley.com/journal/10.1002/(ISSN)2198-3844.

Li, W., Brubaker, M. D., Spann, B. T., Bertness, K. A., & Fay, P. (2018). GaN nanowire MOSFET with near-ideal subthreshold slope. *IEEE Electron Device Letters.*, *39*(2), 184–187. Available from https://doi.org/10.1109/LED.2017.2785785.

Li, W., Zhou, J., Cai, S., Yu, Z., Zhang, J., Fang, N., Li, T., Wu, Y., Chen, T., Xie, X., Ma, H., Yan, K., Dai, N., Wu, X., Zhao, H., Wang, Z., He, D., Pan, L., Shi, Y., ... Wang, X. (2019). Uniform and ultrathin high-κ gate dielectrics for two-dimensional electronic devices. *Nature Electronics.*, *2*(12), 563–571. Available from https://doi.org/10.1038/s41928-019-0334-y, https://www.nature.com/natelectron/.

Li, Y., Tang, J., Pan, D. X., Sun, L. D., Chen, C., Liu, Y., Wang, Y. F., Shi, S., & Yan, C. H. (2016). A versatile imaging and therapeutic platform based on dual-band luminescent lanthanide nanoparticles toward tumor metastasis inhibition. *ACS Nano.*, *10*(2), 2766–2773. Available from https://doi.org/10.1021/acsnano.5b07873, http://pubs.acs.org/journal/ancac3.

Li, Z., Gadipelli, S., Li, H., Howard, C. A., Brett, D. J. L., Shearing, P. R., Guo, Z., Parkin, I. P., & Li, F. (2020). Tuning the interlayer spacing of graphene laminate films for efficient pore utilization towards compact capacitive energy storage. *Nature Energy.*, *5*(2), 160–168. Available from https://doi.org/10.1038/s41560-020-0560-6, http://www.nature.com/nenergy/.

Liang, Y., Xiao, M., Wu, D., Lin, Y., Liu, L., He, J., Zhang, G., Peng, L. M., & Zhang, Z. (2020). Wafer-Scale uniform carbon nanotube transistors for ultrasensitive and label-free detection of disease biomarkers. *ACS Nano.*, *14*(7), 8866–8874. Available from https://doi.org/10.1021/acsnano.0c03523, http://pubs.acs.org/journal/ancac3.

Liao, L., Lin, Y. C., Bao, M., Cheng, R., Bai, J., Liu, Y., Qu, Y., Wang, K. L., Huang, Y., & Duan, X. (2010). High-speed graphene transistors with a self-aligned nanowire gate. *Nature*, *467*(7313), 305–308. Available from https://doi.org/10.1038/nature09405.

Lin, Y. M., Dimitrakopoulos, C., Jenkins, K. A., Farmer, D. B., Chiu, H. Y., Grill, A., & Avouris, P. (2010). 100-GHz transistors from wafer-scale epitaxial graphene. *Science (New York, N.Y.)*, *327*(5966). Available from https://doi.org/10.1126/science.1184289.

Lin, Y. M., Jenkins, K. A., Alberto, V. G., Small, J. P., Farmer, D. B., & Avouris, P. (2009). Operation of graphene transistors at giqahertz frequencies. *Nano Letters*, *9*(1), 422–426. Available from https://doi.org/10.1021/nl803316h.

Liu, G., Wang, L., Sun, C., Yan, X., Wang, X., Chen, Z., Smith, S. C., Cheng, H. M., & Lu, G. Q. (2009). Band-to-band visible-light photon excitation and photoactivity induced by homogeneous nitrogen doping in layered titanates. *Chemistry of Materials.*, *21*(7), 1266–1274. Available from https://doi.org/10.1021/cm802986r, http://pubs.acs.org/doi/pdfplus/10.1021/cm802986r.

Liu, J., Liu, Y., Liu, N., Han, Y., Zhang, X., Huang, H., Lifshitz, Y., Lee, S. T., Zhong, J., & Kang, Z. (2015). Metal-free efficient photocatalyst for stable visible water splitting via a two-electron pathway. *Science (New York, N.Y.)*, *347*(6225), 970–974. Available from https://doi.org/10.1126/science.aaa3145, http://www.sciencemag.org/content/347/6225/970.full.pdf.

Liu, L., Han, J., Xu, L., Zhou, J., Zhao, C., Ding, S., Shi, H., Xiao, M., Ding, L., Ma, Z., Jin, C., Zhang, Z., & Peng, L. M. (2020). Aligned, high-density semiconducting carbon nanotube arrays for high-performance electronics. *Science (New York, N.Y.)*, *368*(6493), 850–856. Available from https://doi.org/10.1126/science.aba5980, https://science.sciencemag.org/content/sci/368/6493/850.full.pdf.

Liu, M., Johnston, M. B., & Snaith, H. J. (2013). Efficient planar heterojunction perovskite solar cells by vapour deposition. *Nature*, *501*(7467), 395–398. Available from https://doi.org/10.1038/nature12509.

Liu, X. H., Zhong, L., Huang, S., Mao, S. X., Zhu, T., & Huang, J. Y. (2012). Size-dependent fracture of silicon nanoparticles during lithiation. *ACS Nano*, *6*(2), 1522–1531. Available from https://doi.org/10.1021/nn204476h.

Liu, Y., Duan, X., Shin, H. J., Park, S., Huang, Y., & Duan, X. (2021). Promises and prospects of two-dimensional transistors. *Nature*, *591*(7848), 43–53. Available from https://doi.org/10.1038/s41586-021-03339-z, http://www.nature.com/nature/index.html.

Liu, Y., Guo, J., Wu, Y., Zhu, E., Weiss, N. O., He, Q., Wu, H., Cheng, H. C., Xu, Y., Shakir, I., Huang, Y., & Duan, X. (2016). Pushing the performance limit of sub-100 nm molybdenum disulfide transistors. *Nano Letters*, *16*(10), 6337–6342. Available from https://doi.org/10.1021/acs.nanolett.6b02713, http://pubs.acs.org/journal/nalefd.

Loth, S., Baumann, S., Lutz, C. P., Eigler, D. M., & Heinrich, A. J. (2012). Bistability in atomic-scale antiferromagnets. *Science (New York, N.Y.)*, *335*(6065), 196–199. Available from https://doi.org/10.1126/science.1214131, http://www.sciencemag.org/content/335/6065/196.full.pdf.

Lu, Y., Tu, Z., & Archer, L. A. (2014). Stable lithium electrodeposition in liquid and nanoporous solid electrolytes. *Nature Materials*, *13*(10), 961–969. Available from https://doi.org/10.1038/nmat4041, http://www.nature.com/nmat/.

Luo, G., & Gao, S. J. (2020). Global health concerns stirred by emerging viral infections. *Journal of Medical Virology*, *92*(4), 399–400. Available from https://doi.org/10.1002/jmv.25683, http://onlinelibrary.wiley.com/journal/10.1002/(ISSN)1096-9071.

Luo, J., Im, J. H., Mayer, M. T., Schreier, M., Nazeeruddin, M. K., Park, N. G., Tilley, S. D., Fan, H. J., & Grätzel, M. (2014). Water photolysis at 12.3% efficiency via perovskite photovoltaics and Earth-abundant catalysts. *Science (New York, N.Y.)*, *345*(6204), 1593–1596. Available from https://doi.org/10.1126/science.1258307, http://www.sciencemag.org/content/345/6204/1593.full.pdf.

Luo, X., & Chung, D. D. L. (1996). Electromagnetic interference shielding reaching 130 dB using flexible graphite. *Carbon*, *34*(10), 1293–1294. Available from https://doi.org/10.1016/0008-6223(96)82798-9.

MacFarlane, D. R., Cherepanov, P. V., Choi, J., Suryanto, B. H. R., Hodgetts, R. Y., Bakker, J. M., Ferrero Vallana, F. M., & Simonov, A. N. (2020). A roadmap to the ammonia economy. *Joule.*, *4*(6), 1186–1205. Available from https://doi.org/10.1016/j.joule.2020.04.004, https://www.journals.elsevier.com/joule.

Maeda, K., Teramura, K., Lu, D., Takata, T., Saito, N., Inoue, Y., & Domen, K. (2006). Photocatalyst releasing hydrogen from water. *Nature*, *440*(7082), 295. Available from https://doi.org/10.1038/440295a.

Maeda, K., Takata, T., Hara, M., Saito, N., Inoue, Y., Kobayashi, H., & Domen, K. (2005). GaN:ZnO solid solution as a photocatalyst for visible-light-driven overall water splitting. *Journal of the American Chemical Society*, *127*(23), 8286–8287. Available from https://doi.org/10.1021/ja0518777.

Mahan, G. D., & Sofo, J. O. (1996). The best thermoelectric. *Proceedings of the National Academy of Sciences*, *93*(15), 7436–7439. Available from https://doi.org/10.1073/pnas.93.15.7436.

Malik, K. (2014). Human development report "*Sustaining human progress: Reducing vulnerabilities and building resilience*." by the United Nations Development Programme 1 UN Plaza, New York, NY 10017, USA

Martel, R., Schmidt, T., Shea, H. R., Hertel, T., & Avouris, P. (1998). Single- and multi-wall carbon nanotube field-effect transistors. *Applied Physics Letters*, *73*(17), 2447–2449. Available from https://doi.org/10.1063/1.122477.

Mcdonald, S. A., Konstantatos, G., Zhang, S., Cyr, P. W., Klem, E. J. D., Levina, L., & Sargent, E. H. (2005). Solution-processed PbS quantum dot infrared photodetectors and photovoltaics. *Nature Materials*, *4*(2), 138–142. Available from https://doi.org/10.1038/nmat1299, http://www.nature.com/nmat/.

McEnaney, J. M., Singh, A. R., Schwalbe, J. A., Kibsgaard, J., Lin, J. C., Cargnello, M., Jaramillo, T. F., & Nørskov, J. K. (2017). Ammonia synthesis from N_2 and H_2O using a lithium cycling electrification strategy at atmospheric pressure. *Energy and*

Environmental Science., *10*(7), 1621−1630. Available from https://doi.org/10.1039/c7ee01126a, http://www.rsc.org/Publishing/Journals/EE/About.asp.

Mei, X., Yoshida, W., Lange, M., Lee, J., Zhou, J., Liu, P.-H., Leong, K., Zamora, A., Padilla, J., Sarkozy, S., Lai, R., & Deal, W. R. (2015). First demonstration of amplification at 1THz using 25-nm InP high electron mobility transistor process. *IEEE Electron Device Letters.*, *36*(4), 327−329. Available from https://doi.org/10.1109/LED.2015.2407193.

Mennel, L., Symonowicz, J., Wachter, S., Polyushkin, D. K., Molina-Mendoza, A. J., & Mueller, T. (2020). Ultrafast machine vision with 2D material neural network image sensors. *Nature*, *579*(7797), 62−66. Available from https://doi.org/10.1038/s41586-020-2038-x, http://www.nature.com/nature/index.html.

Miao, X., Chabak, K., Zhang, C., Mohseni, P. K., Walker, D., & Li, X. (2015). High-speed planar GaAs nanowire arrays with fmax > 75GHz by wafer-scale bottom-up growth. *Nano Letters*, *15*(5), 2780−2786. Available from https://doi.org/10.1021/nl503596j, http://pubs.acs.org/journal/nalefd.

Mikhaylik, Y. V., & Akridge, J. R. (2004). Polysulfide shuttle study in the Li/S battery system. *Journal of the Electrochemical Society*, *151*(11), A1969−A1976. Available from https://doi.org/10.1149/1.1806394.

Mirzaeian, M., & Hall, P. J. (2009). Preparation of controlled porosity carbon aerogels for energy storage in rechargeable lithium oxygen batteries. *Electrochimica Acta*, *54*(28), 7444−7451. Available from https://doi.org/10.1016/j.electacta.2009.07.079.

Mizuno, K., Ishii, J., Kishida, H., Hayamizu, Y., Yasuda, S., Futaba, D. N., Yumura, M., & Hata, K. (2009). A black body absorber from vertically aligned single-walled carbon nanotubes. *Proceedings of the National Academy of Sciences of the United States of America.*, *106*(15), 6044−6047. Available from https://doi.org/10.1073/pnas.0900155106Japan, http://www.pnas.org/content/106/15/6044.full.pdf.

Monroe, C., & Newman, J. (2003). Dendrite growth in lithium/polymer systems. *Journal of the Electrochemical Society*, *150*(10). Available from https://doi.org/10.1149/1.1606686.

Moser, J., Eichler, A., Güttinger, J., Dykman, M. I., & Bachtold, A. (2014). Nanotube mechanical resonators with quality factors of up to 5 million. *Nature Nanotechnology.*, *9*(12), 1007−1011. Available from https://doi.org/10.1038/nnano.2014.234.

Mouli, S. K., Zhao, L. C., Omary, R. A., & Thaxton, C. S. (2010). Lymphotropic nanoparticle enhanced MRI for the staging of genitourinary tumors. *Nature Reviews Urology.*, 7(2), 84−93. Available from https://doi.org/10.1038/nrurol.2009.254.

Mu, L., Chang, Y., Sawtelle, S. D., Wipf, M., Duan, X., & Reed, M. A. (2015). Silicon nanowire field-effect transistors—A versatile class of potentiometric nanobiosensors. *IEEE Access.*, *3*, 287−302. Available from https://doi.org/10.1109/ACCESS.2015.2422842, http://ieeexplore.ieee.org/xpl/RecentIssue.jsp?punumber = 6287639.

Mu, L., Zhao, Y., Li, A., Wang, S., Wang, Z., Yang, J., Wang, Y., Liu, T., Chen, R., Zhu, J., Fan, F., Li, R., & Li, C. (2016). Enhancing charge separation on high symmetry $SrTiO_3$ exposed with anisotropic facets for photocatalytic water splitting. *Energy and Environmental Science.*, *9*(7), 2463−2469. Available from https://doi.org/10.1039/c6ee00526h, http://www.rsc.org/Publishing/Journals/EE/About.asp.

Nakada, A., Uchiyama, T., Kawakami, N., Sahara, G., Nishioka, S., Kamata, R., Kumagai, H., Ishitani, O., Uchimoto, Y., & Maeda, K. (2019). Solar water oxidation by a visible-light-responsive tantalum/nitrogen-codoped rutile titania anode for photoelectrochemical water splitting and carbon dioxide fixation. *ChemPhotoChem.*, *3*(1), 37−45. Available from https://doi.org/10.1002/cptc.201800157, onlinelibrary.wiley.com/journal/23670932.

Nan, C. W., Liu, G., Lin, Y., & Li, M. (2004). Interface effect on thermal conductivity of carbon nanotube composites. *Applied Physics Letters, 85*(16), 3549–3551. Available from https://doi.org/10.1063/1.1808874.

Nel, A. E., & Miller, J. F. (2021). Nano-enabled COVID-19 vaccines: Meeting the challenges of durable antibody plus cellular immunity and immune escape. *ACS Nano., 15*(4), 5793–5818. Available from https://doi.org/10.1021/acsnano.1c01845, http://pubs.acs.org/journal/ancac3.

Nishiyama, H., Yamada, T., Nakabayashi, M., Maehara, Y., Yamaguchi, M., Kuromiya, Y., Nagatsuma, Y., Tokudome, H., Akiyama, S., Watanabe, T., Narushima, R., Okunaka, S., Shibata, N., Takata, T., Hisatomi, T., & Domen, K. (2021). Photocatalytic solar hydrogen production from water on a 100-m^2 scale. *Nature, 598* (7880), 304–307. Available from https://doi.org/10.1038/s41586-021-03907-3, http://www.nature.com/nature/index.html.

Ohta, H., Kim, S., Mune, Y., Mizoguchi, T., Nomura, K., Ohta, S., Nomura, T., Nakanishi, Y., Ikuhara, Y., Hirano, M., Hosono, H., & Koumoto, K. (2007). Giant thermoelectric Seebeck coefficient of a two-dimensional electron gas in $SrTiO_3$. *Nature Materials, 6*(2), 129–134. Available from https://doi.org/10.1038/nmat1821, http://www.nature.com/nmat/.

O'Regan, B., & Grätzel, M. (1991). A low-cost, high-efficiency solar cell based on dye-sensitized colloidal TiO_2 films. *Nature, 353*(6346), 737–740. Available from https://doi.org/10.1038/353737a0, http://www.nature.com/nature/index.html.

Organick, L., Ang, S. D., Chen, Y. J., Lopez, R., Yekhanin, S., Makarychev, K., Racz, M. Z., Kamath, G., Gopalan, P., Nguyen, B., Takahashi, C. N., Newman, S., Parker, H. Y., Rashtchian, C., Stewart, K., Gupta, G., Carlson, R., Mulligan, J., Carmean, D., ... Strauss, K. (2018). Random access in large-scale DNA data storage. *Nature Biotechnology, 36*(3), 242–248. Available from https://doi.org/10.1038/nbt.4079, http://www.nature.com/nbt/index.html.

Paithankar, D., Hwang, B. H., Munavalli, G., Kauvar, A., Lloyd, J., Blomgren, R., Faupel, L., Meyer, T., & Mitragotri, S. (2015). Ultrasonic delivery of silica-gold nanoshells for photothermolysis of sebaceous glands in humans: Nanotechnology from the bench to clinic. *Journal of Controlled Release., 206*, 30–36. Available from https://doi.org/10.1016/j.jconrel.2015.03.004, http://www.elsevier.com/locate/jconrel.

Pardi, N., Hogan, M. J., Porter, F. W., & Weissman, D. (2018). mRNA vaccines-a new era in vaccinology. *Nature Reviews. Drug Discovery, 17*(4), 261–279. Available from https://doi.org/10.1038/nrd.2017.243, http://www.nature.com/nrd/index.html.

Patolsky, F., Timko, B. P., Yu, G., Fang, Y., Greytak, A. B., Zheng, G., & Lieber, C. M. (2006). Detection, stimulation, and inhibition of neuronal signals with high-density nanowire transistor arrays. *Science (New York, N.Y.), 313*(5790), 1100–1104. Available from https://doi.org/10.1126/science.1128640.

Peharz, G., Dimroth, F., & Wittstadt, U. (2007). Solar hydrogen production by water splitting with a conversion efficiency of 18%. *International Journal of Hydrogen Energy, 32*(15), 3248–3252. Available from https://doi.org/10.1016/j.ijhydene.2007.04.036, http://www.journals.elsevier.com/international-journal-of-hydrogen-energy/.

Pekarik, J. J., Adkisson, J., Gray, P., Liu, Q., Camillo-Castillo, R., Khater, M., Jain, V., Zetterlund, B., DiVergilio, A., Tian, X., Vallett, A., Ellis-Monaghan, J., Gross, B. J., Cheng, P., Kaushal, V., He, Z., Lukaitis, J., Newton, K., Kerbaugh, M., Cahoon, N., Vera, L., Zhao, Y., Long, J. R., Valdes-Garcia, A., Reynolds, S., Lee, W., Sadhu, B., & Harame, D. (2014). A 90nm SiGe BiCMOS technology for mm-wave and high-performance analog applications. In *Bipolar/BiCMOS Circuits and Technology Meeting*, (pp. 92–95). Coronado, CA, USA: Proceedings of the IEEE.

Peng, L. M., Zhang, Z., & Wang, S. (2014). Carbon nanotube electronics: Recent advances. *Materials Today.*, *17*(9), 433–442. Available from https://doi.org/10.1016/j.mattod.2014.07.008, http://www.journals.elsevier.com/materials-today/.

Peng, Z., Freunberger, S. A., Chen, Y., & Bruce, P. G. (2012). A reversible and higher-rate Li-O_2 battery. *Science (New York, N.Y.)*, *337*(6094), 563–566. Available from https://doi.org/10.1126/science.1223985, http://www.sciencemag.org/content/337/6094/563.full.pdf.

Pi, S., Li, C., Jiang, H., Xia, W., Xin, H., Yang, J. J., & Xia, Q. (2019). Memristor crossbar arrays with 6-nm half-pitch and 2-nm critical dimension. *Nature Nanotechnology.*, *14*(1), 35–39. Available from https://doi.org/10.1038/s41565-018-0302-0, http://www.nature.com/nnano/index.html.

Ping, L., Hou, P. X., Liu, C., Li, J., Zhao, Y., Zhang, F., Ma, C., Tai, K., Cong, H., & Cheng, H. M. (2017). Surface-restrained growth of vertically aligned carbon nanotube arrays with excellent thermal transport performance. *Nanoscale*, *9*(24), 8213–8219. Available from https://doi.org/10.1039/c7nr01529a, http://www.rsc.org/publishing/journals/NR/Index.asp.

Polking, M. J., Han, M. G., Yourdkhani, A., Petkov, V., Kisielowski, C. F., Volkov, V. V., Zhu, Y., Caruntu, G., Paul Alivisatos, A., & Ramesh, R. (2012). Ferroelectric order in individual nanometre-scale crystals. *Nature Materials*, *11*(8), 700–709. Available from https://doi.org/10.1038/nmat3371, http://www.nature.com/nmat/.

Polman, A., Knight, M., Garnett, E. C., Ehrler, B., & Sinke, W. C. (2016). Photovoltaic materials: Present efficiencies and future challenges. *Netherlands Science.*, *352*(6283). Available from https://doi.org/10.1126/science.aad4424, http://science.sciencemag.org/content/352/6283/aad4424.full.pdf.

Prezioso, M., Merrikh-Bayat, F., Hoskins, B. D., Adam, G. C., Likharev, K. K., & Strukov, D. B. (2015). Training and operation of an integrated neuromorphic network based on metal–oxide memristors. *Nature*, *521*(7550), 61–64. Available from https://doi.org/10.1038/nature14441.

Pugno, N. M. (2006). On the strength of the carbon nanotube-based space elevator cable: From nanomechanics to megamechanics. *Journal of Physics Condensed Matter*, *18*(33), S1971–S1990. Available from https://doi.org/10.1088/0953-8984/18/33/S14.

Qiu, C., Zhang, Z., Xiao, M., Yang, Y., Zhong, D., & Peng, L. M. (2017). Scaling carbon nanotube complementary transistors to 5-nm gate lengths. *Science (New York, N.Y.)*, *355*(6322), 271–276. Available from https://doi.org/10.1126/science.aaj1628, http://science.sciencemag.org/content/sci/355/6322/271.full.pdf.

Raji, A. R. O., Villegas Salvatierra, R., Kim, N. D., Fan, X., Li, Y., Silva, G. A. L., Sha, J., & Tour, J. M. (2017). Lithium batteries with nearly maximum metal storage. *ACS Nano.*, *11*(6), 6362–6369. Available from https://doi.org/10.1021/acsnano.7b02731, http://pubs.acs.org/journal/ancac3.

Rastinehad, A. R., Anastos, H., Wajswol, E., Winoker, J. S., Sfakianos, J. P., Doppalapudi, S. K., Carrick, M. R., Knauer, C. J., Taouli, B., Lewis, S. C., Tewari, A. K., Schwartz, J. A., Canfield, S. E., George, A. K., West, J. L., & Halas, N. J. (2019). Gold nanoshell-localized photothermal ablation of prostate tumors in a clinical pilot device study. *Proceedings of the National Academy of Sciences*, *116*(37), 18590–18596. Available from https://doi.org/10.1073/pnas.1906929116.

Razavi, P., & Fagas, G. (2013). Electrical performance of III–V gate-all-around nanowire transistors. *Applied Physics Letters*, *103*(6). Available from https://doi.org/10.1063/1.4817997.

Reece, S. Y., Hamel, J. A., Sung, K., Jarvi, T. D., Esswein, A. J., Pijpers, J. J. H., & Nocera, D. G. (2011). Wireless solar water splitting using silicon-based semiconductors and earth-abundant catalysts. *Science (New York, N.Y.)*, *334*(6056), 645–648. Available from https://doi.org/10.1126/science.1209816, http://www.sciencemag.org/content/334/6056/645.full.pdf.

Reinsel, D., Gantz, J., Rydning, J. (2018).The digitization of the world from edge to core. An IDC White Paper – #US44413318.

Ren, Y., Zhang, D., Suo, J., Cao, Y., Eickemeyer, F. T., Vlachopoulos, N., Zakeeruddin, S. M., Hagfeldt, A., & Grätzel, M. (2023). Hydroxamic acid pre-adsorption raises the efficiency of cosensitized solar cells. *Nature*, *613*(7942), 60–65. Available from https://doi.org/10.1038/s41586-022-05460-z, https://www.nature.com/nature/.

Ross, R. T., & Hsiao, T. L. (1977). Limits on the yield of photochemical solar energy conversion. *Journal of Applied Physics*, *48*(11), 4783–4785. Available from https://doi.org/10.1063/1.323494.

Ruggiero, A., Villa, C. H., Bander, E., Rey, D. A., Bergkvist, M., Batt, C. A., Manova-Todorova, K., Deen, W. M., Scheinberg, D. A., & McDevitt, M. R. (2010). Paradoxical glomerular filtration of carbon nanotubes. *Proceedings of the National Academy of Sciences*, *107*(27), 12369–12374. Available from https://doi.org/10.1073/pnas.0913667107.

Rutherglen, C., Jain, D., & Burke, P. (2009). Nanotube electronics for radio frequency applications. *Nature Nanotechnology.*, *4*(12), 811–819. Available from https://doi.org/10.1038/nnano.2009.355, http://www.nature.com/nnano/index.html.

Sales, B. C. (1998). Electron crystals and phonon glasses: A new path to improved thermoelectric materials. *MRS Bulletin.*, *23*(1), 15–21. Available from https://doi.org/10.1557/S0883769400031419, https://www.springer.com/journal/43577.

Sánchez-López, E., Gomes, D., Esteruelas, G., Bonilla, L., Lopez-Machado, A. L., Galindo, R., Cano, A., Espina, M., Ettcheto, M., Camins, A., Silva, A. M., Durazzo, A., Santini, A., Garcia, M. L., & Souto, E. B. (2020). Metal-based nanoparticles as antimicrobial agents: An overview. *Nanomaterials.*, *10*(2). Available from https://doi.org/10.3390/nano10020292, https://www.mdpi.com/2079-4991/10/2/292/pdf.

Sasaki, T., Watanabe, M., Hashizume, H., Yamada, H., & Nakazawa, H. (1996). Macromolecule-like aspects for a colloidal suspension of an exfoliated titanate. Pairwise association of nanosheets and dynamic reassembling process initiated from it. *Journal of the American Chemical Society*, *118*(35), 8329–8335. Available from https://doi.org/10.1021/ja960073b.

Schreier, M., Curvat, L., Giordano, F., Steier, L., Abate, A., Zakeeruddin, S. M., Luo, J., Mayer, M. T., & Grätzel, M. (2015). Efficient photosynthesis of carbon monoxide from CO_2 using perovskite photovoltaics. *Nature Communications.*, *6*(1). Available from https://doi.org/10.1038/ncomms8326.

Schreier, M., Héroguel, F., Steier, L., Ahmad, S., Luterbacher, J. S., Mayer, M. T., Luo, J., & Grätzel, M. (2017). Solar conversion of CO_2 to CO using earth-abundant electrocatalysts prepared by atomic layer modification of CuO. *Nature Energy.*, *2*(7). Available from https://doi.org/10.1038/nenergy.2017.87, http://www.nature.com/nenergy/.

Sekiguchi, Y., Yao, Y., Ohko, Y., Tanaka, K., Ishido, T., Fujishima, A., & Kubota, Y. (2007). Self-sterilizing catheters with titanium dioxide photocatalyst thin films for clean intermittent catheterization: Basis and study of clinical use. *International Journal of Urology*, *14*(5), 426–430. Available from https://doi.org/10.1111/j.1442-2042.2007.01743.x.

Seliuta, D., Subačius, L., Kašalynas, I., Shuba, M., Paddubskaya, A., Ksenevich, V., Kuzhir, P., Maksimenko, S., & Valušis, G. (2013). Electrical conductivity of single-wall carbon nanotube films in strong electric field. *Journal of Applied Physics*, *113*(18). Available from https://doi.org/10.1063/1.4804658.

Shannon, C. E. (1949). Communication in the presence of noise. *Proceedings of the IRE*, *37*(1), 10–21. Available from https://doi.org/10.1109/JRPROC.1949.232969.

Shehada, N., Cancilla, J. C., Torrecilla, J. S., Pariente, E. S., Brönstrup, G., Christiansen, S., Johnson, D. W., Leja, M., Davies, M. P. A., Liran, O., Peled, N., & Haick, H. (2016). Silicon nanowire sensors enable diagnosis of patients via exhaled breath. *ACS Nano.*, *10*(7), 7047–7057. Available from https://doi.org/10.1021/acsnano.6b03127, http://pubs.acs.org/journal/ancac3.

Shi, X., Zuo, Y., Zhai, P., Shen, J., Yang, Y., Gao, Z., Liao, M., Wu, J., Wang, J., Xu, X., Tong, Q., Zhang, B., Wang, B., Sun, X., Zhang, L., Pei, Q., Jin, D., Chen, P., & Peng, H. (2021). Large-area display textiles integrated with functional systems. *Nature*, *591*(7849), 240–245. Available from https://doi.org/10.1038/s41586-021-03295-8, http://www.nature.com/nature/index.html.

Shih, C. F., Zhang, T., Li, J., & Bai, C. (2018). Powering the future with liquid sunshine. *Joule.*, *2*(10), 1925–1949. Available from https://doi.org/10.1016/j.joule.2018.08.016, https://www.journals.elsevier.com/joule.

Shin, M. D., Shukla, S., Chung, Y. H., Beiss, V., Chan, S. K., Ortega-Rivera, O. A., Wirth, D. M., Chen, A., Sack, M., Pokorski, J. K., & Steinmetz, N. F. (2020). COVID-19 vaccine development and a potential nanomaterial path forward. *Nature Nanotechnology.*, *15*(8), 646–655. Available from https://doi.org/10.1038/s41565-020-0737-y, http://www.nature.com/nnano/index.html.

Shin, T. H., & Cheon, J. (2017). Synergism of nanomaterials with physical stimuli for biology and medicine. *Accounts of Chemical Research*, *50*(3), 567–572. Available from https://doi.org/10.1021/acs.accounts.6b00559, http://pubs.acs.org/journal/achre4.

Shin, T. H., Choi, Y., Kim, S., & Cheon, J. (2015). Recent advances in magnetic nanoparticle-based multi-modal imaging. *Chemical Society Reviews*, *44*(14), 4501–4516. Available from https://doi.org/10.1039/c4cs00345d, http://pubs.rsc.org/en/journals/journal/cs.

Shockley, W., & Queisser, H. J. (1961). Detailed balance limit of efficiency of p-n junction solar cells. *Journal of Applied Physics*, *32*(3), 510–519. Available from https://doi.org/10.1063/1.1736034.

Shulaker, M. M., Hills, G., Park, R. S., Howe, R. T., Saraswat, K., Wong, H. S. P., & Mitra, S. (2017). Three-dimensional integration of nanotechnologies for computing and data storage on a single chip. *Nature*, *547*(7661), 74–78. Available from https://doi.org/10.1038/nature22994, http://www.nature.com/nature/index.html.

Shulaker, M. M., Hills, G., Patil, N., Wei, H., Chen, H. Y., Wong, H. S. P., & Mitra, S. (2013). Carbon nanotube computer. *Nature*, *501*(7468), 526–530. Available from https://doi.org/10.1038/nature12502.

Singh, N., Agarwal, A., Bera, L. K., Liow, T. Y., Yang, R., Rustagi, S. C., Tung, C. H., Kumar, R., Lo, G. Q., Balasubramanian, N., & Kwong, D.-L. (2006). High-performance fully depleted silicon nanowire (diameter/spl les/5nm) gate-all-around CMOS devices. *IEEE Electron Device Letters.*, *27*(5), 383–386. Available from https://doi.org/10.1109/LED.2006.873381.

Soo Choi, H., Liu, W., Misra, P., Tanaka, E., Zimmer, J. P., Itty Ipe, B., Bawendi, M. G., & Frangioni, J. V. (2007). Renal clearance of quantum dots. *Nature Biotechnology*, *25*(10), 1165–1170. Available from https://doi.org/10.1038/nbt1340.

Sorcar, S., Hwang, Y., Lee, J., Kim, H., Grimes, K. M., Grimes, C. A., Jung, J. W., Cho, C. H., Majima, T., Hoffmann, M. R., & In, S. I. (2019). CO_2, water, and sunlight to hydrocarbon fuels: A sustained sunlight to fuel (Joule-to-Joule) photoconversion efficiency of 1%. *Energy and Environmental Science*, *12*(9), 2685–2696. Available from https://doi.org/10.1039/c9ee00734b, http://pubs.rsc.org/en/journals/journal/ee.

Stamenkovic, V. R., Mun, B. S., Arenz, M., Mayrhofer, K. J. J., Lucas, C. A., Wang, G., Ross, P. N., & Markovic, N. M. (2007). Trends in electrocatalysis on extended and nanoscale Pt-bimetallic alloy surfaces. *Nature Materials*, *6*(3), 241–247. Available from https://doi.org/10.1038/nmat1840, http://www.nature.com/nmat/.

Steiner, M., Engel, M., Lin, Y. M., Wu, Y., Jenkins, K., Farmer, D. B., Humes, J. J., Yoder, N. L., Seo, J. W. T., Green, A. A., Hersam, M. C., Krupke, R., & Avouris, P. (2012). High-frequency performance of scaled carbon nanotube array field-effect transistors. *Applied Physics Letters*, *101*(5). Available from https://doi.org/10.1063/1.4742325.

Stern, E., Klemic, J. F., Routenberg, D. A., Wyrembak, P. N., Turner-Evans, D. B., Hamilton, A. D., LaVan, D. A., Fahmy, T. M., & Reed, M. A. (2007). Label-free immunodetection with CMOS-compatible semiconducting nanowires. *Nature*, *445* (7127), 519–522. Available from https://doi.org/10.1038/nature05498, http://www.nature.com/nature/index.html.

Stranks, S. D., Eperon, G. E., Grancini, G., Menelaou, C., Alcocer, M. J. P., Leijtens, T., Herz, L. M., Petrozza, A., & Snaith, H. J. (2013). Electron-hole diffusion lengths exceeding 1 micrometer in an organometal trihalide perovskite absorber. *Science (New York, N.Y.)*, *342*(6156), 341–344. Available from https://doi.org/10.1126/science.1243982, http://www.sciencemag.org/content/342/6156/341.full.pdf.

Strasser, P., Koh, S., Anniyev, T., Greeley, J., More, K., Yu, C., Liu, Z., Kaya, S., Nordlund, D., Ogasawara, H., Toney, M. F., & Nilsson, A. (2010). Lattice-strain control of the activity in dealloyed core-shell fuel cell catalysts. *Nature Chemistry.*, *2*(6), 454–460. Available from https://doi.org/10.1038/nchem.623.

Suhr, J., Koratkar, N., Keblinski, P., & Ajayan, P. (2005). Viscoelasticity in carbon nanotube composites. *Nature Materials*, *4*(2), 134–137. Available from https://doi.org/10.1038/nmat1293, http://www.nature.com/nmat/.

Sun, D. M., Timmermans, M. Y., Tian, Y., Nasibulin, A. G., Kauppinen, E. I., Kishimoto, S., Mizutani, T., & Ohno, Y. (2011). Flexible high-performance carbon nanotube integrated circuits. *Nature Nanotechnology.*, *6*(3), 156–161. Available from https://doi.org/10.1038/nnano.2011.1, http://www.nature.com/nnano/index.html.

Suryanto, B. H. R., Du, H. L., Wang, D., Chen, J., Simonov, A. N., & MacFarlane, D. R. (2019). Challenges and prospects in the catalysis of electroreduction of nitrogen to ammonia. *Nature Catalysis.*, *2*(4), 290–296. Available from https://doi.org/10.1038/s41929-019-0252-4, https://www.nature.com/natcatal/.

Takata, T., Jiang, J., Sakata, Y., Nakabayashi, M., Shibata, N., Nandal, V., Seki, K., Hisatomi, T., & Domen, K. (2020). Photocatalytic water splitting with a quantum efficiency of almost unity. *Nature*, *581*(7809), 411–414. Available from https://doi.org/10.1038/s41586-020-2278-9, http://www.nature.com/nature/index.html.

Tamelen, E. E. V., & Akermark, B. (1968). Electrolytic reduction of molecular nitrogen. *Journal of the American Chemical Society*, *90*(16), 4492–4493. Available from https://doi.org/10.1021/ja01018a074.

Tang, C., & Qiao, S. Z. (2019). How to explore ambient electrocatalytic nitrogen reduction reliably and insightfully. *Chemical Society Reviews*, *48*(12), 3166–3180. Available from https://doi.org/10.1039/c9cs00280d, http://pubs.rsc.org/en/journals/journal/cs.

Tang, Z., Kotov, N. A., Magonov, S., & Ozturk, B. (2003). Nanostructured artificial nacre. *Nature Materials*, *2*(6), 413–418. Available from https://doi.org/10.1038/nmat906, http://www.nature.com/nmat/.

Tans, S. J., Verschueren, A. R. M., & Dekker, C. (1998). Room-temperature transistor based on a single carbon nanotube. *Nature*, *393*(6680), 49–52. Available from https://doi.org/10.1038/29954.

Tao, Y., Xie, X., Lv, W., Tang, D. M., Kong, D., Huang, Z., Nishihara, H., Ishii, T., Li, B., Golberg, D., Kang, F., Kyotani, T., & Yang, Q. H. (2013). Towards ultrahigh volumetric capacitance: Graphene derived highly dense but porous carbons for supercapacitors. *Scientific Reports.*, *3*. Available from https://doi.org/10.1038/srep02975.

Tee, B. C. K., Wang, C., Allen, R., & Bao, Z. (2012). An electrically and mechanically self-healing composite with pressure- and flexion-sensitive properties for electronic skin applications. *Nature Nanotechnology.*, 7(12), 825–832. Available from https://doi.org/10.1038/nnano.2012.192, http://www.nature.com/nnano/index.html.

Tellier, R., Li, Y., Cowling, B. J., & Tang, J. W. (2019). Recognition of aerosol transmission of infectious agents: A commentary. *BMC Infectious Diseases, 19*(1). Available from https://doi.org/10.1186/s12879-019-3707-y, http://www.biomedcentral.com/bmcinfectdis/.

Tian, B., Cohen-Karni, T., Qing, Q., Duan, X., Xie, P., & Lieber, C. M. (2010). Three-dimensional, flexible nanoscale field-effect transistors as localized bioprobes. *Science (New York, N.Y.), 329*(5993), 830–834. Available from https://doi.org/10.1126/science.1192033, http://www.sciencemag.org/cgi/reprint/329/5993/830.pdf, United States.

Tian, B., Liu, J., Dvir, T., Jin, L., Tsui, J. H., Qing, Q., Suo, Z., Langer, R., Kohane, D. S., & Lieber, C. M. (2012). Macroporous nanowire nanoelectronic scaffolds for synthetic tissues. *Nature Materials, 11*(11), 986–994. Available from https://doi.org/10.1038/nmat3404, http://www.nature.com/nmat/.

Tsoi, K. M., Macparland, S. A., Ma, X. Z., Spetzler, V. N., Echeverri, J., Ouyang, B., Fadel, S. M., Sykes, E. A., Goldaracena, N., Kaths, J. M., Conneely, J. B., Alman, B. A., Selzner, M., Ostrowski, M. A., Adeyi, O. A., Zilman, A., McGilvray, I. D., & Chan, W. C. W. (2016). Mechanism of hard-nanomaterial clearance by the liver. *Nature Materials, 15*(11), 1212–1221. Available from https://doi.org/10.1038/nmat4718, http://www.nature.com/nmat/.

Van Tamelen, E. E., & Seeley, D. A. (1969). Catalytic fixation of molecular nitrogen by electrolytic and chemical reduction. *Journal of the American Chemical Society, 91*(18), 5194. Available from https://doi.org/10.1021/ja01046a063.

Ventola, C. L. (2017). Progress in nanomedicine: Approved and investigational nanodrugs. *P and T., 42*(12), 742–755. Available from https://www.ptcommunity.com/system/files/pdf/ptj4212742.pdf.

Viola, K. L., Sbarboro, J., Sureka, R., De, M., Bicca, M. A., Wang, J., Vasavada, S., Satpathy, S., Wu, S., Joshi, H., Velasco, P. T., Macrenaris, K., Waters, E. A., Lu, C., Phan, J., Lacor, P., Prasad, P., Dravid, V. P., & Klein, W. L. (2015). Towards non-invasive diagnostic imaging of early-stage Alzheimer's disease. *Nature Nanotechnology., 10*(1), 91–98. Available from https://doi.org/10.1038/nnano.2014.254, http://www.nature.com/nnano/index.html.

De Volder, M. F. L., Tawfick, S. H., Baughman, R. H., & Hart, A. J. (2013). Carbon nanotubes: Present and future commercial applications. *Science (New York, N.Y.), 339* (6119), 535–539. Available from https://doi.org/10.1126/science.1222453, http://www.sciencemag.org/content/339/6119/535.full.pdf.

Wallentin, J., Anttu, N., Asoli, D., Huffman, M., Åberg, I., Magnusson, M. H., Siefer, G., Fuss-Kailuweit, P., Dimroth, F., Witzigmann, B., Xu, H. Q., Samuelson, L., Deppert, K., & Borgström, M. T. (2013). InP nanowire array solar cells achieving 13.8% efficiency by exceeding the ray optics limit. *Science (New York, N.Y.), 339* (6123), 1057–1060. Available from https://doi.org/10.1126/science.1230969, http://www.sciencemag.org/content/339/6123/1057.full.pdf.

Wang, B. W., Jiang, S., Zhu, Q. B., Sun, Y., Luan, J., Hou, P. X., Qiu, S., Li, Q. W., Liu, C., Sun, D. M., & Cheng, H. M. (2018). Continuous fabrication of meter-scale single-wall carbon nanotube films and their use in flexible and transparent integrated circuits. *Advanced Materials., 30*(32). Available from https://doi.org/10.1002/adma.201802057, http://onlinelibrary.wiley.com/journal/10.1002/(ISSN)1521-4095.

Wang, D. W., Li, F., Liu, M., Lu, G. Q., & Cheng, H. M. (2008). 3D aperiodic hierarchical porous graphitic carbon material for high-rate electrochemical capacitive energy storage. *Angewandte Chemie—International Edition.*, *47*(2), 373–376. Available from https://doi.org/10.1002/anie.200702721.

Wang, H., He, Y., Li, Y. H., Su, Z. E., Li, B., Huang, H. L., Ding, X., Chen, M. C., Liu, C., Qin, J., Li, J. P., He, Y. M., Schneider, C., Kamp, M., Peng, C. Z., Höfling, S., Lu, C. Y., & Pan, J. W. (2017). High-efficiency multiphoton boson sampling. *Nature Photonics.*, *11*(6), 361–365. Available from https://doi.org/10.1038/nphoton.2017.63, http://www.nature.com/nphoton/index.html.

Wang, Q., Hisatomi, T., Jia, Q., Tokudome, H., Zhong, M., Wang, C., Pan, Z., Takata, T., Nakabayashi, M., Shibata, N., Li, Y., Sharp, I. D., Kudo, A., Yamada, T., & Domen, K. (2016). Scalable water splitting on particulate photocatalyst sheets with a solar-to-hydrogen energy conversion efficiency exceeding 1%. *Nature Materials*, *15*(6), 611–615. Available from https://doi.org/10.1038/nmat4589, http://www.nature.com/nmat/.

Wang, Q., Warnan, J., Rodríguez-Jiménez, S., Leung, J. J., Kalathil, S., Andrei, V., Domen, K., & Reisner, E. (2020). Molecularly engineered photocatalyst sheet for scalable solar formate production from carbon dioxide and water. *Nature Energy.*, *5*(9), 703–710. Available from https://doi.org/10.1038/s41560-020-0678-6, http://www.nature.com/nenergy/.

Wang, X., Maeda, K., Thomas, A., Takanabe, K., Xin, G., Carlsson, J. M., Domen, K., & Antonietti, M. (2009). A metal-free polymeric photocatalyst for hydrogen production from water under visible light. *Nature Materials*, *8*(1), 76–80. Available from https://doi.org/10.1038/nmat2317, http://www.nature.com/nmat/.

Wang, Y., Wu, T., Barbaud, J., Kong, W., Cui, D., Chen, H., Yang, X., & Han, L. (2019). Stabilizing heterostructures of soft perovskite semiconductors. *Science (New York, N.Y.)*, *365*(6454), 687–691. Available from https://doi.org/10.1126/science.aax8018, https://science.sciencemag.org/content/365/6454/687/tab-pdf.

Wang, Y. F., Liu, G. Y., Sun, L. D., Xiao, J. W., Zhou, J. C., & Yan, C. H. (2013). Nd^{3+}-sensitized upconversion nanophosphors: Efficient in vivo bioimaging probes with minimized heating effect. *ACS Nano*, 7(8), 7200–7206. Available from https://doi.org/10.1021/nn402601d.

Wang, Z., Wu, H., Burr, G. W., Hwang, C. S., Wang, K. L., Xia, Q., & Yang, J. J. (2020). Resistive switching materials for information processing. *Nature Reviews Materials.*, *5*(3), 173–195. Available from https://doi.org/10.1038/s41578-019-0159-3, http://www.nature.com/natrevmats/.

Wang, Z. L., Xu, D., Xu, J. J., & Zhang, X. B. (2014). Oxygen electrocatalysts in metal-air batteries: From aqueous to nonaqueous electrolytes. *Chemical Society Reviews*, *43* (22), 7746–7786. Available from https://doi.org/10.1039/c3cs60248f, http://pubs.rsc.org/en/journals/journal/cs.

Wee, K. W., Kang, G. Y., Park, J., Kang, J. Y., Yoon, D. S., Park, J. H., & Kim, T. S. (2005). Novel electrical detection of label-free disease marker proteins using piezoresistive self-sensing micro-cantilevers. *Biosensors and Bioelectronics*, *20*(10), 1932–1938. Available from https://doi.org/10.1016/j.bios.2004.09.023, http://www.elsevier.com/locate/bios.

Winter, M., & Brodd, R. J. (2004). What are batteries, fuel cells, and supercapacitors? *Chemical Reviews*, *104*(10), 4245–4270. Available from https://doi.org/10.1021/cr020730k.

Wright, K., Beck, K. M., Debnath, S., Amini, J. M., Nam, Y., Grzesiak, N., Chen, J.-S., Pisenti, N. C., Chmielewski, M., Collins, C., Hudek, K. M., Mizrahi, J., Wong-Campos, J. D., Allen, S., Apisdorf, J., Solomon, P., Williams, M., Ducore, A. M., Blinov, A., . . . Kim, J. (2019). Benchmarking an 11-qubit quantum computer. *Nature Communications.*, *10*(1). Available from https://doi.org/10.1038/s41467-019-13534-2.

Wu, J., Qi, L., You, H., Gross, A., Li, J., & Yang, H. (2012). Icosahedral platinum alloy nanocrystals with enhanced electrocatalytic activities. *Journal of the American Chemical Society*, *134*(29), 11880–11883. Available from https://doi.org/10.1021/ja303950v.

Wu, Y., Jenkins, K. A., Valdes-Garcia, A., Farmer, D. B., Zhu, Y., Bol, A. A., Dimitrakopoulos, C., Zhu, W., Xia, F., Avouris, P., & Lin, Y. M. (2012). State-of-the-art graphene high-frequency electronics. *Nano Letters*, *12*(6), 3062–3067. Available from https://doi.org/10.1021/nl300904k.

Wunderbaldinger, P., Josephson, L., Bremer, C., Moore, A., & Weissleder, R. (2002). Detection of lymph node metastases by contrast-enhanced MRI in an experimental model. *Magnetic Resonance in Medicine.*, *47*(2), 292–297. Available from https://doi.org/10.1002/mrm.10068, http://onlinelibrary.wiley.com/journal/10.1002/(ISSN)1522-2594.

Wuttig, M., & Yamada, N. (2007). Phase-change materials for rewriteable data storage. *Nature Materials*, *6*(11), 824–832. Available from https://doi.org/10.1038/nmat2009, http://www.nature.com/nmat/.

Xin, G., Yao, T., Sun, H., Scott, S. M., Shao, D., Wang, G., & Lian, J. (2015). Highly thermally conductive and mechanically strong graphene fibers. *Science (New York, N. Y.)*, *349*(6252), 1083–1087. Available from https://doi.org/10.1126/science.aaa6502, http://www.sciencemag.org/content/349/6252/1083.full.pdf.

Xu, Z., & Gao, C. (2011). Graphene chiral liquid crystals and macroscopic assembled fibres. *Nature Communications.*, *2*(1). Available from https://doi.org/10.1038/ncomms1583.

Xu, Z., & Gao, C. (2014). Graphene in macroscopic order: Liquid crystals and wet-spun fibers. *Accounts of Chemical Research*, *47*(4), 1267–1276. Available from https://doi.org/10.1021/ar4002813, http://pubs.acs.org/journal/achre4.

Yamaguchi, M., Ito, A., Ono, A., Kawabe, Y., & Kamihira, M. (2014). Heat-inducible gene expression system by applying alternating magnetic field to magnetic nanoparticles. *ACS Synthetic Biology.*, *3*(5), 273–279. Available from https://doi.org/10.1021/sb4000838, http://pubs.acs.org/journal/asbcd6.

Yang, H. G., Sun, C. H., Qiao, S. Z., Zou, J., Liu, G., Smith, S. C., Cheng, H. M., & Lu, G. Q. (2008). Anatase TiO_2 single crystals with a large percentage of reactive facets. *Nature*, *453*(7195), 638–641. Available from https://doi.org/10.1038/nature06964.

Yang, Y., Yin, L. C., Gong, Y., Niu, P., Wang, J. Q., Gu, L., Chen, X., Liu, G., Wang, L., & Cheng, H. M. (2018). An unusual strong visible-light absorption band in red anatase TiO_2 photocatalyst induced by atomic hydrogen-occupied oxygen vacancies. *Advanced Materials.*, *30*(6). Available from https://doi.org/10.1002/adma.201704479, http://www3.interscience.wiley.com/journal/119030556/issue.

Yin, W. J., Shi, T., & Yan, Y. (2014). Unusual defect physics in $CH_3NH_3PbI_3$ perovskite solar cell absorber. *Applied Physics Letters*, *104*(6). Available from https://doi.org/10.1063/1.4864778, http://scitation.aip.org/content/aip/journal/apl.

Yoshida, T., & Kojima, K. (2015). Toyota MIRAI fuel cell vehicle and progress toward a future hydrogen society. *Interface Magazine.*, *24*(2), 45–49. Available from https://doi.org/10.1149/2.f03152if.

Yoshikawa, G., Akiyama, T., Gautsch, S., Vettiger, P., & Rohrer, H. (2011). Nanomechanical membrane-type surface stress sensor. *Nano Letters*, *11*(3), 1044–1048. Available from https://doi.org/10.1021/nl103901a.

Yu, J.-W., Yeh, P.-C., Wang, S.-L., Wu, Y.-R., Mao, M.-H., Lin, H.-H., & Peng, L.-H. (2012). Short channel effects on gallium nitride/gallium oxide nanowire transistors. *Applied Physics Letters*, *101*(18). Available from https://doi.org/10.1063/1.4764554.

Yu, S., & Chen, P. Y. (2016). Emerging memory technologies: Recent trends and prospects. *IEEE Solid-State Circuits Magazine*, *8*(2), 43–56. Available from https://doi.org/10.1109/MSSC.2016.2546199, http://ewh.ieee.org/soc/sscs/index.php?option = com_content&task = view&id = 32&Itemid = 22.

Zhang, J., Lang, H. P., Huber, F., Bietsch, A., Grange, W., Certa, U., Mckendry, R., Güntherodt, H.-J., Hegner, M., & Gerber, C. (2006). Rapid and label-free nanomechanical detection of biomarker transcripts in human RNA. *Nature Nanotechnology*, *1* (3), 214–220. Available from https://doi.org/10.1038/nnano.2006.134.

Zhang, L., Liu, D., Muhammad, Z., Wan, F., Xie, W., Wang, Y., Song, L., Niu, Z., & Chen, J. (2019). Single nickel atoms on nitrogen-doped graphene enabling enhanced kinetics of lithium–sulfur batteries. *Advanced Materials.*, *31*(40). Available from https://doi.org/10.1002/adma.201903955, http://onlinelibrary.wiley.com/journal/10.1002/(ISSN)1521-4095.

Zhang, Z., Liang, X., Wang, S., Yao, K., Hu, Y., Zhu, Y., Chen, Q., Zhou, W., Li, Y., Yao, Y., Zhang, J., & Peng, L. M. (2007). Doping-free fabrication of carbon nanotube based ballistic CMOS devices and circuits. *Nano Letters*, 7(12), 3603–3607. Available from https://doi.org/10.1021/nl0717107.

Zhao, L. D., Dravid, V. P., & Kanatzidis, M. G. (2014). The panoscopic approach to high performance thermoelectrics. *Energy and Environmental Science*, 7(1), 251–268. Available from https://doi.org/10.1039/c3ee43099e, http://www.rsc.org/Publishing/Journals/EE/About.asp.

Zheng, G., Gao, X. P. A., & Lieber, C. M. (2010). Frequency domain detection of biomolecules using silicon nanowire biosensors. *Nano Letters*, *10*(8), 3179–3183. Available from https://doi.org/10.1021/nl1020975.

Zhong, D., Shi, H., Ding, L., Zhao, C., Liu, J., Zhou, J., Zhang, Z., & Peng, L. M. (2019). Carbon nanotube film-based radio frequency transistors with maximum oscillation frequency above 100GHz. *ACS Applied Materials and Interfaces*, *11*(45), 42496–42503. Available from https://doi.org/10.1021/acsami.9b15334, http://pubs.acs.org/journal/aamick.

Zhong, M., Tran, K., Min, Y., Wang, C., Wang, Z., Dinh, C. T., De Luna, P., Yu, Z., Rasouli, A. S., Brodersen, P., Sun, S., Voznyy, O., Tan, C. S., Askerka, M., Che, F., Liu, M., Seifitokaldani, A., Pang, Y., Lo, S. C., ... Sargent, E. H. (2020). Accelerated discovery of CO_2 electrocatalysts using active machine learning. *Nature*, *581*(7807), 178–183. Available from https://doi.org/10.1038/s41586-020-2242-8, http://www.nature.com/nature/index.html.

Zhou, J., Ren, L., Li, H., Cheng, X., Pan, Z., Zhang, Z., Ding, L., & Peng, L. M. (2023). Carbon nanotube radiofrequency transistors with fT/fMAXof 376/318 GHz. *IEEE Electron Device Letters.*, *44*(2), 329–332. Available from https://doi.org/10.1109/LED.2022.3227133, https://ieeexplore.ieee.org/servlet/opac?punumber = 55.

Zhu, Q. B., Li, B., Yang, D. D., Liu, C., Feng, S., Chen, M. L., Sun, Y., Tian, Y. N., Su, X., Wang, X. M., Qiu, S., Li, Q. W., Li, X. M., Zeng, H. B., Cheng, H. M., & Sun, D. M. (2021). A flexible ultrasensitive optoelectronic sensor array for neuromorphic vision systems. *Nature Communications.*, *12*(1). Available from https://doi.org/10.1038/s41467-021-22047-w, http://www.nature.com/ncomms/index.html.

Zidan, M. A., Strachan, J. P., & Lu, W. D. (2018). The future of electronics based on memristive systems. *Nature Electronics.*, *1*(1), 22–29. Available from https://doi.org/10.1038/s41928-017-0006-8, https://www.nature.com/natelectron/.

Zota, C. B., Roll, G., Wernersson, L. E., & Lind, E. (2014). Radio-frequency characterization of selectively regrown InGaAs lateral nanowire MOSFETs. *IEEE Transactions on Electron Devices.*, *61*(12), 4078–4083. Available from https://doi.org/10.1109/TED.2014.2363732.

CHAPTER SEVEN

The past, present, and future of low-dimensional materials

7.1 Achievements in fundamental research

7.1.1 Development of fundamental theories

Technology revolutions often originate from breakthroughs in basic science and discoveries of new fundamental principles. Low-dimensional theory is the basis for understanding and designing low-dimensional materials and devices. Modern science has two important directions. One is about the interactions of fundamental particles and fields, from Newtonian mechanics at macroscopic scale to atomic theories, and to the discovery of subatomic elementary particles. The other direction is about unification, such as Maxwell's unification of light and electromagnetic waves, Einstein's unification of time and space, energy and matter, and unification of particles and waves in quantum mechanics.

Theoretical basis of low-dimensional materials is mainly the many-body quantum mechanics in low-dimensional systems, including interactions of electrons, phonons, and photons. In Chapter 1, we have discussed energy distributions at ground states, excitation arising from interactions with light, and nonequilibrium states in transport processes. We have seen fascinating mathematical correspondence between electrons in a low-dimensional material and the high-energy particles. A famous example is the electronic structure of two-dimensional graphene. Due to the lattice symmetry, electrons near the K-points are characterized by a linear dispersion relation, manifesting zero-mass Dirac fermions described by relativistic quantum mechanics (Novoselov et al.; Geim et al., 2005; Novoselov, Jiang, et al., 2005).

Quantum mechanics is often called the first principles, however, because of the scale and complexity of low-dimensional materials, theories with suitable and necessary assumptions and simplifications have been developed to catch the essential interactions and properties. For small

Low-Dimensional Materials.
DOI: https://doi.org/10.1016/B978-0-443-29046-6.00013-4

molecules, their properties can be investigated and completely described by solving the Schrödinger's equation accurately by first principles. At larger length scales, density functional theory is used to obtain the ground state properties; Green's function method to study the excited states; and nonequilibrium Green's function method to calculate nonequilibrium transport properties (Louie et al., 2021).

In this book, we have mentioned that many seemingly different properties share common theoretical ground, such as the symmetry principles in exploring topological phases of electron and photon energy band structures (Fu & Kane, 2007; He et al., 2016), the reciprocal correspondences between crystalline lattice and reciprocal space, between energy and momentum space, and between diffraction and imaging methods. There have been many surprising discoveries in fundamental research of low-dimensional materials, such as stacked bilayer graphene with a "magic" twisting angle to have quantum phase transitions and superconducting properties similar to those in high-temperature superconducting materials (Cao et al., 2018).

7.1.2 Discovery of low-dimensional structures

If we consider all possible combinations of elements as a space and each stable composition is represented as a vector in this space, we will find that the space is largely empty, similar to the starry sky, where the stars shine in a vast vacuum. Following the analogy, recently discovered low-dimensional materials are the most dazzling "supernova." On the one hand, the discoveries are the results of development of advanced technology. On the other hand, serendipity and imagination have been indispensable. C60, known as the most perfect molecule, was discovered by Smalley, Curl, and Kroto, driven by their curiosity about organic molecules in deep interstellar space (Kroto et al., 1985). Carbon nanotubes, on the other hand, owe the discovery to the development and application of high-resolution transmission electron microscopy (TEM) by Iijima (1991). And surprisingly, two-dimensional materials such as graphene, which were thought to be instable in classical statistical mechanics, were isolated by Novoselov and Geim simply using a Scotch tape (Novoselov et al., 2004; Novoselov et al., 2005; Novoselov, Jiang, et al., 2005).

The "supernovas" of the newly discovered low-dimensional structures have gradually developed into "galaxies" that are the low-dimensional material systems, including zero-dimensional nanocrystals, one-dimensional nanowires and nanotubes, and two-dimensional materials families. With in-depth understanding of the growth mechanism and optimization of the growth conditions, scientists have been able to precisely control low-dimensional structures such as size, shape, and surface facets of nanocrystals (Jing et al., 2016). Through structure-control growth, it becomes possible to obtain carbon nanotubes with desired structures, including wall number, diameter, location, alignment, electrical conductivity type, and even specific chirality (Yang et al., 2020). For graphene and other two-dimensional materials, it is now possible to control the number of layers, size of grains, edge structure, degree of cleanliness, even interlayer stacking order and orientation (Lin et al., 2019; Ma et al., 2013, 2017).

Low-dimensional materials have distinct properties from macroscopic materials of the same composition due to their low dimension, small length scale, and high specific surface area. For example, gold, one of the most inert elements, is an excellent catalyst at nanometer scale (Haruta et al., 1989). Due to the absence of dangling bonds on surface, electrons in low-dimensional carbon materials exhibit ballistic transport at micrometer scale (Mayorov et al., 2011). In addition, assemblies of low-dimensional structures may have properties that are absent from those of the constituent units, as an emergent phenomenon. An example is the superior toughness of laminar structures with soft and hard phases in alternative layers (Wegst et al., 2015).

7.1.3 Invention of advanced technologies

Discoveries of low-dimensional materials and invention of novel technologies are like intertwined helical lines that grow simultaneously, promoting each other. On the one hand, invention of new tools such as scanning tunneling microscopy (STM) and TEM enabled the discovery, characterization, and measurement of individual low-dimensional structures. On the other hand, application of low-dimensional materials have resulted in more and more new techniques (Jain et al., 2016).

Scanning probe microscopy (SPM), represented by STM and atomic force microscope, uses the "touch" method to visualize nanostructures with atomic resolution, including chemical bonds, to simultaneously characterize the electronic structures, and to manipulate single atoms and molecules to assemble artificial structures with quantum effects.

TEM, on the other hand, uses a "seeing" approach to explore the internal structure of low-dimensional materials with an electron beam as the "light." With the invention of spherical aberration correctors and drastically improvement of resolution, not only can we directly see the atomic structure and defects, but also obtain the information of local composition and chemical state by electron spectroscopy. TEM was indispensable for the discovery of carbon nanotubes (Iijima, 1991). Recently, electron emitters made from LaB_6 nanowires have been developed with superior performance regarding the energy dispersion that could be applied for the next-generation analytical electron microscope (Tang et al., 2022; Zhang et al., 2016, 2022).

By integration SPM with TEM, in situ TEM probing technique combines the high efficiency of TEM's "seeing" and SPM's "touching" manipulation and measurement functions, allowing direct structure—process—property correlation at atomic level. Recently, we have applied in situ TEM probing technique to modify the chirality of an individual single-walled carbon nanotube, realizing a metal-semiconductor transition to construct a field-effect transistor with a channel length of 2.8 nm that exhibits quantum coherent transport at room temperature (Tang et al., 2021).

7.1.4 New paradigm of materials development

Materials science is to understand structure—property relations so as to design materials with desired properties for various applications. Usually, there is a clear application requirement and development goal, a general direction, and some clues, but no complete information. Therefore development of new materials is somewhat like solving a puzzle. If we see material design as chess game, and the periodic table of elements as a chessboard, there are almost infinite compositions. The vast composition space is the greatest challenges screening the right combination. Traditionally, material design is limited to a small composition space, relying on experience, inspiration, trial-and-error optimization. Similarly, computational and theoretical approaches are also limited by computational accuracy, scale, complexity, and time (Curtarolo et al., 2013).

Discovery, development, and application of new materials have a long cycle, generally taking 10 years or more. In 2011 Materials Genome Initiative was started in the United States to adopt a strategy similar to that of biogenetic engineering, to apply advanced information and data science and technology, to shorten the cycle of materials development. Over the last decade, machine learning and artificial intelligence—assisted research have become a new paradigm of materials science. Large amounts of data are collected by high-throughput experiments and computations to establish a large-scale database. Machine learning is used to construct a model of the training data, to reveal hidden patterns, and to predict new compositions for material with desired microstructures and properties (Warren, 2018). Machine learning has widely been applied to the study of low-dimensional materials, including robotic automated experimentation and optimization (King et al., 2004), literature mining (Tshitoyan et al., 2019), and structural characterization (Spurgeon et al., 2021), which has led to the discovery of novel metal–organic framework structures (Raccuglia et al., 2016), energy storage materials (Zhang et al., 2019), and thermoelectric materials (Tshitoyan et al., 2019).

7.2 Achievements and challenges in practical applications

7.2.1 Application achievements

In 2000 National Nanotechnology Initiative was launched with a great vision of nanotechnology applications, such as lightweight, superstrong materials with a strength 10 times higher than steel; high-density memory devices that can store all of the books in Library of Congress within a compact size of a sugar cube; and early detection of cancerous lesions with the size of a few cells (Downey et al., 2000). After two decades of efforts by scientists and continuous investment in nanotechnology by governments and enterprises, low-dimensional materials have started to be applied in a wide range of fields.

According to the statistics by Nanotechnology Products Database, there have been more than 9000 products containing nanomaterials, produced by more than 2600 companies in more than 60 countries, including catalysts, transistors, vaccines, therapeutics, batteries, clean water filters, quantum dot light-emitting diodes (QLEDs)-based displays, and environmental sensors. According to "Research and Markets," the global market reached approximately $54.2 billion in 2020, with a growing rate averaged at 17% during 2018–2024 (Global Nanotechnology Market Outlook, 2024).

The great potential of low-dimensional materials has already been explored by the semiconductor industry. The theoretical foundation of semiconductors is quantum theory that is applied to crystals with periodic lattices and potentials, leading to the electron energy band theory. For more than half a century, along with the miniature of semiconductor devices scaling from micrometer to nanometer, the performance and functions were greatly enhanced, while the cost was dramatically reduced. Nowadays, mobile electronics have been popularized in every corner of the world in the hands of almost every person, profoundly changing lifestyle and business models. Along with the increase in device integration and power density, thermal management is becoming critical for high performance processors. In recent years, graphene-based thin films with a high flexibility and a high thermal conductivity have been applied for efficient heat dissipation in high-end smartphones.

For display and lighting applications, nanocrystalline quantum dots have shown advantages of high energy conversion efficiency, high brightness, and tunable wavelength. OSRAM has already applied CdSe quantum dots as fluorescent layer for GaN-based light-emitting diode (LED). Samsung electronics launched an 85-in. "QLED" display consisting of quantum dots as green and red light—emitting pixel units, demonstrating advantages of high brightness and low power consumption (Efros & Brus, 2021).

For applications in clean energy, nanostructured zeolite catalysts have been widely used in oil refining on a global scale (Dusselier & Davis, 2018). And carbon nanotubes are being used as conductive additives, effectively improving the cycling life of lithium-ion batteries (Wen et al., 2016). Based on carbon nanotube growth technology developed by Tsinghua University, Jiangsu Cnano Technology announced in 2022 that it has reached an annual production capacity of 120,000 tons of conductive paste.

For applications in healthcare, low-dimensional materials have played important roles for development of nanovaccines and gene sequencing technology. Lipid nanoparticles were critical to the stability of m-RNA and protein vaccines in human body, effectively improving the delivery efficiency (Corbett et al., 2020; Sahin et al., 2020). Oxford nanopore was the key technology that enabled complete sequencing of human genome by telomere-to-telomere consortium (Logsdon et al., 2021; Nurk et al., 2022), because of the advantage in sequencing long and complex gene fragments.

7.2.2 Best practice, standards, and ethics

Nanotechnology is developing rapidly, with up to tens of thousands of publications every year, and higher and higher device performance reported. However, there have been serious concerns about the credibility and reproducibility of research results.

In 2018 Novoselov et al. conducted a survey of 60 commercially available "graphene" products on the market and found that most of them contained less than 10% of well-defined graphene (Kauling et al., 2018). In 2019 Chorkendorff et al. at the Technical University of Denmark repeated experiments reported in the literature on catalysts for electrochemical synthesis of ammonia, and only one of the findings was verified successfully (Andersen et al., 2019). Schmidt-Mende et al. analyzed 375 papers on organic and organic–inorganic solar cells published in 13 journals during 2011–2012 and found that over 37% of the papers, the short-circuit currents did not agree with the external quantum efficiency (Zimmermann et al., 2014). Grätzel pointed out problems of perovskite solar cells regarding the accuracy and reliability of the reported conversion efficiency, particularly the artifacts caused by current–voltage scanning hysteresis (Grätzel, 2014).

There are many possible causes for the crisis of credibility and reproducibility. One of the reasons is the over emphasis on positive results. Other reasons include quantitative management of the research projects and evaluation of the researchers. Just like any challenging work, there could be success and failure for research. And success can be achieved only after years of repeated failures, in most of the cases. However, the evaluation system typically only counts the success in a short term.

Another reason for the difficulty in reproducing reported results is the sensitivity of experimental conditions and lack of standards. For example, it was reported that many factors could influence the synthesis results of gold nanorods, such as water quality and small change of pH value. Korgel et al. obtained dramatically different morphologies of gold nanocrystals by simply changing surfactants from different manufacturers (Smith & Korgel, 2008).

The key to solving the trust crisis in research is to establish standards and ethics for best practice, which have already begun in many fields, including synthesis of inorganic nanocrystals (Murphy & Buriak, 2015), transistor mobility (Choi et al., 2017), solar cells (Jeong et al., 2020; Wang et al., 2019), lithium-ion batteries (Betz et al., 2019; Li, 2019), supercapacitors (Li, Qi, et al., 2019; Zhang & Pan, 2015), LEDs (Forrest et al., 2003), catalysis (Bligaard et al., 2016), electrocatalysis (Appel & Helm, 2014), photocatalysis (Rajeshwar et al., 2015), and nano-bio experiments (Faria et al., 2018).

A good example is the standard certification for reporting performance of solar cells, provided by three institutions: National Renewable Energy Laboratory in the United States, Advanced Institute of Science and Technology in Japan, and the Fraunhofer Institute for Solar Energy in Germany. In the National Center for Photovoltaics database, only certified data could be included. Starting in 2015, Nature Photonics has required all papers on solar cells to fill out a checklist that includes device area, current–voltage scanning ranges and sequences, conversion efficiencies versus external quantum efficiencies, light source correction methods, mask dimensions, third-party certificates, performance statistics, and long-term stability (Editorial, 2015).

Similarly, for reliably measuring and reporting the performance of lithium-ion batteries, Li (2019) of Institute of Physics, Chinese Academy of Sciences, suggested to provide detailed information such as the size, mass, density, thickness, and active substance ratio of cathode and anode, as well as test conditions such as voltage range, current density, Coulombic efficiency, and energy density. Winter and Schmuch et al. at University of Münster called for more transparency in calculations of batteries and proposed a standard for the estimation of theoretical and practical capacity in six levels, starting from the active materials in electrode, electrode-plus-electrolyte system, then considering inactive substances, and finally reporting the structure and capacity at whole cell level (Betz et al., 2019).

7.3 Limit and breakthrough at low dimensions

Great achievements have been made for the low-dimensional materials in both basic research and industrialization. How will low-dimensional materials develop in the future? What challenges still need to be overcome? Where is the limit of development? How to break the limits? There are many questions worthy of further thinking and exploration:

Is it possible to achieve material simulations in full scales of space and time?

Is it possible to achieve atomic-level control of material structure on a large scale?

Is it possible to assemble atom-by-atom on a large scale and create heterostructures and devices for practical applications?

Is it possible to synthesize biological macromolecules such as DNA and proteins to create new life forms?

How to extract the information contained in DNA sequences? Can personalized precision medicine be achieved?

How to understand the interaction between drugs and organs at molecular level?

Is it possible to simulate, understand, or even realize brain functions through inorganic materials and devices?

A fundamental question is whether there is a limit to the development of low-dimensional materials or generally human civilization. According to the growth theory (Smil, 2019), growth of a resource-constrained system (N) typically follows a logistic pattern:$\frac{dN}{dt} = r\left(\frac{K-N}{K}\right)N$, where K is the upper limit. This is an S-shaped curve, with exponential growth at the beginning; as N approaches the K limit, the growth rate declines and approaches zero growth.

Human civilization is a resource-constrained system on the planet Earth in the solar system. Through the industrial age, our civilization has experienced exponential growth in almost all aspects, such as the population, information, productivity, device size, and performance, but also consumption fossil fuel and CO_2 emissions.

Along with tremendous progress, human activity has caused enormous damage on the environment. Human beings have already utilized more energy and matter than all other lives combined. There have been many alarms about the growth limit, such as shrinkage of Antarctic ice sheet, accelerated melting of Greenland glaciers, weakening of Atlantic circulation, and the irreversible deforestation of Amazon rainforest (Human Development Report; United Nations Development Programme, 2020).

If material recycling and energy regeneration can be fully realized to break the precondition of "limited resources," it is possible to break the growth limit. Opportunities for disruptive scientific and technological innovations to realize renewable energy and material cycles lie in low-dimensional materials. The limit of low-dimensional materials is to fully utilize every atom. The size of commercial transistors is now approaching sub-10 nm. Single-atom transistors (Fuechsle et al., 2012) and single-molecule switches (Terabe et al., 2005) have been reported. Single-atom catalysts are the theoretically achievable limit of catalytic activity (Qiao et al., 2011). The capacity of many energy-storage systems is approaching or has reached the theoretical capacity (Dunn et al., 2011; Larcher & Tarascon, 2015). The mechanical strengths of many low-dimensional materials are approaching the theoretical strength (Zhu & Li, 2010). In addition, sequences of human DNA have been completely mapped, paving the way to explore all secret of genetics (Miga et al., 2020).

Furthermore, we can envision that single atom is not the physical limit of low-dimensional materials. Is it possible to explore, control, and utilize "lower, deeper and hidden dimensions"? The development of quantum information may give some indication and inspiration. For traditional information bit 0 or 1, usually represented by charge, voltage, and spin, the limit density of storage is the atomic density. In quantum mechanics, a quantum state is represented by a wave function in the Schrödinger equation or by a vector in Hilbert space, which can be viewed as a linear superposition of vectors $|0\rangle$ and $|1\rangle$, with infinity embedded in the finite system. By fully exploiting quantum information, human beings may establish a "quantum civilization." Low-dimensional materials with well-defined atomic structures, quantum levels, and long coherent time will become the material basis for the storage, transportation, and transformation of information, matter, and energy at the quantum level.

7.4 Concluding remarks

Low-dimensional materials are one of the most active scientific frontiers in the 21st century. In this book we had a tour in this rapidly developing and wide field, starting from theoretical foundations, structural control and characterization, and properties and applications, to the limits and breakthroughs. We have seen marvelous scientific discoveries, amazing technologies, emerging applications, as well as numerous difficulties and challenges. To fulfill the full potential of low-dimensional materials, there is still a long way to go.

Among all the possible paths, we believe that the most fundamental approach is most important. By exploring the intrinsic and unique properties of low-dimensional materials, they will be the key solutions to the grand problems such as clean energy, information technology, and sustainable environment. We hope that this book will inspire young researchers and students in this wonderland and will stimulate discussions in the research community and public society.

References

Andersen, S. Z., Čolić, V., Yang, S., Schwalbe, J. A., Nielander, A. C., McEnaney, J. M., Enemark-Rasmussen, K., Baker, J. G., Singh, A. R., Rohr, B. A., Statt, M. J., Blair, S. J., Mezzavilla, S., Kibsgaard, J., Vesborg, P. C. K., Cargnello, M., Bent, S. F., Jaramillo, T. F., Stephens, I. E. L., ... Chorkendorff, I. (2019). A rigorous electrochemical ammonia synthesis protocol with quantitative isotope measurements. *Nature Publishing Group, Denmark Nature*, *570*(7762), 504–508. Available from https://doi.org/10.1038/s41586-019-1260-x, http://www.nature.com/nature/index.html.

Appel, A. M., & Helm, M. L. (2014). Determining the overpotential for a molecular electrocatalyst. *ACS Catalysis*, *4*(2), 630–633. Available from https://doi.org/10.1021/cs401013v.

Betz, J., Bieker, G., Meister, P., Placke, T., Winter, M., & Schmuch, R. (2019). Theoretical versus practical energy: A plea for more transparency in the energy calculation of different rechargeable battery systems. *Advanced Energy Materials*, *9*(6). Available from https://doi.org/10.1002/aenm.201803170, http://onlinelibrary.wiley.com/journal/10.1002/(ISSN)1614-6840.

Bligaard, T., Bullock, R. M., Campbell, C. T., Chen, J. G., Gates, B. C., Gorte, R. J., Jones, C. W., Jones, W. D., Kitchin, J. R., & Scott, S. L. (2016). Toward benchmarking in catalysis science: Best practices, challenges, and opportunities. *ACS Catalysis*, *6*(4), 2590–2602. Available from https://doi.org/10.1021/acscatal.6b00183, http://pubs.acs.org/page/accacs/about.html.

Cao, Y., Fatemi, V., Fang, S., Watanabe, K., Taniguchi, T., Kaxiras, E., & Jarillo-Herrero, P. (2018). Unconventional superconductivity in magic-angle graphene superlattices. *Nature*, *556*(7699), 43–50. Available from https://doi.org/10.1038/nature26160, http://www.nature.com/nature/index.html.

Choi, H. H., Cho, K., Frisbie, C. D., Sirringhaus, H., & Podzorov, V. (2017). Critical assessment of charge mobility extraction in FETs. *Nature Materials*, *17*(1), 2–7. Available from https://doi.org/10.1038/nmat5035, http://www.nature.com/nmat/.

Corbett, K. S., Flynn, B., Foulds, K. E., Francica, J. R., Boyoglu-Barnum, S., Werner, A. P., Flach, B., O'Connell, S., Bock, K. W., Minai, M., Nagata, B. M., Andersen, H., Martinez, D. R., Noe, A. T., Douek, N., Donaldson, M. M., Nji, N. N., Alvarado, G. S., Edwards, D. K., ... Graham, B. S. (2020). Evaluation of the mRNA-1273 vaccine against SARS-COV-2 in nonhuman primates. *New England Journal of Medicine*, *383*(16), 1544–1555. Available from https://doi.org/10.1056/NEJMoa2024671, http://www.nejm.org/medical-index.

Curtarolo, S., Hart, G. L. W., Nardelli, M. B., Mingo, N., Sanvito, S., & Levy, O. (2013). The high-throughput highway to computational materials design. *Nature Materials*, *12*(3), 191–201. Available from https://doi.org/10.1038/nmat3568.

Downey,M.L., Moore,D.T., Bachula,G.R., EtterD.M., Carey,E.F., & Perine, L. (2000). *National Nanotechnology Initiative (NNI)—Leading to the next industrial revolution*. Committee on Technology National Science and Technology Council.

Dunn, B., Kamath, H., & Tarascon, J. M. (2011). Electrical energy storage for the grid: A battery of choices. *Science (New York, N.Y.)*, *334*(6058), 928–935. Available from https://doi.org/10.1126/science.1212741, http://www.sciencemag.org/content/334/6058/928.full.pdf.

Dusselier, M., & Davis, M. E. (2018). Small-pore zeolites: Synthesis and catalysis. *Chemical Reviews*, *118*(11), 5265–5329. Available from https://doi.org/10.1021/acs.chemrev.7b00738, http://pubs.acs.org/journal/chreay.

Editorial. (2015). A solar checklist. *Nature Photonics.*, *9*.

Efros, A. L., & Brus, L. E. (2021). Nanocrystal quantum dots: From discovery to modern development. *ACS Nano, 15*(4), 6192–6210. Available from https://doi.org/10.1021/acsnano.1c01399, http://pubs.acs.org/journal/ancac3.

Faria, M., Björnmalm, M., Thurecht, K. J., Kent, S. J., Parton, R. G., Kavallaris, M., Johnston, A. P. R., Gooding, J. J., Corrie, S. R., Boyd, B. J., Thordarson, P., Whittaker, A. K., Stevens, M. M., Prestidge, C. A., Porter, C. J. H., Parak, W. J., Davis, T. P., Crampin, E. J., & Caruso, F. (2018). Minimum information reporting in bio–nano experimental literature. *Nature Nanotechnology, 13*(9), 777–785. Available from https://doi.org/10.1038/s41565-018-0246-4, http://www.nature.com/nnano/index.html.

Forrest, S. R., Bradley, D. D. C., & Thompson, M. E. (2003). Measuring the efficiency of organic light-emitting devices. *Advanced Materials., 15*(13), 1043–1048. Available from https://doi.org/10.1002/adma.200302151.

Fu, L., & Kane, C. L. (2007). Topological insulators with inversion symmetry. *Physical Review B., 76*(4). Available from https://doi.org/10.1103/physrevb.76.045302.

Fuechsle, M., Miwa, J. A., Mahapatra, S., Ryu, H., Lee, S., Warschkow, O., Hollenberg, L. C. L., Klimeck, G., & Simmons, M. Y. (2012). A single-atom transistor. *Nature Nanotechnology,* 7(4), 242–246. Available from https://doi.org/10.1038/nnano.2012.21, http://www.nature.com/nnano/index.html.

Global Nanotechnology Market Outlook. (2024).

Grätzel, M. (2014). The light and shade of perovskite solar cells. *Nature Materials, 13*(9), 838–842. Available from https://doi.org/10.1038/nmat4065, http://www.nature.com/nmat/.

Haruta, M., Yamada, N., Kobayashi, T., & Iijima, S. (1989). Gold catalysts prepared by coprecipitation for low-temperature oxidation of hydrogen and of carbon monoxide. *Journal of Catalysis, 115*(2), 301–309. Available from https://doi.org/10.1016/0021-9517(89)90034-1.

He, C., Sun, X.-C., Liu, X.-P., Lu, M.-H., Chen, Y., Feng, L., & Chen, Y.-F. (2016). Photonic topological insulator with broken time-reversal symmetry. *Proceedings of the National Academy of Sciences, 113*(18), 4924–4928. Available from https://doi.org/10.1073/pnas.1525502113.

Human Development Report. (2020). United Nations Development Programme.

Iijima, S. (1991). Helical microtubules of graphitic carbon. *Nature, 354*(6348), 56–58. Available from https://doi.org/10.1038/354056a0.

Jain, M., Olsen, H. E., Paten, B., & Akeson, M. (2016). The Oxford nanopore MinION: Delivery of nanopore sequencing to the genomics community. *Genome Biology, 17*(1). Available from https://doi.org/10.1186/s13059-016-1103-0l, http://genomebiology.com/.

Jeong, S. H., Park, J., Han, T. H., Zhang, F., Zhu, K., Kim, J. S., Park, M. H., Reese, M. O., Yoo, S., & Lee, T. W. (2020). Characterizing the efficiency of perovskite solar cells and light-emitting diodes. *Joule, 4*(6), 1206–1235. Available from https://doi.org/10.1016/j.joule.2020.04.007, https://www.journals.elsevier.com/joule.

Jing, L., Kershaw, S. V., Li, Y., Huang, X., Li, Y., Rogach, A. L., & Gao, M. (2016). Aqueous based semiconductor nanocrystals. *Chemical Reviews, 116*(18), 10623–10730. Available from https://doi.org/10.1021/acs.chemrev.6b00041, http://pubs.acs.org/journal/chreay.

Kauling, A. P., Seefeldt, A. T., Pisoni, D. P., Pradeep, R. C., Bentini, R., Oliveira, R. V. B., Novoselov, K. S., & Castro Neto, A. H. (2018). The worldwide graphene flake production. *Advanced Materials, 30*(44). Available from https://doi.org/10.1002/adma.201803784, http://onlinelibrary.wiley.com/journal/10.1002/(ISSN)1521-4095.

King, R. D., Whelan, K. E., Jones, F. M., Reiser, P. G. K., Bryant, C. H., Muggleton, S. H., Kell, D. B., & Oliver, S. G. (2004). Functional genomic hypothesis generation and experimentation by a robot scientist. *Nature*, *427*(6971), 247–252. Available from https://doi.org/10.1038/nature02236.

Kroto, H. W., Heath, J. R., O'Brien, S. C., Curl, R. F., & Smalley, R. E. (1985). C60: Buckminsterfullerene. *Nature*, *318*(6042), 162–163. Available from https://doi.org/10.1038/318162a0.

Larcher, D., & Tarascon, J.-M. (2015). Towards greener and more sustainable batteries for electrical energy storage. *Nature Chemistry*, 7(1), 19–29. Available from https://doi.org/10.1038/nchem.2085.

Li, H. (2019). Practical evaluation of Li-ion batteries. *Joule*, *3*(4), 911–914. Available from https://doi.org/10.1016/j.joule.2019.03.028, https://www.journals.elsevier.com/joule.

Li, H., Qi, C., Tao, Y., Liu, H., Wang, D. W., Li, F., Yang, Q. H., & Cheng, H. M. (2019). Quantifying the volumetric performance metrics of supercapacitors. *Advanced Energy Materials*, *9*(21). Available from https://doi.org/10.1002/aenm.201900079, http://onlinelibrary.wiley.com/journal/10.1002/(ISSN)1614-6840.

Lin, L., Zhang, J., Su, H., Li, J., Sun, L., Wang, Z., Xu, F., Liu, C., Lopatin, S., Zhu, Y., Jia, K., Chen, S., Rui, D., Sun, J., Xue, R., Gao, P., Kang, N., Han, Y., Xu, H. Q., ... Liu, Z. (2019). Towards super-clean graphene. *Nature Communications*, *10*(1). Available from https://doi.org/10.1038/s41467-019-09565-4, http://www.nature.com/ncomms/index.html.

Logsdon, G. A., Vollger, M. R., Hsieh, P. H., Mao, Y., Liskovykh, M. A., Koren, S., Nurk, S., Mercuri, L., Dishuck, P. C., Rhie, A., de Lima, L. G., Dvorkina, T., Porubsky, D., Harvey, W. T., Mikheenko, A., Bzikadze, A. V., Kremitzki, M., Graves-Lindsay, T. A., Jain, C., ... Eichler, E. E. (2021). The structure, function and evolution of a complete human chromosome 8. *Nature*, *593*(7857), 101–107. Available from https://doi.org/10.1038/s41586-021-03420-7, http://www.nature.com/nature/index.html.

Louie, S. G., Chan, Y. H., da Jornada, F. H., Li, Z., & Qiu, D. Y. (2021). Discovering and understanding materials through computation. *Nature Materials*, *20*(6), 728–735. Available from https://doi.org/10.1038/s41563-021-01015-1, http://www.nature.com/nmat/.

Ma, T., Liu, Z., Wen, J., Gao, Y., Ren, X., Chen, H., Jin, C., Ma, X. L., Xu, N., Cheng, H. M., & Ren, W. (2017). Tailoring the thermal and electrical transport properties of graphene films by grain size engineering. *Nature Communications*, *8*. Available from https://doi.org/10.1038/ncomms14486, http://www.nature.com/ncomms/index.html.

Ma, T., Ren, W., Zhang, X., Liu, Z., Gao, Y., Yin, L.-C., Ma, X.-L., Ding, F., & Cheng, H.-M. (2013). Edge-controlled growth and kinetics of single-crystal graphene domains by chemical vapor deposition. *Proceedings of the National Academy of Sciences*, *110*(51), 20386–20391. Available from https://doi.org/10.1073/pnas.1312802110.

Mayorov, A. S., Gorbachev, R. V., Morozov, S. V., Britnell, L., Jalil, R., Ponomarenko, L. A., Blake, P., Novoselov, K. S., Watanabe, K., Taniguchi, T., & Geim, A. K. (2011). Micrometer-scale ballistic transport in encapsulated graphene at room temperature. *Nano Letters*, *11*(6), 2396–2399. Available from https://doi.org/10.1021/nl200758b.

Miga, K. H., Koren, S., Rhie, A., Vollger, M. R., Gershman, A., Bzikadze, A., Brooks, S., Howe, E., Porubsky, D., Logsdon, G. A., Schneider, V. A., Potapova, T., Wood, J., Chow, W., Armstrong, J., Fredrickson, J., Pak, E., Tigyi, K., Kremitzki, M., ... Phillippy, A. M. (2020). Telomere-to-telomere assembly of a complete human X chromosome. *Nature*, *585*(7823), 79–84. Available from https://doi.org/10.1038/s41586-020-2547-7, http://www.nature.com/nature/index.html.

Murphy, C. J., & Buriak, J. M. (2015). Best practices for the reporting of colloidal inorganic nanomaterials, American Chemical Society, undefined*Chemistry of Materials, 27* (14), 4911–4913. Available from https://doi.org/10.1021/acs.chemmater.5b02323, http://pubs.acs.org/journal/cmatex.

Novoselov, K. S., Geim, A. K., Morozov, S. V., Jiang, D., Katsnelson, M. I., Grigorieva, I. V., Dubonos, S. V., & Firsov, A. A. (2005). Two-dimensional gas of massless Dirac fermions in graphene. *Nature, 438*(7065), 197–200. Available from https://doi.org/10.1038/nature04233.

Novoselov, K. S., Geim, A. K., Morozov, S. V., Jiang, D., Zhang, Y., Dubonos, S. V., Grigorieva, I. V., & Firsov, A. A. (2004). Electric field effect in atomically thin carbon films. *Science (New York, N.Y.), 306*(5696), 666–669. Available from https://doi.org/10.1126/science.1102896.

Novoselov, K. S., Jiang, D., Schedin, F., Booth, T. J., Khotkevich, V. V., Morozov, S. V., & Geim, A. K. (2005). Two-dimensional atomic crystals. *Proceedings of the National Academy of Sciences of the United States of America., 102*(30), 10451–10453. Available from https://doi.org/10.1073/pnas.0502848102.

Nurk, S., Koren, S., Rhie, A., Rautiainen, M., Bzikadze, A. V., Mikheenko, A., Vollger, M. R., Altemose, N., Uralsky, L., Gershman, A., Aganezov, S., Hoyt, S. J., Diekhans, M., Logsdon, G. A., Alonge, M., Antonarakis, S. E., Borchers, M., Bouffard, G. G., Brooks, S. Y., … Phillippy, A. M. (2022). The complete sequence of a human genome. *Science (New York, N.Y.), 376*(6588), 44–53. Available from https://doi.org/10.1126/science.abj6987, https://www.science.org/doi/10.1126/science.abj6987.

Qiao, B., Wang, A., Yang, X., Allard, L. F., Jiang, Z., Cui, Y., Liu, J., Li, J., & Zhang, T. (2011). Single-atom catalysis of CO oxidation using Pt1/FeO_x. *Nature Chemistry, 3*(8), 634–641. Available from https://doi.org/10.1038/nchem.1095.

Raccuglia, P., Elbert, K. C., Adler, P. D. F., Falk, C., Wenny, M. B., Mollo, A., Zeller, M., Friedler, S. A., Schrier, J., & Norquist, A. J. (2016). Machine-learning-assisted materials discovery using failed experiments. *Nature, 533*(7601), 73–76. Available from https://doi.org/10.1038/nature17439, http://www.nature.com/nature/index.html.

Rajeshwar, K., Thomas, A., & Janáky, C. (2015). Photocatalytic activity of inorganic semiconductor surfaces: Myths, hype, and reality. *The Journal of Physical Chemistry Letters, 6*(1), 139–147. Available from https://doi.org/10.1021/jz502586p.

Sahin, U., Muik, A., Derhovanessian, E., Vogler, I., Kranz, L. M., Vormehr, M., Baum, A., Pascal, K., Quandt, J., Maurus, D., Brachtendorf, S., Lörks, V., Sikorski, J., Hilker, R., Becker, D., Eller, A. K., Grützner, J., Boesler, C., Rosenbaum, C., … Türeci, Ö. (2020). COVID-19 vaccine BNT162b1 elicits human antibody and TH1 T cell responses. *Nature Research, Germany Nature, 586*(7830), 594–599. Available from https://doi.org/10.1038/s41586-020-2814-7, http://www.nature.com/nature/index.html.

Smil, V. (2019). *Growth: From microorganisms to megacities*. The MIT Press.

Smith, D. K., & Korgel, B. A. (2008). The importance of the CTAB surfactant on the colloidal seed-mediated synthesis of gold nanorods. *Langmuir: The ACS Journal of Surfaces and Colloids, 24*(3), 644–649. Available from https://doi.org/10.1021/la703625a.

Spurgeon, S. R., Ophus, C., Jones, L., Petford-Long, A., Kalinin, S. V., Olszta, M. J., Dunin-Borkowski, R. E., Salmon, N., Hattar, K., Yang, W. C. D., Sharma, R., Du, Y., Chiaramonti, A., Zheng, H., Buck, E. C., Kovarik, L., Penn, R. L., Li, D., Zhang, X., … Taheri, M. L. (2021). Towards data-driven next-generation transmission electron microscopy. *Nature Materials, 20*(3), 274–279. Available from https://doi.org/10.1038/s41563-020-00833-z, http://www.nature.com/nmat/.

Tang, D. M., Erohin, S. V., Kvashnin, D. G., Demin, V. A., Cretu, O., Jiang, S., Zhang, L., Hou, P. X., Chen, G., Futaba, D. N., Zheng, Y., Xiang, R., Zhou, X., Hsia, F. C., Kawamoto, N., Mitome, M., Nemoto, Y., Uesugi, F., Takeguchi, M., ... Golberg, D. (2021). Semiconductor nanochannels in metallic carbon nanotubes by thermomechanical chirality alteration. *Science (New York, N.Y.)*, *374*(6575), 1616–1620. Available from https://doi.org/10.1126/science.abi8884, https://www.science.org/doi/10.1126/science.abi8884.

Tang, S., Tang, J., Okunishi, E., Ninota, Y., Yasuhara, A., Uzuhashi, J., Ohkubo, T., Takeguchi, M., Yuan, J., & Qin, L. C. (2022). A stable LaB6 nanoneedle field-emission electron source for atomic resolution imaging with a transmission electron microscope. *Materials Today*, *57*, 35–42. Available from https://doi.org/10.1016/j.mattod.2022.06.003, http://www.journals.elsevier.com/materials-today/.

Terabe, K., Hasegawa, T., Nakayama, T., & Aono, M. (2005). Quantized conductance atomic switch. *Nature*, *433*(7021), 47–50. Available from https://doi.org/10.1038/nature03190.

Tshitoyan, V., Dagdelen, J., Weston, L., Dunn, A., Rong, Z., Kononova, O., Persson, K. A., Ceder, G., & Jain, A. (2019). Unsupervised word embeddings capture latent knowledge from materials science literature. *Nature*, *571*(7763), 95–98. Available from https://doi.org/10.1038/s41586-019-1335-8, http://www.nature.com/nature/index.html.

Wang, Y., Liu, X., Zhou, Z., Ru, P., Chen, H., Yang, X., & Han, L. (2019). Reliable measurement of perovskite solar cells. *Advanced Materials*, *31*(47). Available from https://doi.org/10.1002/adma.201803231, http://onlinelibrary.wiley.com/journal/10.1002/(ISSN)1521-4095.

Warren, J. A. (2018). The materials genome initiative and artificial intelligence. *MRS Bulletin*, *43*(6), 452–457. Available from https://doi.org/10.1557/mrs.2018.122, http://journals.cambridge.org/MRS.

Wegst, U. G. K., Bai, H., Saiz, E., Tomsia, A. P., & Ritchie, R. O. (2015). Bioinspired structural materials. *Nature Materials*, *14*(1), 23–36. Available from https://doi.org/10.1038/nmat4089, http://www.nature.com/nmat/.

Wen, L., Li, F., & Cheng, H. M. (2016). Carbon nanotubes and graphene for flexible electrochemical energy storage: From materials to devices. *Advanced Materials*, *28*(22), 4306–4337. Available from https://doi.org/10.1002/adma.201504225, http://www3.interscience.wiley.com/journal/119030556/issue.

Yang, F., Wang, M., Zhang, D., Yang, J., Zheng, M., & Li, Y. (2020). Chirality pure carbon nanotubes: Growth, sorting, and characterization. *Chemical Reviews*, *120*(5), 2693–2758. Available from https://doi.org/10.1021/acs.chemrev.9b00835, http://pubs.acs.org/journal/chreay.

Zhang, H., Jimbo, Y., Niwata, A., Ikeda, A., Yasuhara, A., Ovidiu, C., Kimoto, K., Kasaya, T., Miyazaki, H. T., Tsujii, N., Wang, H., Yamauchi, Y., Fujita, D., Kitamura, Si, & Manabe, H. (2022). High-endurance micro-engineered LaB6 nanowire electron source for high-resolution electron microscopy. *Nature Nanotechnology*, *17*(1), 21–26. Available from https://doi.org/10.1038/s41565-021-00999-w, http://www.nature.com/nnano/index.html.

Zhang, H., Tang, J., Yuan, J., Yamauchi, Y., Suzuki, T. T., Shinya, N., Nakajima, K., & Qin, L. C. (2016). An ultrabright and monochromatic electron point source made of a LaB 6 nanowire. *Nature Nanotechnology*, *11*(3), 273–279. Available from https://doi.org/10.1038/nnano.2015.276, http://www.nature.com/nnano/index.html.

Zhang, S., & Pan, N. (2015). Supercapacitors performance evaluation. *Advanced Energy Materials*, *5*(6). Available from https://doi.org/10.1002/aenm.201401401, http://onlinelibrary.wiley.com/journal/10.1002/(ISSN)1614-6840.

Zhang, Y., He, X., Chen, Z., Bai, Q., Nolan, A. M., Roberts, C. A., Banerjee, D., Matsunaga, T., Mo, Y., & Ling, C. (2019). Unsupervised discovery of solid-state lithium ion conductors. *Nature Communications*, *10*(1). Available from https://doi.org/10.1038/s41467-019-13214-1, http://www.nature.com/ncomms/index.html.

Zhu, T., & Li, J. (2010). Ultra-strength materials. *Progress in Materials Science*, *55*(7), 710–757. Available from https://doi.org/10.1016/j.pmatsci.2010.04.001.

Zimmermann, E., Ehrenreich, P., Pfadler, T., Dorman, J. A., Weickert, J., & Schmidt-Mende, L. (2014). Erroneous efficiency reports harm organic solar cell research. *Nature Photonics*, *8*(9), 669–672. Available from https://doi.org/10.1038/nphoton.2014.210, http://www.nature.com/nphoton/index.html.

Index

Note: Page numbers followed by "*f*" refer to figures.

R

S

X

Y

Z